save buk

12

ADVANCED MATHEMATICS

PNG UPPER SECONDARY

Christine McRae
Esther Gawac Ehava
Rory Barrett
Tom Sidebotham

OXFORD

OXFORD
UNIVERSITY PRESS

Oxford University Press is a department of the University of Oxford. It furthers the University's objective of excellence in research, scholarship, and education by publishing worldwide. Oxford is a registered trademark of Oxford University Press in the UK and in certain other countries.

Published in Australia by
Oxford University Press
Level 8, 737 Bourke Street, Docklands, Victoria 3008, Australia

First published 2013
Reprinted 2014 (twice), 2017 (twice), 2018, 2019, 2021, 2022, 2023, 2024, 2025

This book was originally published by ESA Publications, Auckland, New Zealand. This edition, specially adapted for the Grade 12 syllabus in Papua New Guinea, is published by arrangement with ESA Publications. Authors of the original work were Phyl Haydock, John Watson and Rory Barrett and this adaptation has been produced by Christine McRae with the assistance of Esther Gawac Ehava, Mac Dandava and John Gesa Junior.

ISBN 978 0 19 557872 0

Illustrated by diacriTech, Chennai, India
Typeset by diacriTech, Chennai, India
Printed in China by Golden Cup Printing Co. Ltd

Oxford University Press Australia & New Zealand is committed to sourcing paper responsibly.

FSC www.fsc.org MIX Paper | Supporting responsible forestry FSC™ C110497

Contents

Unit 12.1 Patterns and Algebra

Unit 12.2 Trigonometry and Vectors

Unit 12.3 Calculus

Example B

If A = {2, 4, 6, 8, 10, 12, 14, 16, 18}, B = {4, 8, 12, 16} and C = {1, 2, 3, 4, 5, 6}

Set B is a subset of set A ($B \subset A$) because all the elements in set B are also in set A.

Set C is not a subset of set A ($C \not\subset A$) because set C contains some elements that are not in set A.

- If a set has p elements then there are 2^p subsets of the set. This is because each element is either in the subset or not in the subset so there is a choice of 2 possibilities for each element.
- The null set is a subset of all sets. The set itself is also a subset of a set.

Example C

The set D = {2, 4, 6, 8}.

There are four elements in D so there should be a total of $2^4 = 16$ possible subsets of D.

Writing down all the possible subsets of D:

∅, {2}, {4}, {6}, {8}, {2, 4}, {2, 6}, {2, 8}, {4, 6}, {4, 8}, {6, 8}, {2, 4, 6}, {2, 4, 8}, {2, 6, 8}, {4, 6, 8}, {2, 4, 6, 8}

Unit 12.1 Activity 1A: Definition and notation of sets

1. a. List the following sets:

i. P, the set of prime numbers less than 20.

ii. Q, the set of people in your immediate family (parents and siblings).

iii. R, the set of square integers between 100 and 200 inclusive.

b. State whether each of the above sets is finite or infinite.

c. Write down the measure of each of the finite sets.

2. If A = {1, 3, 5, 7, 9, 11, 13}; B = { 1, 2, 4, 9} and C = {1, 2, 3, 4, 6, 8, 9}, state whether each of the following is true or false:

a. $n(A) = 8$

b. $5 \in A$

c. $B \not\subset A$

d. A is an infinite set

e. $n(C) = 7$

f. $n(\varnothing) = 1$

g. B has 8 possible subsets

h. $B \in C$

i. $7 \in C$

j. $C \subset B$

k. $\varnothing \subset C$

l. $A \subset A$

m. $2 \notin A$

n. A and C have the same number of possible subsets

3. Write down all the possible subsets of X = {a, b, c}.

4. For the set Y = {p, q, r, s, t}

a. State the number of possible subsets.

b. Write down all the subsets that have 4 elements.

Unit 12.1 Patterns and Algebra
Topic 2: Sets – Venn diagrams

This first Unit in the Grade 12 Syllabus focuses on different methods of solving real-life problems. Having summarised the basic definitions and notation in Topic 1, Topic 2 looks at Venn diagrams.

Venn diagrams are named after John Venn (1834–1923), who was the first person to use these types of diagrams in formal logic. Venn diagrams are used in solving problems relating to probability, statistics and computer science.

Venn diagrams

- The universal set is the set containing all the elements under consideration. The universal set is denoted by the Greek letter ξ.
- The **complement** of a set A, written as A', is the set that contains all the elements that are in the universal set but are *not* in set A.

Example A

If the universal set is the set of **natural numbers** 1 to 10, and A is the set containing the natural numbers that are multiples of 3, then:

ξ = {1, 2, 3, 4, 5, 6, 7, 8, 9, 10}

A = {3, 6, 9}

The complement of A, A' = {1, 2, 4, 5, 7, 8, 10}

A **Venn diagram** is a diagrammatic way of representing sets. On a Venn diagram the universal set is shown as a rectangle and subsets are shown as circles.

A Venn diagram, with the set A shaded, for Example A above, would look like:

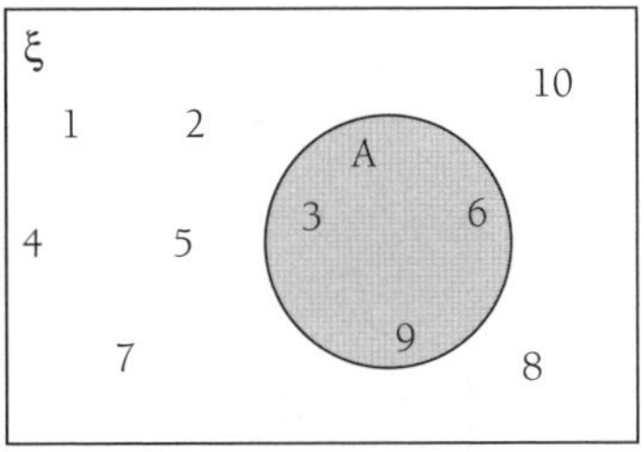

A Venn diagram with A' shaded would look like:

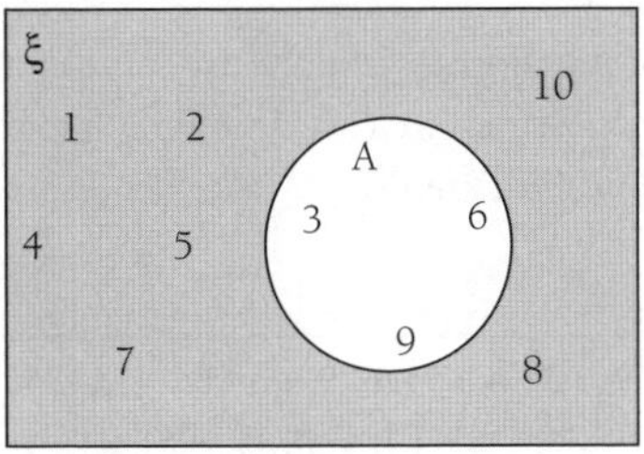

If set B = {3, 9} then set B is a subset of set A.

The Venn diagram below shows the relationship between set A and set B:

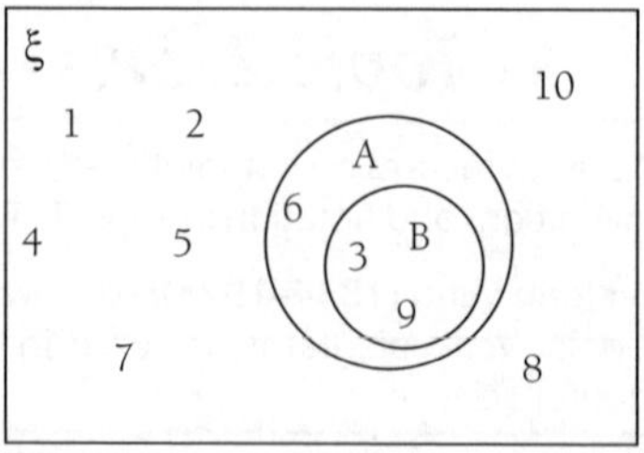

- The **intersection** of two sets is the set of elements that the two sets have in common. The intersection of sets A and B is written as **A ∩ B**.

On the Venn diagram below, the intersection of sets A and B is shaded:

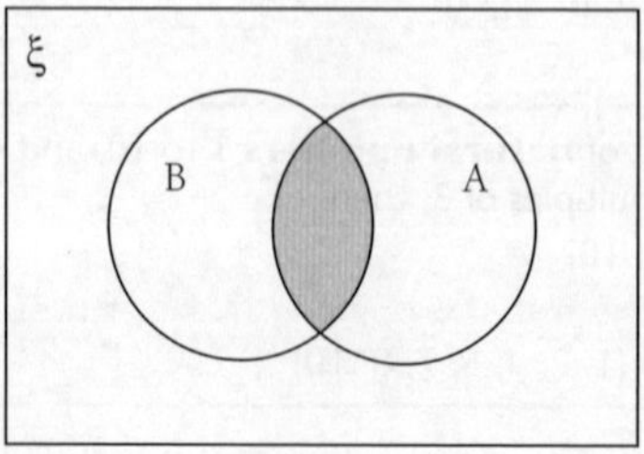

- The **union** of two sets is the set of elements that belong to both A and B or both. The union of sets A and B is written as **A ∪ B**.

On the Venn diagram below the union of A and B is shaded:

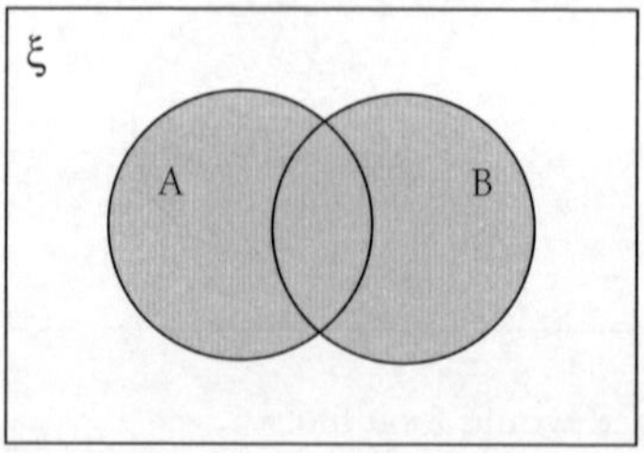

- Two sets are equal if, and only if, they contain the same elements. The order that elements are written in a set does not change the set. {red, yellow, blue} = {yellow, blue, red}

Example B

If ξ = {1, 2, 3, 4, 5, 6, 7, 8, 9, 10}, A = {3, 6, 9} and B = {3, 4, 5, 6, 7}
Then A' = {1, 2, 4, 5, 7, 8, 10} and B' = { 1, 2, 8, 9, 10}
The intersection of sets A and B, A ∩ B = {3, 6}
The union of sets A and B, A ∪ B = {3, 4, 5, 6, 7, 9}

On a Venn diagram these sets could be shown:

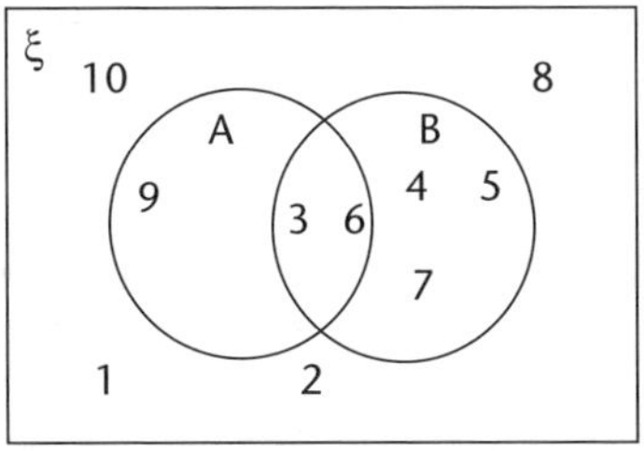

Unit 12.1 Activity 2A: Universal sets, complements and Venn diagrams

1. If ξ = {integers from 1 to 10 inclusive}
A = {even numbers}
B = {prime numbers}
C = {multiples of 3}

a. List each of the sets A, B and C.

b. List the sets:

i. $A \cap B$
ii. $A \cap C$
iii. $B \cap C$
iv. $B \cap \xi$
v. $C \cap \varnothing$
vi. $(A \cap B) \cap C$ (do bracket first)

c. List the sets:

i. $A \cup B$
ii. $A \cup C$
iii. $B \cup C$
iv. $C \cup \xi$
v. $B \cup \varnothing$

d. Draw a Venn diagram to illustrate the relationship between the sets ξ, A, B and C.

e. List the sets A', B' and C'.

f. List the sets:

i. $A' \cup B$
ii. $A \cup C'$
iii. $B' \cap C'$
iv. $A' \cap B$
v. $B' \cap \xi$

2. If ξ = {letters of the alphabet}
U = {u, n, i, t, e}
V = {the vowels}
W = {letters of the word 'consonant'}

a. List each of the sets V and W.

b. List the sets:

i. $V \cap W$

ii. $V \cap U$

iii. $W \cap U$

iv. $W \cap \xi$

v. $U \cap \varnothing$

c. List the sets:

i. $V \cup W$

ii. $V \cup U$

iii. $W \cup U$

iv. $U \cup \xi$

v. $W \cup \varnothing$

d. List the following sets (*Hint:* Do the brackets first):

i. $U \cup (V \cup W)$

ii. $U \cap (V \cap W)$

iii. $(U \cap V) \cap W$

iv. $U \cup (V \cap W)$

v. $U \cap (V \cup W)$

vi. $U \cap (V \cap \xi)$

vii. $U \cup (\varnothing \cap W)$

viii. $(U \cap V) \cap \xi$

ix. $(U \cap V) \cap (V \cap W)$

x. $(U \cup V) \cup (V \cup W)$

xi. $(U \cap V) \cup (U \cap W)$

xii. $(U \cup V) \cap (U \cup W)$

3. If ξ = {integers from 1 to 20 inclusive}

P = {multiples of 3}

Q = {multiples of 4}

R = {multiples of 6}

a. List the following sets:

i. $P \cap Q$

ii. $P \cap R$

iii. $Q \cap R$

iv. $P \cap Q \cap R$

b. What relationship is there between set P and set R?

c. Construct a Venn diagram to illustrate the relationship between sets P, Q and R. Include all the elements of ξ on the diagram.

Unit 12.1 Patterns and Algebra

Topic 3: Sets – field laws

In Topic 3 we look at some of the laws of sets before we move on to exploring how set theory is used in problem solving, in the next Topic.

The field laws for sets under intersection

1. Commutative law: $A \cap B = B \cap A$

The commutative law means that you can swap the sets and still get the same answer. For example, working with numbers: $2 \times 4 = 4 \times 2$

2. Associative law: $(A \cap B) \cap C = A \cap (B \cap C)$

The associative law means that whatever way you group the sets, you still get the same answer. For example, working with numbers: $(2 \times 3) \times 4 = 2 \times (3 \times 4)$

3. Identity law: $A \cap \xi = A$

4. **Distributive law:** $A \cap (B \cup C) = (A \cap B) \cup (A \cap C)$

The distributive law, working with numbers: $2 \times (3 + 5) = 2 \times 3 + 2 \times 5$. The 2 can be 'distributed' across 3 + 5, into 2×3 and 2×5.

The distributive law can be demonstrated using Venn diagrams:

Left-hand side of the equation: $A \cap (B \cup C)$

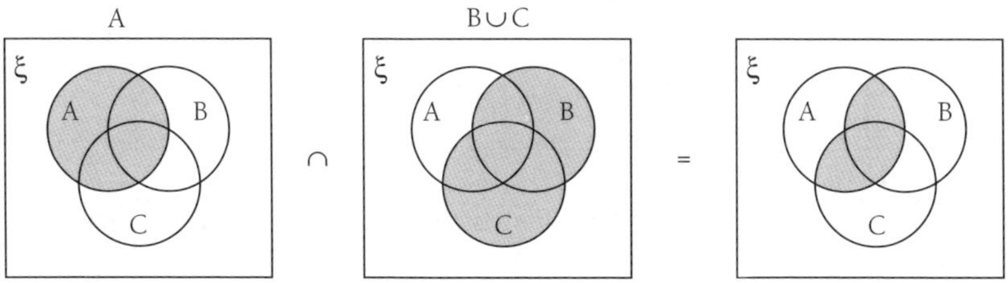

Right-hand side of the equation: $(A \cap B) \cup (A \cap C)$

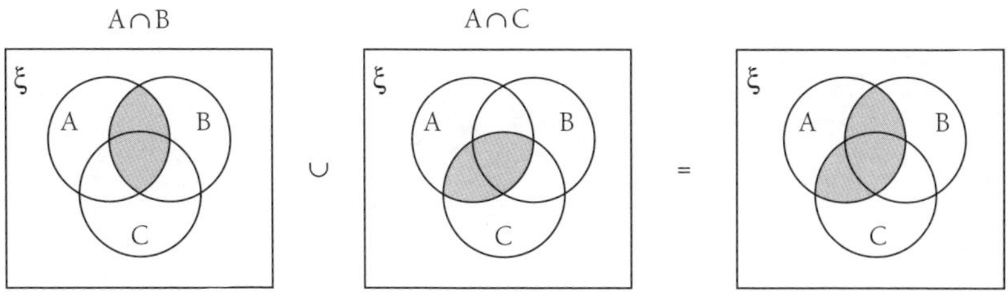

The two Venn diagrams for both sides of the equation are the same. Therefore we have shown that the distributive law for sets under intersection $A \cap (B \cup C) = (A \cap B) \cup (A \cap C)$ is true.

The field laws for sets under union

1. Commutative law: $A \cup B = B \cup A$
2. Associative law: $(A \cup B) \cup C = A \cup (B \cup C)$
3. Identity law: $A \cup \varnothing = A$
4. Distributive law: $A \cup (B \cap C) = (A \cup B) \cap (A \cup C)$

The distributive law can be demonstrated using Venn diagrams:

Left-hand side of the equation: $A \cup (B \cap C)$

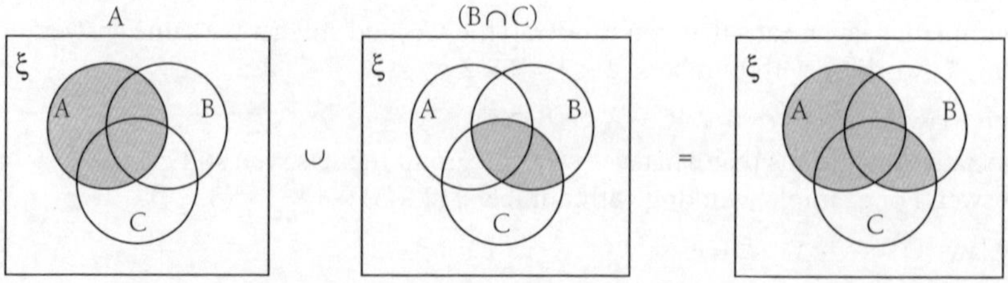

Right-hand side of the equation: $(A \cup B) \cap (A \cup C)$

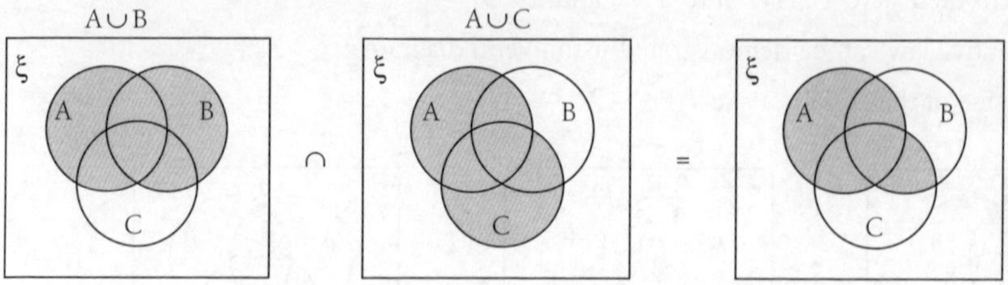

The two Venn diagrams for both sides of the equation are the same. Therefore we have shown that the distributive law for sets under union, $A \cup (B \cap C) = (A \cup B) \cap (A \cup C)$, is true.

- **Disjoint sets** have no elements in common.
  If A and B are disjoint sets then on a Venn diagram they would look like this:

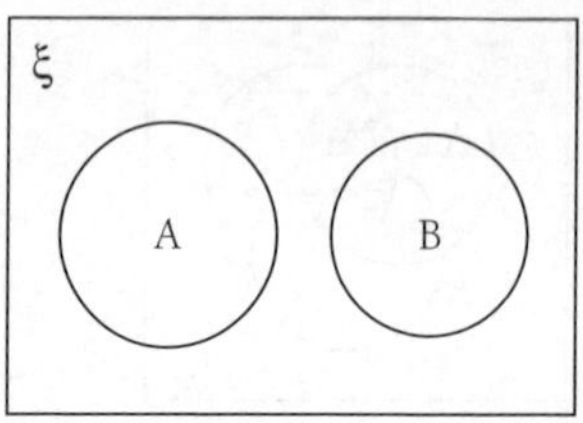

$A \cap B = \varnothing$

Unit 12.1 Activity 3A: Field laws for sets

1. Shade each of the following sets on a Venn diagram. Assume there is intersection between all mentioned sets:

a. $A \cup B \cup C$ **b.** $A \cap B \cap C$ **c.** $A \cap B'$ **d.** $A \cup B'$

e. $A' \cup B'$ **f.** $(A \cap B)'$ **g.** $(A \cup B)'$ **h.** $A' \cap B'$

i. Using your answers to parts **e**. to **h**. above, complete the following:

$(A \cup B)' = A' \ldots B'$

$(A \cap B)' = A' \ldots B'$

2. If the sets A and B are disjoint sets, shade each of the following sets on a Venn diagram:

a. $A \cup B$ **b.** $A \cap B$ **c.** $A' \cup B'$

3. If the set A is a subset of set B, shade each of the following sets on a Venn diagram:

a. $A \cup B$ **b.** $A \cap B$ **c.** $A' \cap B$ **d.** $A \cap B'$

4. If the sets A, B and C are as shown in the Venn diagram below, shade each of the following sets:

a. $A \cup B \cup C$ **b.** $(A \cap B) \cup C$ **c.** $B' \cap C$ **d.** $A \cap B' \cap C$

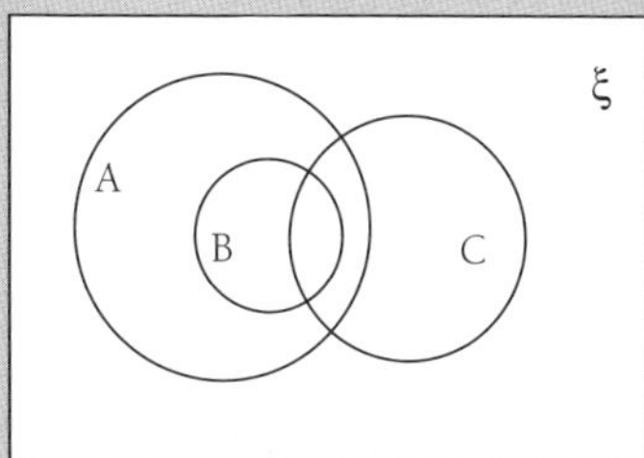

5. If the universal set is the set of all triangles, T, the set I is the set of isosceles triangles, the set R is the set of right-angled triangles and the set E is the set of equilateral triangles, show these sets on a Venn diagram.

6. If the universal set Q is the set of all quadrilaterals, the set R is the set of rectangles, the set S is the set of squares and the set P is the set of parallelograms, show these sets on a Venn diagram.

7. State the sets that are shaded in the following Venn diagrams:

a.

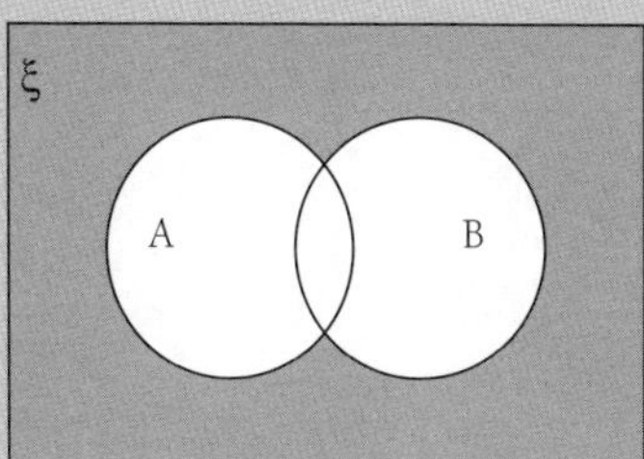

b.

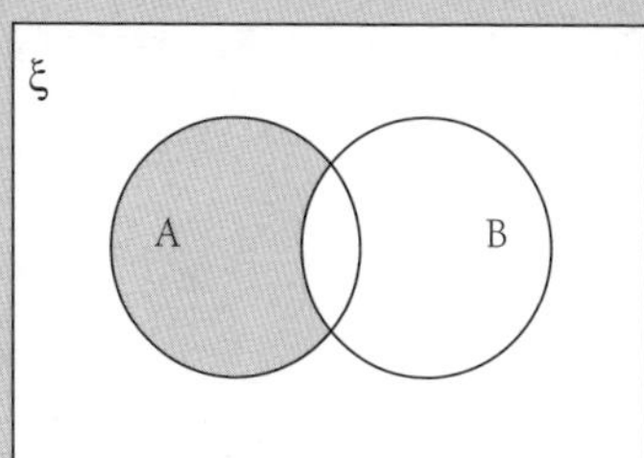

c.

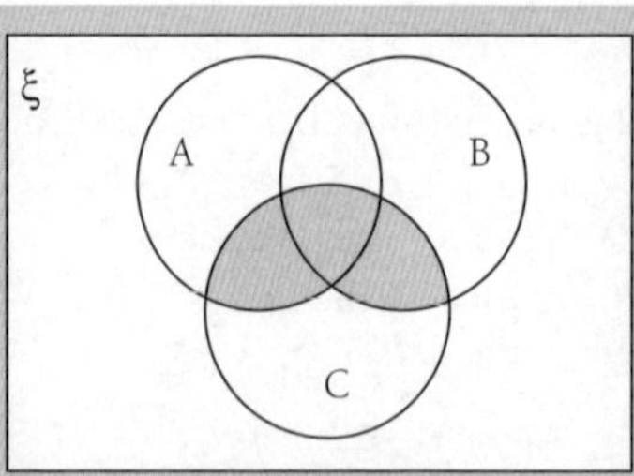

d.

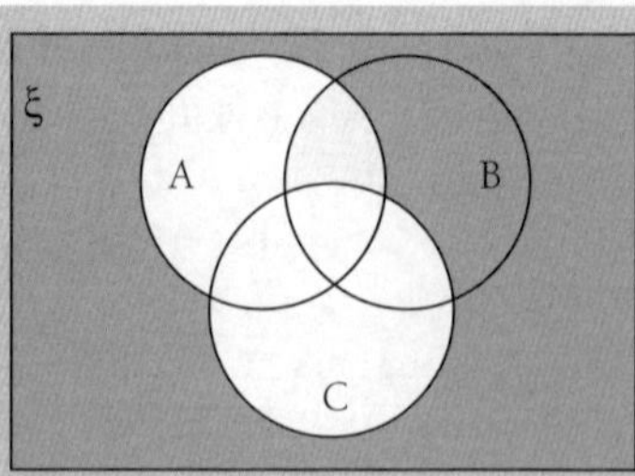

e.

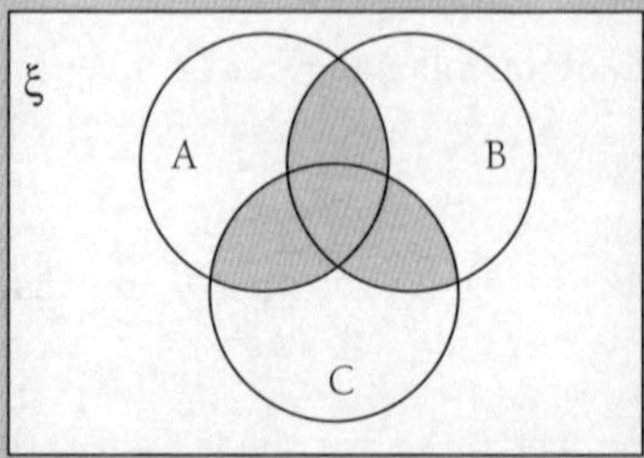

f.

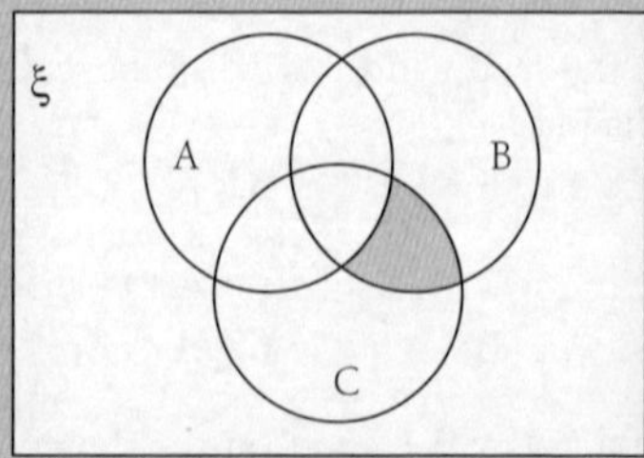

Unit 12.1 Patterns and Algebra

Topic 4: Using sets to solve problems

The first three Topics in Unit 12.1 examined basic definitions and notation relating to sets and set theory. In Topic 4 we focus on how sets can be used to solve problems. As stated in the Syllabus, set theory is used in many real-life situations relating to social, economic, biological, physical and mathematical contexts.

Sets represented on Venn diagrams can be used to solve problems. In most cases the *number* of elements in a set is shown on the Venn diagram, rather than the elements themselves.

For sets that intersect, a **rule** that is often used is:

$$n(A \cup B) = n(A) + n(B) - n(A \cap B)$$

For example:

The rule $n(A \cup B) = n(A) + n(B) - n(A \cap B)$ can be demonstrated:

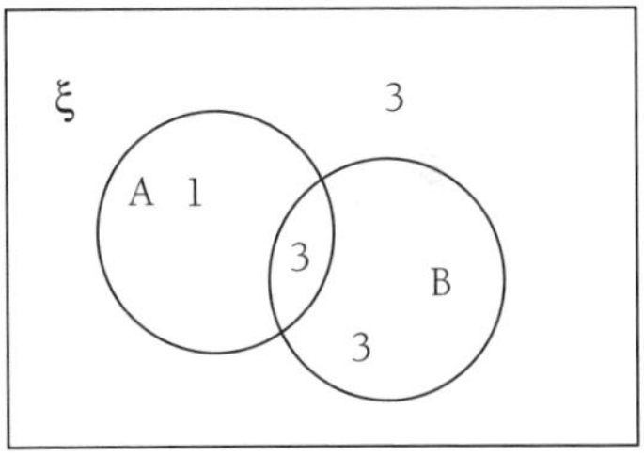

$\xi = \{1, 2, 3, 4, 5, 6, 7, 8, 9, 10\}$

$A = \{2, 4, 6, 8\}$

$B = \{3, 4, 5, 6, 7, 8\}$

$A \cup B = \{2, 3, 4, 5, 6, 7, 8\}$

$A \cap B = \{4, 6, 8\}$

$n(A) = 4$; $n(B) = 6$; $n(A \cup B) = 7$; $n(A \cap B) = 3$

$n(A) + n(B) - n(A \cap B) = 4 + 6 - 3 = 7 = n(A \cup B)$

The numbers on the Venn diagram represent the number of elements in each of the sets.

When drawing a Venn diagram containing the number of elements, it is good to start with the number of elements in the intersections first, then fill in all the other possible areas.

Example A

Q. In a class of 27 students at a school, 22 study mathematics (M), 17 study physics (P) and 18 study chemistry (C). 14 of the students study all three subjects, 16 study mathematics and chemistry, while 17 study mathematics and physics.

If there are no students who study physics and chemistry but not mathematics, how many students:

a. do not study any of the three subjects?
b. study chemistry only?
c. study mathematics only?

A. Construct a Venn diagram and start with the intersection of all three subjects (14 students).

For the intersection of M and C:

$n(M \cap C) = 16$, but 14 study all three subjects leaving only 2 who study mathematics and chemistry (but not physics).

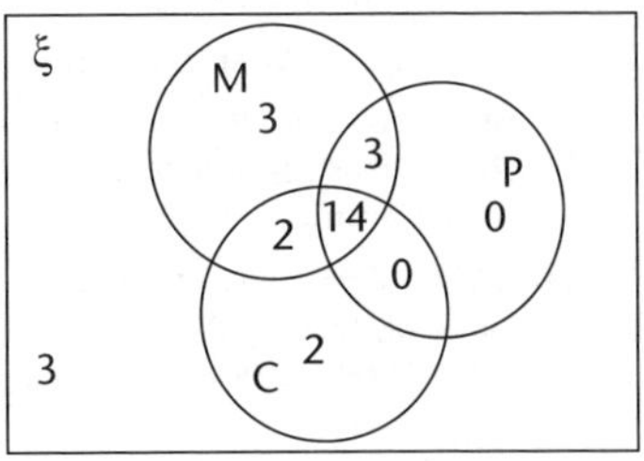

The numbers of all 4 **regions** that are in:

- the mathematics set have to sum to 22
- the physics set have to sum to 17
- the chemistry set have to sum to 18

Summing all the regions in M ∪ P ∪ C gives 24, so that leaves 27 – 24 = 3 outside this area.

Having drawn the diagram we can answer the questions:

a. The number of students who do not study any of the three subjects is 3 (region shaded on the Venn diagram):

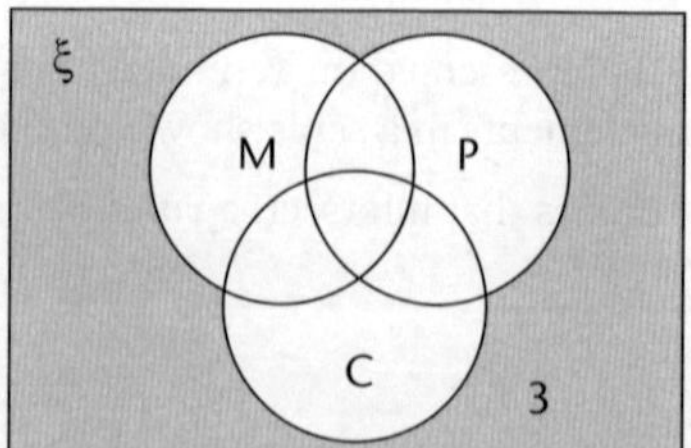

b. The number of students who study chemistry only is 2 (region shaded on the Venn diagram):

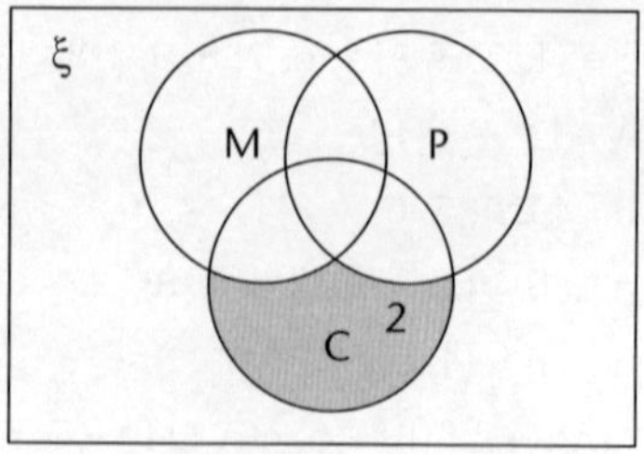

c. The number of students who study mathematics only is 3 (region shaded on the Venn diagram):

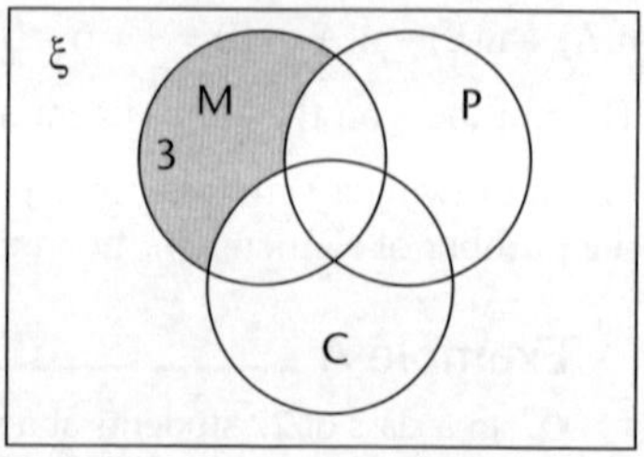

Example B

Q. In a group of 44 students, 20 study economics (E), 28 study accounting (A) and 5 study neither of these subjects.

Construct a Venn diagram for this situation and find how many students study:

a. both accounting and economics
b. at least one of accounting and economics
c. accounting only

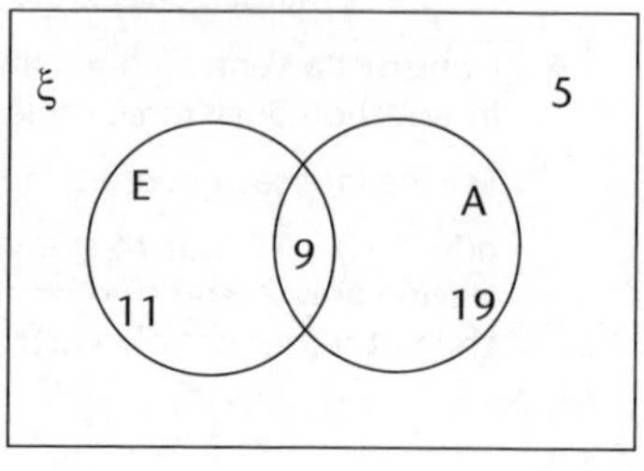

A. In this case there are 5 students who study neither subject, leaving 44 – 5 = 39 who study at least one of economics or accounting.

Using:

n(E ∪ A) = n(E) + n(A) – n(E ∩ A)

$39 = 20 + 28 - n(E \cap A)$

$n(E \cap A) = 48 - 39 = 9$

Having found the number in the intersection of E and A, the other numbers are found by subtracting this number from the totals ($20 - 9 = 11$ for $n(E \cap A')$ and $28 - 9 = 19$ for $n(A \cap E')$).

a. 9 students study both accounting and economics.
b. This is the set $E \cup A$ so 39 students study at least one of the subjects.
c. 19 students study accounting only.

Unit 12.1 Activity 4A: Using sets to solve problems

1. The Venn diagram shows the number of elements in each section of three sets A, B and C.

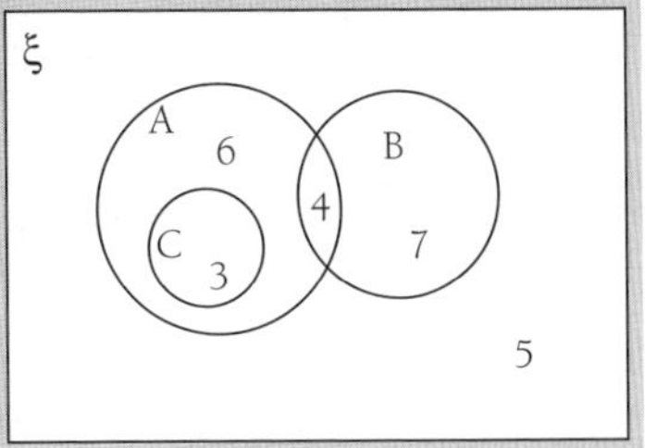

Find:

a. $n(A)$
b. $n(B)$
c. $n(A \cap B)$
d. $n(A \cup B)$
e. $n(A \cap C)$
f. $n(A \cup C)$
g. $n(A')$
h. $n(B')$
i. $n(C')$
j. $n(C \cap B)$
k. $n((A \cup B)')$
l. $n(A' \cap B)$
m. $n(A \cap C')$

2. At a sporting club members can play volleyball (V) and/or either netball (N) or basketball (B).
50 members of the club were asked which sports they had played and the following Venn diagram was constructed from the results:

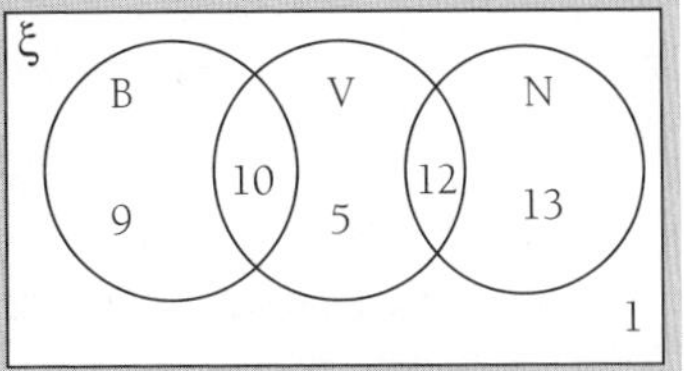

How many members play:

a. Volleyball only.
b. Volleyball and basketball.
c. None of the sports.
d. Netball.
e. Basketball and/or volleyball.
f. Netball only.
g. Basketball or netball.

3. Of 80 senior students, 70 had access to a mobile phone and/or a computer.
55 had access to a computer and 54 had access to a mobile phone. Construct a Venn diagram for this situation and find the number of students who have access to:

a. Both a computer and a mobile phone.
b. A computer only.
c. A mobile phone only.
d. Neither a computer nor a mobile phone.

4. Students doing a particular subject at university have a choice of two textbooks, A or B, to buy. Of the 83 students doing the subject, 51 have chosen textbook A and of these 51 students, 30 have chosen to buy an update of the textbook, C. Seven students have not bought a textbook. Construct a Venn diagram showing the number of elements in sets A, B and C. How many students have bought textbook B?

5. A store sells food, drinks and newspapers. The owner has collected the following information about her 140 customers on one day:

- 2 didn't buy anything.
- 26 bought drink only.
- 38 bought food only.
- 21 bought newspapers only.
- 24 bought food and drink but not newspapers.
- 7 bought drink and newspapers but not food.
- 14 bought food and newspapers but not drink.

Construct a Venn diagram for this situation and find the number of customers who bought:

a. Food, drink and newspapers.

b. Food.

c. Drink.

d. Newspapers.

6. In a village, 30 of the 33 families grow some vegetables to sell. 13 of the families grow only kaukau to sell. All the 10 families who grow taro also grow kaukau. There are 4 families who grow kaukau, taro and cassava. No families grow taro only but 3 families grow cassava only. Construct a Venn diagram for this situation. How many families grow:

a. Kaukau and cassava.

b. Kaukau and/or cassava.

c. Cassava.

d. Taro and/or cassava.

Unit 12.1 Patterns and Algebra
Topic 5: Sequences and series

Sequences and series is another method of solving problems. Topic 5 introduces the concepts of a 'sequence' and a 'series' and how they can be used to solve problems involving arithmetic and geometric sequences; in particular:

- Using the Σ notation.
- Exploring problems involving sequences.

Introduction

A **sequence** is an ordered list of numbers. Unless otherwise stated, it is always taken to be infinite (it goes on forever). A sequence associates a **real number** with each natural number.

The real number associated with the natural number 1 is called the *first* **term**.

The number associated with 2 is called the *second* term, and so on.

The number associated with n is called the **nth term** or **general term**.

Example A

The infinite set of numbers: 1, 3, 5, 7, … is a sequence with a 1st term of 1, 2nd term of 3, and 3rd term of 5. The general or nth term is $2n - 1$. The dots at the end of the sequence indicate that the sequence is infinite.

Example B

Q. Consider the following pattern of rectangles:

Write down the first four terms and the general term of the sequences for the following:

1. The number of rectangles.

2. The number of vertices (corners).

3. The number of sides.

A. **1.** 1, 2, 3, 4, …, n, …

2. 4, 6, 8, 10, …, $2n + 2$, …

3. 4, 7, 10, 13, …, $3n + 1$, …

Unit 12.1 Activity 5A: Finding terms of a sequence

1. For each of the following sequences, write down the next two terms and the general term:

a. 2, 3, 4, 5, … **b.** 0, 1, 2, 3, … **c.** 3, 6, 9, 12, …

d. 3, 5, 7, 9, 11, … **e.** –3, –1, 1, 3, … **f.** 1, 2, 4, 8, …

g. 1, 3, 9, 27, … **h.** 2, 6, 18, 54, … **i.** $\frac{1}{2}, \frac{2}{3}, \frac{3}{4}, \frac{4}{5}, \ldots$

j. $0, \frac{1}{2}, \frac{2}{3}, \frac{3}{4}, \ldots$ **k.** 10, 8, 6, … **l.** 16, 8, 4, 2,…

m. $1.00, 0.50, 0.\dot{3}, 0.25, \ldots$ **n.** 1, 4, 9, 16, … **o.** $1, \frac{1}{4}, \frac{1}{9}, \frac{1}{16}, \ldots$

p. 1, 3, 5, 7, 9, … **q.** 2, 5, 10, 17, … **r.** 1.00, 1.41, 1.73, 2.00, …

2. Consider the following pattern of triangles:

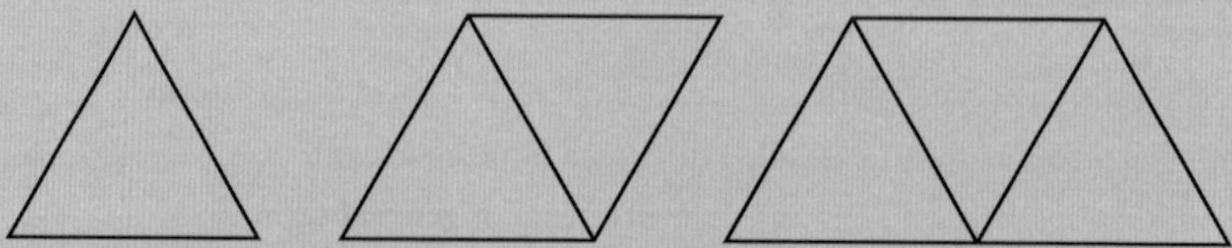

I II III

Write down the first four terms and the general term of the sequences of:

a. The number of triangles.

b. The number of vertices.

c. The number of sides.

3. Consider this pattern of shapes:

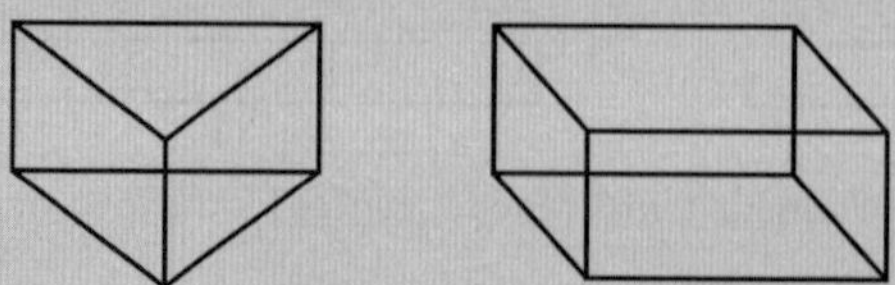

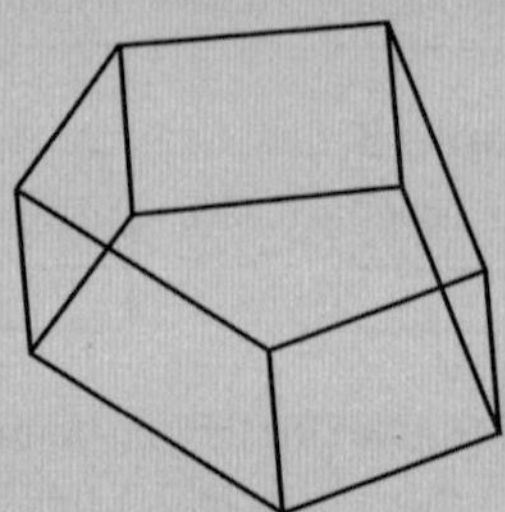

I II III

Write down the first four terms and the general term of the sequences of:

a. The number of vertices (or corners).

b. The number of faces.

c. The number of edges.

Listing sequences

Sequences can be listed in the following ways:

a. **Enumeration** or **listing** the terms:
1, 3, 5, 7, …, $2n - 1$, …

b. Writing down the rule for the nth term:
in **triangular brackets** $<2n - 1>$ or by **formula**, $T_n = 2n - 1$.

Example C

Q. List the first four terms of each of the following sequences:

1. $<n^3 - n^2>$ **2.** $T_n = 3n^2 - n$

A. In each case the terms are worked out by **substitution**.

1. The first term is $1^3 - 1^2 = 0$ [substituting $n = 1$ in $n^3 - n^2$]

The second term is $2^3 - 2^2 = 4$ [substituting $n = 2$ in $n^3 - n^2$]

The third term is $3^3 - 3^2 = 18$ [substituting $n = 3$]

The fourth term is $4^3 - 4^2 = 48$ [substituting $n = 4$]

$\therefore$ the first four terms are 0, 4, 18 and 48.

2. The first four terms are: $3 \times 1^2 - 1$, $3 \times 2^2 - 2$, $3 \times 3^2 - 3$ and $3 \times 4^2 - 4$

$\therefore$ the terms are 2, 10, 24 and 44.

Unit 12.1 Activity 5B: Listing sequences from formulae

1. For each of the sequences given by the following formulae, write down the first 4 terms:

a. $<n^2 + 1>$ **b.** $<2n + 3>$ **c.** $A_n = n^3$

d. $A_n = n^2 + n + 1$ **e.** $<\dfrac{n^2 + 1}{n}>$ **f.** $A_n = (1 + \dfrac{1}{n})^n$

g. $A_n = \dfrac{(-1)^n n}{2^n}$ **h.** $<n^2 + (-1)^n>$

2. Write down the first four terms of these sequences:

a. First term is 30, each term thereafter is 40% larger than its predecessor.

b. First term is 12, each term is 10% larger than its predecessor.

c. Second term is 48, each term is bigger by 7 than its predecessor.

d. Each term is the square of the one before it. The fourth term is 16.

e. Each term is double the one before it. First term is 6.

Limit of a sequence

A sequence $<T_n>$ has a **limit** if there is a number L which the terms of the sequence get closer and closer to as n tends to infinity (ie as n gets very large).
If $<T_n>$ has a limit and the limit is L, then this is written:

$\lim_{n \to \infty} T_n = L$ or $T_n \to L$ as $n \to \infty$ which is read 'T_n tends to L as n tends to infinity.'

Example D

Q. Show that the sequence $T_n = \dfrac{3n + 1}{2n - 2}$ has a limit as n tends to infinity, and find the limit.

A. Using a calculator and substituting gives:

$T_{100} = 1.520\,202$, $T_{1000} = 1.502\,002$ and $T_{10\,000} = 1.500\,200$

[eg $T_{100} = \dfrac{3 \times 100 + 1}{2 \times 100 - 2} = \dfrac{301}{198} = 1.520\,202$]

As n increases, the terms get closer and closer to 1.5.

The limit is thus 1.5 and can be written $T_n \to 1.5$ as $n \to \infty$.

Example E

Q. A person on a diet loses 5 kg the first week, 2.5 kg the next week, 1.25 kg the next week, etc. His weight was initially 105 kg.

1. Write down his body weight at the end of the first 4 weeks.

2. His body weight (in kg) after n weeks on the diet is $95 + \frac{10}{2^n}$. What weight does he get closer and closer to?

A. 1. $105 - 5 = 100$ [subtracting weekly losses from previous weights]

$100 - 2.5 = 97.5$

$97.5 - 1.25 = 96.25$

$96.25 - 0.625 = 95.625$

Body weight at the end of the fourth week is 95.625 kg.

2. Substituting $n = 50$ and $n = 100$ shows that his body weight gets closer and closer to 95 kg. (This is the limit of the sequence.)

Unit 12.1 Activity 5C: Limits of sequences

1. Find the limits as $n \to \infty$, if they exist, of the following sequences:

a. $<2n>$ **b.** $<\frac{2}{n}>$ **c.** $<1 + \frac{1}{n}>$ **d.** $<\frac{2^n}{3^n}>$

e. $<\frac{n}{2^n}>$ **f.** $A_n = (n)^{\frac{1}{2}}$ **g.** $A_n = (n)^{\frac{1}{n}}$

h. $A_n = \frac{3^{n+1}}{2n + 1}$ **i.** $T_n = \frac{2 - n}{1 - 2n}$ **j.** $L_n = \frac{n^2 + n + 1}{2n^2 - n + 2}$

2. A person on a diet loses 9 kg the first month, 6 kg the next, 4 kg the next. Each month she loses 2/3 of the amount she lost in the previous month. Initially her weight was 100 kg.

a. Write down her weight at the end of each of the first 6 months.

b. What is the limit of the sequence of monthly weights?

c. What does this limit represent in practical terms?

3. A company employs 2 workers. It pays them a total of K50 an hour. Each new worker added to the payroll thereafter, gets paid K8 an hour.

a. What is the total number of workers if n new workers have been added?

b. What is the total hourly wage bill if n new workers have been added?

c. What is the average hourly wage if n new workers have been added?

d. Explain why the sequence of average hourly wage rates has a limit of K8 an hour, in practical terms.

Series

A **series** is obtained by *adding the successive terms of a sequence*.

Example F

1 + 2 + 3 + 4 + … is an infinite series

1 + 2 + 3 + 4 + 5 is a finite series

Sigma notation

The Greek letter Σ (sigma) is used to indicate a summation of terms. Numbers below and above the symbol Σ indicate which terms are to be added.

For example,

$\sum_{i=1}^{5} a_i$ means $a_1 + a_2 + a_3 + a_4 + a_5$ and $\sum_{n=3}^{7} u_n$ means $u_3 + u_4 + u_5 + u_6 + u_7$

Example G

$$\sum_{n=1}^{4}(2n+1) = (2 \times 1 + 1) + (2 \times 2 + 1) + (2 \times 3 + 1) + (2 \times 4 + 1)$$
$$= 3 + 5 + 7 + 9$$
$$= 24$$

By establishing a sequential pattern in a series, the sum can be written using **sigma notation**.

Example H

Q. Express 2 + 4 + 6 + 8 + 10 in sigma notation.

A. $2 + 4 + 6 + 8 + 10 = 2 \times 1 + 2 \times 2 + 2 \times 3 + 2 \times 4 + 2 \times 5$

$= \sum_{j=1}^{5} 2j$ [any letter can be used instead of *j*]

Unit 12.1 Activity 5D: Sigma notation

1. a. $\langle a_n \rangle$ is the sequence 1, 3, 1, 3, … Find $\sum_{n=1}^{7} a_n$

b. $\langle b_n \rangle$ is the sequence 1, 1.1, 1.11, 1.111, … Find $\sum_{n=1}^{5} b_n$

c. Find $\sum_{n=3}^{5}\left(\frac{1}{n}\right)$

d. Evaluate $\sum_{n=1}^{16}(-1)^n$

e. $T_n = n^2 + 1$. Evaluate $\sum_{j=1}^{4} T_j$

f. $L_n = (n + 1)^2 - n^2$. Calculate $\sum_{n=2}^{6} L_n$

g. Evaluate $\sum_{n=1}^{7} n^2 - \sum_{n=1}^{6} n^2$

h. Evaluate $\sum_{n=1}^{50} n^2 - \sum_{n=1}^{49} n^2$

2. Write the following using sigma notation:
   a. $3 + 6 + 9$ b. $4 + 6 + 8$ c. $5 + 6 + 7 + 8$
   d. $1 + 4 + 9 + 16$ e. $4 + 8 + 16 + 32$

3. Find the value of:
   a. $\sum_{n=1}^{4}(2 - n)$ b. $\sum_{n=1}^{3} n^3$ c. $\sum_{k=1}^{5} 2k$
   d. $\sum_{j=4}^{7}(2j + 5)$ e. $\sum_{n=1}^{11} 5$

4. Find a if $\sum_{n=1}^{4}(an - 1) = 16$

5. Find k if $\sum_{n=1}^{5} kn^2 = 55$

Unit 12.1 Patterns and Algebra

Topic 6: Sequences and series – arithmetic sequences

Topic 6 continues the exploration of sequences and solving straightforward problems involving arithmetic and geometric sequences:

- General terms of an arithmetic sequence.
- Partial sums of an arithmetic sequence.
- Applications of arithmetic sequences.
- Explore situations and interpret results of problems involving arithmetic sequences.

Arithmetic sequences

An **arithmetic sequence**, often called an **arithmetic progression** (AP), is a sequence in which each term after the first differs from the previous term by a **constant** amount. This amount is called the **common difference** or simply the **difference** and is represented by the letter d.

Example A

2, 5, 8, 11 and 14 are the first five terms of an arithmetic sequence because each term after the first differs by 3 from the term before it ($5 - 2 = 3$, $8 - 5 = 3$, $11 - 8 = 3$ and $14 - 11 = 3$).

Thus $d = 3$ and the sequence is arithmetic.

The general term of an arithmetic sequence

If the first term of an arithmetic sequence is a and the common difference is d, then the sequence is: $a, a + d, a + 2d, a + 3d, \ldots$

Thus the first term has no d's, the second term has 1 d, the third term has 2 d's and so on.

Since the number of the d's is always *one less* than the number of the term, the nth term or general term, t_n, of an arithmetic sequence has the formula:

$$\boxed{t_n = a + (n - 1)d}$$

Example B

Q. Janette is saving money and aims to increase the amount of money she deposits in her savings account by K10 each month. The first month she starts with a deposit of K25.

1. Write down the amounts she deposits during the first five months.

2. How much will Janette be depositing exactly two years after she started?

A. 1. Amounts deposited are K25, K35, K45, K55, K65, …

2. The sequence is arithmetic with $a = 25$ and $d = 10$

$\therefore\ t_n = 25 + (n - 1)10$ [substituting $a = 25$ and $d = 10$]

$\therefore$ Amount deposited $= 25 + (24 - 1)10$ [after 2 years, $n = 24$]

$= \text{K}255$

Example C

Q. The fifteenth term of an arithmetic sequence is 50 and the sixtieth term is 185. Find the first term and the common difference.

A. Let the first term be a and the difference be d. Substituting into the formula $t_n = a + (n - 1)d$ gives:

$a + 14d = 50$	… (1)
$a + 59d = 185$	… (2)
$\therefore\ 45d = 135$	[solving simultaneously, (2) – (1)]
$\therefore\ d = 3$	[dividing by 45]
$a + 42 = 50$	[substituting for d in (1)]
$\therefore\ a = 8$	[subtracting 42]

Unit 12.1 Activity 6A: Terms of arithmetic sequences

The following sequences are arithmetic. Find the indicated terms:

1. a. 2, 5, 8, 11, … t_6, t_{20}, t_n

b. 11, 15, 19, 23, … t_7, t_{11}, t_n

c. 4, 2, 0, –2, … t_6, t_{20}, t_n

d. 1, 1.5 , 2, … t_6, t_{12}, t_n

e. 2, 2.1, 2.2, 2.3, … t_6, t_{12}, t_n

2. a. In an arithmetic sequence, $t_3 = 4$, $t_4 = 8$. Write down the first 6 terms and the nth term.

b. In an arithmetic sequence, $t_5 = 6$, $t_6 = 11$. Write down the first 8 terms and the nth term.

c. In an arithmetic sequence, $t_4 = 5$, $t_7 = 11$. Write down the first 8 terms and the nth term.

d. In an arithmetic sequence, $t_3 = 6$, $t_6 = -3$. Write down the first 8 terms and the nth term.

e. 11 is the 5th term of an arithmetic sequence and –17 is the 12th term. Find the 58th term.

f. $t_7 = 210$ in an arithmetic sequence with a difference of 7. Which term is 448?

g. In an arithmetic sequence with difference of 8, the 4th term is 35. What term is 115?

h. The 12th term of an arithmetic sequence is 29. If the 3rd term is subtracted from the 7th term the result is –8. Find the 52nd term.

i. The 15th term of an arithmetic sequence is 219 and the difference is 3. How many terms are there before the terms of the sequence exceed 1 000?

j. What is the common difference of an arithmetic sequence whose difference is half the first term and whose 12th term is 169?

3. Stephen starts with 100 phonecards in his collection in January 2012 and increases the number by 3 each month. In January 2012 Stephanie starts with 200 and increases the number by 2 each month.

a. How much time will elapse before Stephen has more than 500 cards?

b. i. Will Stephanie ever have 240 cards?

ii. If so, at the end of what month?

c. When will Stephen first have more than Stephanie?

4.

This diagram shows a sequence of pictures made by successively adding layers of triangles.

L_n is the number of triangles in the bottom layer of the nth picture.

J_n is the number of triangles in the nth picture.

S_n is the number of sides in the bottom layer of the nth picture.

P_n is the number of points or vertices in the bottom layer of the nth picture.

W_n is the total number of points or vertices in the nth picture.

a. Draw the next diagram in the sequence.

b. Enumerate the first four terms of each of the sequences $<L_n>$, $<J_n>$, $<S_n>$, $<P_n>$, $<W_n>$. Find the general term for each.

c. Which sequences are arithmetic?

Partial sums of an arithmetic sequence

The sum of the first n terms of a sequence is called the **nth partial sum**. The nth partial sum of an arithmetic progression with first term a and difference d has the following formula:

$$S_n = \frac{n}{2}[2a + (n - 1)d] \text{ or } S_n = \frac{n}{2}(t_1 + t_n)$$

Example D

Q. A company increases the value of goods it exports annually by K86 000 per year. In 2007 it exported K975 000 worth of goods. What will be the total value of all goods exported during the ten years beginning in 2007?

A. Annual exports are an arithmetic sequence with $a = 975\ 000$ and $d = 86\ 000$.

Using the formula $S_n = \frac{n}{2}[2a + (n - 1)d]$ for $n = 10$ years gives:

$$S_{10} = 5[2 \times 975\ 000 + (10 - 1)\ 86\ 000]$$
$$= \text{K}13\ 620\ 000$$

Example E

Q. How many terms of the arithmetic sequence with first term 1 and common difference of 4 should be added together to give a sum of 153?

A. Let the number of terms be n.

$$S_n = \frac{n}{2}[2a + (n - 1)d]$$

$$153 = \frac{n}{2}[2 + (n - 1) \times 4] \quad \text{[substituting } a = 1,\ d = 4,\ S_n = 153]$$

$$306 = n[2 + (n - 1)4] \quad \text{[multiplying by 2]}$$

$= 2n + 4n(n - 1)$ [removing [ ] by expansion]

$= 2n + 4n^2 - 4n$ [expanding]

$\therefore\ 4n^2 - 2n - 306 = 0$ [**rearranging** and simplifying]

$2n^2 - n - 153 = 0$ [dividing by 2]

$\therefore\ (2n + 17)(n - 9) = 0$ [factorising]

$\therefore\ n = 9$ [disregarding $n = -8\frac{1}{2}$ since n is a whole number]

Note: The terms of the sequence could be added until 153 is obtained. The number of terms added to get 153 is then the required answer (the nth term is $4n - 3$).

Unit 12.1 Activity 6B: Partial sums of arithmetic sequences

1. What is the sum of the first 20 terms of the sequence 2, 5, 8, . . .?

2. What is the sum of the first 200 terms of the sequence 3, 6, 9, . . .?

3. An arithmetic sequence has first term of 7 and difference of 5. Find the sum of the first 50 terms.

4. Find the sum of the first ten terms of the arithmetic series 3 + 5 + 7 + . . .

5. Find the sum of the sixteen terms of the series 5 + 7 + . . . + 35.

6. What is the sum of the 20 terms 7 + 10 + 13 + . . . + 64?

7. The sum of the first n terms of an arithmetic series is $50(2 \times 7 + 99 \times 3)$.
- **a.** What is the first term?
- **b.** What is the common difference?
- **c.** How many terms are being added (ie find n)?
- **d.** What is this sum?

8. $25(48 + 49 \times 5)$ is the sum of a number of terms of an arithmetic sequence.
- **a.** How many terms are being added?
- **b.** What is the sum?
- **c.** What is the difference?
- **d.** What is the first term?

9. An arithmetic sequence has first term of 4 and difference of –2.5. Find the sum of the first 30 terms.

10. An arithmetic sequence has 3rd term of 45 and difference of 1.1. Find the sum of the first 10 terms.

11. An arithmetic sequence has 5th term of 56 and difference of –1.1. Find the sum of the first 20 terms.

12. An arithmetic sequence has $t_1 = 6$, $t_4 = 21$. Find the sum of the first 24 terms.

13. An arithmetic sequence has $t_3 = 14$ and $t_{11} = 34$. Find the sum of the first 90 terms.

14. Find $\sum_{n=1}^{50}(2 + 3n)$

15. Find $\sum_{n=1}^{60}(4 - 7n)$

16. Find $\sum_{n=1}^{60}(3n + 1)$

17. Find $\sum_{n=15}^{35}(2n + 3)$

18. The first term of a finite arithmetic sequence is 20 and the last is 50. The sum of this series is 385. How many terms does it have?

19. An arithmetic sequence has first term 8 and difference of 3. How many terms must be added in order that the sum exceeds 65?

20. An arithmetic sequence has first term 5. The sum of its first 5 terms is 45. Find the common difference.

21. A very successful sales representative makes 35 sales in the first month of employment and for the next two years increases the number of sales each month by 4. Over the two-year period, how many sales were made?

22. After a scandal involving senior members of a political party, the number of people who resign from the party each week increases by 26, on average, per week. The first week after the scandal first broke there were 134 resignations. This pattern continued for 14 weeks. How many people resigned during that period?

23. A runner on a special program seeks to increase the distance run by 12 km each week. During the fourth week 61 km are run. This program is followed for 12 weeks. What is the total distance run during the period?

Unit 12.1 Patterns and Algebra

Topic 7: Sequences and series – geometric sequences

Topic 7 continues the exploration of sequences and solving straightforward problems involving arithmetic and geometric sequences:

- General terms of a geometric sequence.
- Partial sums of a geometric sequence.
- The sum to infinity of a geometric sequence.
- Applications of geometric sequences such as compound interest.
- Explore situations and interpret results of problems involving geometric sequences.

Geometric sequences

A **geometric sequence** or **geometric progression** (GP) is a sequence where each term after the first gives a constant value when *divided* by the previous term. The value obtained when dividing is called the **common ratio** (or simply the **ratio**) of the GP and is usually represented by the letter r.

Example A

The sequence $-2, 1, \frac{-1}{2}, \frac{1}{4}, \ldots$ is a geometric sequence because each term, when divided by the previous term, is equal to $\frac{-1}{2}$ $\left(1 \div -2 = \frac{-1}{2}, \frac{-1}{2} \div 1 = \frac{-1}{2}, \frac{1}{4} \div \frac{-1}{2} = \frac{-1}{2}\right)$.

Thus $r = \frac{-1}{2}$ and the sequence is geometric.

The general term of a geometric sequence

The geometric sequence with first term a and common ratio r is the sequence:

$$a, ar, ar^2, ar^3, \ldots$$

From this it can be seen that the **power** of r is always *one less* than the number of the term (eg the 4th term is ar^3).

Hence, the general term or nth term, t_n, of the geometric sequence with first term a and ratio r is:

$$\boxed{t_n = ar^{n-1}}$$

Example B

Q. In a certain country, **inflation** has been running at 14% for a number of years. After one year of inflation, a pair of shoes cost K30.00.

a. Find the price of this pair of shoes after 2 years and after 3 years.

b. Show that these prices are in geometric progression.

c. Find the common ratio and hence the price after n years.

d. Find how many full years will elapse before the cost of a pair of shoes will exceed K1 000.

A. a. After one year the price is K30.00.

After two years the price will be K30.00 + 14% of K30.00 = K34.20.

After three years the price is K34.20 + 14% of K34.20 = K38.99.

b. $\frac{38.99}{34.20} = 1.14$ and $\frac{34.20}{30} = 1.14$, thus the ratios are constant and therefore prices are in geometric progression.

c. $r = 1.14$. So after n years the price will be $30(1.14)^{n-1}$

d. Letting n be the number of years that have elapsed when the price reaches K1 000.00, then:

$$30(1.14)^{n-1} = 1\,000 \qquad \text{[since } t_n = ar^{n-1}\text{]}$$

$$\therefore\ (1.14)^{n-1} = \frac{100}{3} \qquad \text{[dividing by 30]}$$

$$\therefore\ \log(1.14^{n-1}) = \log\left(\frac{100}{3}\right) \qquad \text{[taking logs]}$$

$$\therefore\ (n-1)\log(1.14) = \log\left(\frac{100}{3}\right) \qquad [\log A^n = n \log A]$$

$$\therefore\ n - 1 = \frac{\log\left(\frac{100}{3}\right)}{\log(1.14)} \qquad \text{[dividing by log (1.14)]}$$

$$\therefore\ n = \frac{\log\left(\frac{100}{3}\right)}{\log(1.14)} + 1 \qquad \text{[adding 1]}$$

$$= 27.76 \text{ years}$$

∴ after 28 full years the price will exceed K1 000.

Note: Using the formula, price = $30\,(1.14)^{n-1}$, a **trial-and-error** method (substituting likely values for n) could also be used to solve the problem.

Unit 12.1 Activity 7A: Terms of a geometric sequence

1. Which of the following sequences are geometric? State the common ratio of each geometric sequence.

a. 5, 10, 20, 40, … **b.** 5, 7, 9, 11, 13, … **c.** 10, 5, 2.5, 1.25, …
d. 10, 12, 14.4, 17.28, … **e.** –1, 1, –1, 1, –1, 1, … **f.** –2, 3, –4, 5, –6, …
g. –2, 4, –8, 16, –32, … **h.** –2.2, 4.4, –8.8, 16.16, –32.32, …

2. For each of these geometric sequences, find the ratio and the next term of the sequence.

a. 5, 15, 45, 135, … **b.** –5, 15, –45, 135, … **c.** 135, 45, 15, 5, …
d. 300, 90, 27, 8.1, 2.43, … **e.** 2.43, –8.1, 27, –90, 300, …

3. For the following geometric sequences, find the next two terms and the general term.

a. 4, 16, 64, … **b.** –3, 15, –75, … **c.** 2, 2.4, 2.88, 3.456, …
d. 100, 25, 6.25, … **e.** –100, 50, –25, 12.5, …

4. The following sequences are geometric. Find the indicated terms and the nth term:

a. 4, 2, 1, … t_6, t_{10}, t_n **b.** 8, 24, 72, … t_6, t_9, t_n
c. –1, 2, –4, 8, … t_{10}, t_{13}, t_n **d.** $3, \frac{-2}{3}, \frac{4}{27}, \ldots$ t_4, t_8, t_n
e. $\frac{3}{4}, 3\frac{3}{4}, 18\frac{3}{4}, \ldots$ t_4, t_6, t_n

5. The following questions refer to geometric sequences:

a. $t_3 = 5$, $t_4 = 15$. Write down the first 6 terms and give the nth term.

b. $t_2 = 6$, $t_3 = 12$. Write down the first 6 terms and the nth term.

c. $t_5 = 6$, $t_6 = 9$. Find t_2, t_8 and t_n

d. $t_4 = 2$, $t_5 = 3\frac{1}{3}$. Find t_1 and the nth term.

e. $t_1 = 4$, $t_3 = \frac{4}{9}$. Find t_7 and t_n

f. $t_4 = 11$, $t_8 = 176$. Find t_1 and t_n

g. $t_4 = -3$, $t_7 = 0.375$. Find t_2 and t_n

h. $t = 50$, $t = 75$. How many terms of this GP are less than 500?

i. $t_4 = 12$, $t_6 = 21\frac{1}{3}$. What is the last term less than 50?

j. $t_5 = 7$ and the ratio is 1.5. What term is 35.4375?

k. $t_4 = 8$, $t_7 = 125$. What term is 1 953.125?

l. A GP has first term 6 and fourth term 279.936. Find the common ratio.

6. Money placed in an investment fund increases in value by 11% a year. Initially K5 000 was invested.

a. Write out the value of the investment at the beginning of each of the first five years.

b. How long will it take for this investment to reach or exceed K1 000 000? (Give answer to the nearest year.)

7. The population in a country increases on average by 1.3% a year. In 1991 the population is 2 150 000. Find what the population will be in 2021 if this rate of growth is maintained.

8. A business depreciates the value of a computer by 23% per year. The computer originally cost K10 000.

a. If the business owning the computer sold it after 10 years, what would be its value in the records of the business?

b. If the business had a policy of selling computers once their depreciated value was less than a quarter of the cost price, how many years would the business retain the computer before selling it?

Sum of the first *n* terms of a geometric sequence

The sum to n terms (the nth partial sum) of a geometric sequence with first term a and ratio r is:

$$\boxed{S_n = \frac{a(r^n - 1)}{r - 1}} \quad \text{where } r \neq 1$$

Note: If $-1 < r < 1$, the alternative form, $S_n = \frac{a(1 - r^n)}{1 - r}$ can be easier to use.

Example C

Q. Find the sum of the first 25 terms of the geometric series 3 + 6 + 12 + …

A. The required sum is given by the formula for S_n

$$S_{25} = \frac{a(r^{25} - 1)}{r - 1} \quad \text{where } r = 2,\ a = 3 \text{ and } n = 25$$

$$= \frac{3(2^{25} - 1)}{2 - 1}$$

$$= 100\,663\,293$$

Geometric sequences and series have many practical **applications**.

Example D

Q. In 1996 the number of tourists going to a city was 25 000 per year. The annual number of tourists has been increasing by 16% every year since then. Find the total number of tourists who went to the city in the decade starting from 1996.

A. The number of tourists who visit the city each year is a geometric progression with $t_1 = 25\,000$ and $r = 1.16$. The total number who have visited in ten years is:

$$S_{10} = \frac{a(r^{10} - 1)}{r - 1}$$

$$= \frac{25\,000(1.16^{10} - 1)}{1.16 - 1} \qquad \text{[substituting]}$$

$$= 533\,037$$

Unit 12.1 Activity 7B: Partial sums of geometric sequences

1. Calculate:

a. $\frac{4(3^{12} - 1)}{3 - 1}$ **b.** $\frac{8\left(1 - (0.2)^6\right)}{1 - 0.2}$ **c.** $\frac{-5\left((-2)^{12} - 1\right)}{-2 - 1}$

2. The partial sum of a geometric sequence is given by the calculation $\frac{12\left((1.5)^8 - 1\right)}{(1.5 - 1)}$

a. What is the first term of the GP?

b. What is the common ratio of the GP?

c. How many terms are being added from the GP?

d. Evaluate the sum.

3. What is the sum of the first 10 terms of the sequence 3, 6, 12, … ?

4. What is the sum of the first 10 terms of the sequence 1, 1.5, 2.25, 3.375, … ?

5. How many terms of the sequence 4, 3, $2\frac{1}{4}$, … can you add before the sum exceeds 12?

6. How many terms of the sequence 2, 14, 98, … can you add before the sum exceeds 10 000?

7. A geometric sequence has first term 4 and second term 6. Find the sum of the first 10 terms.

8. A geometric sequence has second term 11 and third term 5.5. Find the sum of the first 50 terms.

9. A house costs on average K15 000 in 1990 and increases in value by 10% every year. A man buys a house in 1990 then another one every year after that. How much money does he spend in total from 1990 to 2006?

10. A man on a diet loses 1.5% of his body weight during each week.

a. If he initially weighs 150 kg, write down his body weight at the end of each of the first 5 weeks.

b. How much does he lose in total during that time?

11. Each week a weightlifter increases the total weight he lifted in the previous week by 0.95%. He keeps up this increase for 10 weeks. In the first week he lifts a total of 50 kg. How much in total does he lift during the 10-week period, to the nearest kg?

12. A PNG company which exports bags of coffee to the USA manages to increase its annual exports by 5% each year. During the first year it sells 1 500 bags of coffee. How many bags of coffee does it export in its first 12 years of operation?

13. Find $\sum_{n=1}^{50} 3(1.1)^n$

14. A geometric sequence with first term 2 and ratio 3 has its terms added. How many terms are added to get a sum of 177 146?

15. A geometric sequence with first term 6 and ratio 4 is summed. How many terms are added to get a sum of 2 097 150?

Sum to infinity of a geometric sequence

If successive terms of an arithmetic progression, such as 1, 3, 5, 7, … are added together, the sum gets larger and larger without limit.

Similarly, in a geometric sequence where the ratio r *exceeds 1 in magnitude*, successive additions of the terms will yield a sum that gets bigger and bigger without limit.

Example E

Successive additions of the terms of the geometric sequence 1, 2, 4, 8, … give a sum 1 + 2 + 4 + 8 + … which tends to infinity. Thus if the sum to n terms is denoted by S_n, then $S_n \to \infty$ as $n \to \infty$.

If $-1 < r < 1$, then the sum to infinity of a geometric sequence exists (is finite).

Example F

Q. Show that the geometric sequence with first term 3 and ratio 0.1 has a sum to infinity and find its value.

A. $S_n = \dfrac{3[1-(0.1)^n]}{1-0.1}$ [using $S_n = \dfrac{a(1-r^n)}{1-r}$ since $-1 < r < 1$]

$= \dfrac{3[1-(0.1)^n]}{\frac{9}{10}}$ [rearranging]

$= \dfrac{10}{3}[1-(0.1)^n]$ [rearranging]

As n increases, $(0.1)^n$ decreases to zero and so S_n gets closer and closer to $\dfrac{10}{3}$. The geometric sequence has a sum to infinity of $\dfrac{10}{3}$ which is written $S_n \to \dfrac{10}{3}$ as $n \to \infty$ or $S_\infty = \dfrac{10}{3}$

Generally the sum to infinity of a sequence $<t_n>$ is denoted by either S_∞ or $\sum_{n=1}^{\infty} t_n$

> If $<t_n>$ is a geometric sequence with $-1 < r < 1$ then $S_\infty = \dfrac{a}{1-r}$

Example G

The sequence $3, -1, \frac{1}{3}, -\frac{1}{9}, \ldots$ is a geometric sequence with ratio $-\frac{1}{3}$.

Since $-1 < -\frac{1}{3} < 1$, the sum to infinity exists and has the value:

$S_\infty = \dfrac{a}{1-r}$

$= \dfrac{3}{1--\frac{1}{3}}$ [substituting $a = 3$, $r = -\frac{1}{3}$]

$= 2.25$

Unit 12.1 Activity 7C: Sum to infinity of a geometric sequence

1. Which of the following geometric sequences have a sum to infinity?

a. 2, 3, 4.5, … **b.** 2, 1, 0.5, 0.25, … **c.** 4, 3, 2.25, …

d. 2.5, 3.5, 4.9, … **e.** 4.9, 3.5, 2.5, …

2. Find the following sums to infinity:

a. $\sum_{n=0}^{\infty} 2\left(\frac{1}{3}\right)^n$ **b.** $\sum_{n=1}^{\infty} 3\left(\frac{2}{3}\right)^n$ **c.** $3 + 0.9 + 0.27 + \ldots$

d. $-4 + 1.5 + -0.5625 + \ldots$ **e.** $\frac{1}{3} + \frac{1}{6} + \frac{1}{12} + \ldots$

3. What is the sum to infinity of a geometric sequence with:

a. First term of 5, ratio of 0.6 **b.** First term of 7, ratio of 0.35

c. 3rd term of 6, ratio of $\frac{2}{3}$ **d.** 4th term of 18, ratio of 0.72

e. First term of 8, 4th term of 3.375

4. A geometric progression with first term 5 and a sum to infinity of 8 has an unknown ratio. Find that ratio.

5. Each year a country exports two-thirds the quantity of the previous year's coffee exports. In 1970 it exported 25 000 tonnes. If it continued to export in this way, what would be the total weight of its coffee exports over all the years from 1970 onwards?

6. A man employs many people. On his birthday he decides to give each employee a gift of money. He gets one of his executives to put the employees in alphabetical order. The first in line he gives K1 000, then the next gets K750, the next K562.50, and so on. If it were possible to pay in this way, what would be the total value of his gifts?

7. A company hires a new accountant who is able to reduce the company's unnecessary expenditure by about 50% every year. When she was first hired, such expenditure was K813 000 per year.

a. Write down the unnecessary expenditure for the first 3 years of her employment.

b. After hiring the accountant, how much can the company expect to spend unnecessarily in the future?

8. A fishing village acquires some new boats and in the first year of use the villagers catch 19 683 tonnes of fish. The next year they catch only one-third of this amount. In each year following they only catch one-third of what they caught the year before. This pattern continues.

a. Find the amount of fish they catch each year for the first 5 years.

b. Calculate the total tonnage caught if they were to go on fishing like this until there were no more fish to catch.

9. The birth rate of a country is falling by 5% per year. In 1961 there were 65 000 births. Find how many babies will be born in the country from the beginning of 1961 if this **trend** continues.

10. When Kate rinses her hair after washing it with a shampoo (or a liquid soap) she removes a quantity of the shampoo. The first rinse removes 5 g of shampoo. Each successive rinse removes 20% of the amount removed by the previous rinse.

a. How much shampoo is removed at the second rinse?

b. Write down the sequence of amounts of shampoo removed at the first 5 rinses.

c. The sequence in **b.** is a geometric sequence. Explain why.

d. Write down an expression, in terms of n, for the amount of shampoo, t_n, removed in the nth rinse.

e. How much shampoo was present in Kate's hair at the beginning of the wash?

Unit 12.1 Patterns and Algebra
Topic 8: Applications of sequences and series

In line with the focus of Unit 12.1, Topic 8 looks at the application of geometric sequences to real-life problems.

There are numerous situations that can be modelled using geometric sequences. Population growth, compound interest and depreciation are some of the more familiar ones.

Example A

Compound interest

Q. K10 000 is invested in an account paying 6.5% compound interest per year.

a. Find the value of the investment after 1, 2 and 3 years.

b. Show that these values form a geometric sequence and find *a*, *r* and an expression for V_n (the value of the investment after *n* years).

c. Find the value of the investment after the 7th year.

d. After how many years is the investment worth at least K20 000?

A. a. After the 1st year: $10\,000 + 6.5/100 \times 10\,000 = 10\,650$

After the 2nd year: $10\,650 + 6.5/100 \times 10\,650 = 11\,342.25$

After the 3rd year: $11\,342.25 + 6.5/100 \times 11\,342.25 = 12\,079.496$

b. $11\,342.25/10625 = 1.065$; $12079.496/11\,342.25 = 1.065$

The ratio of successive terms is the same, hence the terms are for a geometric sequence with $a = 10650$ and $r = 1.065$

c. $V_n = 10650(1.065)^{n-1}$

$V_7 = 10650 \times 1.065^{7-1}$

$= 10650 \times 1.065^6$

$= 15539.865$

The value of the investment after 7 years is K15 539.87

d. Find *n* such that $V_n > 20000$

ie. solve $10650(1.065)^{n-1} > 20000$ for *n*

This can be solved using trial and error, logarithms or on some graphing calculators that have a sequence mode.

i. Trial and error (guess and check):

Trying values of *n* in $10650(1.065)^{n-1}$

$10650(1.065)^{n-1} = 19991.51$; too low $((n - 1) = 10$, so $n = 11)$

$10650(1.065)^{11} = 21\,290.96$; higher $n - 1 = 11$ gives $n = 12$

After 12 years the value of the investment is at least K20 000.

ii. Using **logarithms**:

$10650(1.065)^{n-1} > 20000$

simplifies to

$(1.065)^{n-1} > 1.8779$

Taking $\log_{10}$ of both sides:

$\log_{10} 1.065^{n-1} > \log_{10} 1.8779$

$(n - 1) \log_{10} 1.065 > \log_{10} 1.8779$

$n - 1 > \dfrac{\log_{10} 1.8779}{\log_{10} 1.065}$; $n > 10.006$

$n > 10.006 + 1$

$n > 11.006$ so $n = 12$

After 12 years the value of the investment is at least K20 000.

Example B

Q. The **table** below gives the population in a town for the years 2001 to 2003:

2001	2002	2003
3 620	3 801	3 991

a. Show that the population for the three years represents the terms of a geometric sequence and find a and r and an expression for P_n (the population n years after the year 2000).

Assuming that the town's population grows in the same manner:

b. Calculate the population in 2007.

c. When will the population of the town reach 6 000 people?

A. a. 3 801/3 620 = 1.05; 3 991/3 801 = 1.05

The ratio of successive terms is the same, therefore the terms form a geometric sequence: $a = 3\,620$; $r = 1.05$ and $P_n = 3\,620 \times (1.05)^{n-1}$

b. The population in 2007 is P_7

$P_7 = 3\,620 \times (1.05)^{7-1}$

$= 3\,620 \times (1.05)^6$

$= 4\,851.15$

The population will be 4 851 in 2007.

c. Solve $P_n \geq 6\,000$ for n:

Substituting values of n in $3\,620 \times (1.05)^{n-1}$

$3\,620 \times (1.05)^9 = 5\,615.8$ (below 6 000)

$3\,620 \times (1.05)^{10} = 5\,896.6$ (below 6 000)

$3\,620 \times (1.05)^{11} = 6\,191.4$ (above 6 000); so $n - 1 = 11$, giving $n = 12$

Hence the population reaches 6 000 people twelve years after the year 2000, ie. in the year 2012.

Example C

Q. Maria's grandparents have decided to invest K100 for her on her first birthday and to make a similar investment on the following birthdays. They are going to increase this amount by 3% each year after the first investment, to allow for inflation.

a. Calculate the amounts that will be invested for Maria on her 1st, 2nd and 3rd birthdays.

b. Find a and r for this geometric sequence and a general expression for A_n (the amount invested on the nth birthday).

c. How much will be invested on Maria's 15th birthday?

d. On which birthday will they be first making an investment of K200 or more?

e. How much have the grandparents invested for Maria up to and including her 21st birthday?

A. a. 1st birthday: K100

2nd birthday: 100 + 3/100 × 100 = K103

3rd birthday: 103 + 3/100 × 103 = K106.09

b. $a = 100$, $r = 103/100 = 1.03$ $(106.09/103 = 1.03)$

$A_n = 100(1.03)^{n-1}$

c. $A_{15} = 100(1.03)^{15-1}$

$= 100(1.03)^{14}$

$= 151.26$

They will be investing K1 512.59 on Maria's 15th birthday.

d. Solve $A_n \geq 200$

ie $100(1.03)^{n-1} \geq 200$

This can be solved using trial and error, logarithms or on your calculator.

To use (i) trial and error or (ii) logarithms, it is easier to work with a simplified expression:

$$100(1.03)^{n-1} \geq 200$$

Divide both sides by 100: $n(1.03)^{n-1} \geq 200/100$

$$(1.03)^{n-1} \geq 2$$

i. Try $n = 10$: $1.03^9 = 1.305$ (below)

Try $n = 20$: $1.03^{19} = 1.754$ (below)

Try $n = 24$: $1.03^{23} = 1.974$ (below)

Try $n = 25$: $1.03^{24} = 2.032$ (above)

Hence $n = 25$

ii. Using logarithms:

Solve $1.03^{n-1} = 2$

$\log(1.03)^{n-1} = \log 2$

$(n - 1)\log(1.03) = \log 2$

$n - 1 = \log 2/\log(1.03)$

$n - 1 = 23.44$

hence $n = 24.44$

They will first be making an investment of K200 or more on Maria's 25th birthday.

e. The total amount that the grandparents have invested for Maria in the 21 years is:

$$S_{21} = \frac{100(1.03^{21} - 1)}{1.03 - 1}$$

$= 2\,867.485...$

The grandparents have invested K2 867.65.

Example D

Q. It is estimated that a company's computer equipment will depreciate by 30% per year. The company initially invested K24 000 in the equipment.

a. Calculate the value of the equipment after each of the first three years.

b. Establish a and r for this geometric sequence and find an expression for V_n (the value of the equipment after n years).

c. What is the value of the equipment after 8 years?

d. If the equipment is to be replaced when its value falls below K5 000, when will this happen?

A. a. At the end of the 1st year the value is 24 000 – 30/100 × 24 000 = K16 800

At the end of the 2nd year: 16 800 – 30/100 × 16 800 = K11 760

At the end of the 3rd year: 11 760 – 30/100 × 11 760 = K8 232

b. $a = 16\ 800$, $r = 11\ 760/16\ 800 = 8\ 232/11\ 760 = 0.7$

$V_n = 16\ 800 \times (0.7)^{n-1}$

c. $V_8 = 16\ 800 \times (0.7)^{8-1}$

$= 16\ 800 \times (0.7)^7$

$= 1\ 383.55$

The company's computers are worth K1 383.55 after 8 years.

d. Find n, such that $V_n < 5\ 000$

Solve $16\ 800 \times (0.7)^{n-1} < 5\ 000$ (divide by 16 800)

$(0.7)^{n-1} < 0.2976$

Trial and error gives: $0.7^4 = 0.2401$

$0.7^3 = 0.343$

hence $(0.7)^{n-1} < 0.2976$ when $n - 1 = 4$, ie $n = 5$

The company's computer equipment will be replaced after 5 years.

Example E

Q. The cost of a phone call is K1 for the first minute. The cost per minute decreases by 5% for each minute thereafter.

a. Calculate the cost of a phone call in each of the 2nd, 3rd and 4th minutes. Give answers correct to 3 decimal places.

b. Establish an expression for the cost of a phone call in the nth minute.

c. What is the cost of a phone call in the 10th minute?

d. What is the total cost of a phone call lasting 15 minutes?

e. A phone call costs K12.83. How long was the phone call?

f. What is the **maximum** price of a phone call using this price structure?

A. a. 2nd minute: 1 – 5/100 × 1 = 0.95

3rd minute: 0.95 – 5/100 × 0.95 = 0.903

4th minute: 0.9025 – 5/100 × 0.9025 = 0.857

The terms form a geometric sequence with $a = 1$ and $r = 0.95/1 = 0.9025/.95 = 0.95$

b. Cost of a phone call in the nth minute is C_n, then $C_n = 1 \times (0.95)^{n-1} = (0.95)^{n-1}$

c. $C_{10} = 0.95^{10-1} = 0.95_9 = 0.63$

A phone call costs 63 cents in the 10th minute.

d. The *total cost* of a 15-minute phone call means the sum of the first 15 minutes of the phone call.

$$S_{15} = \frac{1(1 - (0.95)^{15})}{1 - 0.95}$$

$$= \frac{(1 - 0.95^{15})}{0.05}$$

$$= 10.734$$

The cost of a 15-minute phone call is K10.73

e. The total cost is K12.83. Therefore we need to solve $S_n = 12.83$ for n:

$$S_n = 12.83$$

$$12.83 = \frac{1(1 - (0.95)^n)}{1 - 0.95}$$

$$12.83 = \frac{1 - 0.95^n}{0.05}$$

$$12.83 \times 0.05 = 1 - 0.95^n$$

$$0.95^n = 1 - 12.83 \times 0.05$$

$$0.95^n = 0.3585$$

This equation can be solved using:

i. trial and error

ii. logarithms, or on a graphing calculator

i. trial and error

$0.95^{10} = 0.599$

$0.95^{15} = 0.4633$

$0.95^{20} = 0.3585$

$n = 20$

ii. logarithms

$\log(0.95^n) = \log(0.3585)$

$n \log(0.95) = \log(0.3585)$

$n = \log(0.3585)/\log(0.95)$

$= 20$

A phone call costing K12.83 lasted for 20 minutes.

f. The sequence 1, 0.95, 0.9025, ... is a geometric sequence with $r = 0.95$.

Since $-1 < r < 1$ it is possible to sum this sequence to infinity and so S_∞ will be the maximum cost of a phone call.

$$S_\infty = \frac{a}{1 - r}$$

$$= \frac{1}{1 - 0.95}$$

$$= \frac{1}{0.05}$$

$$= 20$$

The maximum cost of a phone call under this pricing structure is K20.

Unit 12.1 Activity 8A: Applications of geometric sequences

1. K6 000 is invested at 8% per annum compound interest.

a. Find the value of the investment after the 1st, 2nd and 3rd years.

b. Show that these values form a geometric sequence and find an expression for V_n, the value of the investment after n years.

c. What is the investment worth after 8 years?
d. When will the initial investment have doubled in value?

2. Find the value of an investment of K15 000 after 7 years if compound interest of 7.5% per annum is paid.

3. The population of a town is decreasing as shown in the table below:

1990	1991	1992
18 125	16 675	15 341

a. Show that these population figures form the terms of a geometric sequence and find a and r for this sequence.
b. If the population continues to decrease in a similar manner, find the population in 1996.
c. In which year will the population fall below 10 000?
d. What is the total decrease in population from 1990 to 1995?

4. The population of a particular type of bacteria has been found to double every 12 hours. 9 of these bacteria have been isolated and are placed in an environment where they can reproduce.
a. Find an expression for B_n, the number of bacteria present at the beginning of the nth 12-hour period.
b. How many bacteria will be present after three days?
c. How long before the colony contains at least 10 000 bacteria?

5. Anthony is saving for an overseas trip in three years' time. He is going to make monthly deposits in an account, starting with K30 in the first month, and increasing these deposits by 10% each month thereafter.
a. Find the amount Anthony deposits in the account in each of the first three months and establish an expression for A_n, the amount he deposits in the account in the nth month.
b. What amount will he be depositing in the 24th month?
c. How much money will he have deposited in the account after three years?

6. A manufacturing company has bought plant equipment for K108 000. The value of this equipment will depreciate (lose value) by 15% each year.
a. Find the value of the equipment after each of the first three years and establish an expression for the value of the equipment after n years.
b. What is the value of the equipment after 5 years?
c. The equipment will be replaced when its value falls below K25 000. When will the equipment be replaced?

7. A company plans to increase its production by 20% each year. If its present production is 5 000 units:
a. Find the company's production after the 1st, 2nd and 3rd years and an expression for the production after n years.

b. After how many years will the production have doubled?
c. How many units has the company produced in the 10 years since it started to increase production?

8. A tree has grown 1 metre, 0.6 metres and 0.36 metres in its first three years. If the tree continues to grow at this rate:
a. How much will it grow in the 5th year?
b. What will be the height of the tree after 5 years?
c. What is the maximum height the tree can be expected to reach?

9. A ball, thrown from a height of 1.4 metres, rebounds to 1.26 metres after its first bounce and then continues to rebound to 90% of its previous height on subsequent bounces.
a. How high will it bounce after it has hit the ground 4 times?
b. How far has the ball travelled when it hits the ground for the 6th bounce?
c. Remember that the ball travels up and down between bounces.
d. What is the total distance the ball can be expected to travel before it comes to rest?

10. Prize money of K10 000 is divided amongst 10 competitors so that 2nd prize is half of 1st prize, 3rd prize is half of 2nd prize and so on. Find the value of 1st and 10th prizes.

Contrasting arithmetic and geometric sequences

A practical comparison of arithmetic and geometric sequences involves simple interest and compound interest:

- *Simple interest*: The same amount, a percentage of the initial investment, is added each time period. The sequence formed by the value of the investment after each time period is an arithmetic sequence and if this sequence is graphed the points are in a straight **line**.
- *Compound interest*: An increasing amount, which is a set percentage of the value of the investment in the previous time period, is added each time period. The sequence formed by the value of the investment after each time period is a geometric sequence and if these points are graphed they follow an exponential curve.

Example F

Q. Stephanie has two options for investing her K1 000 savings:

Option A: An account paying 10% simple interest per year.

Option B: An account paying 8% compound interest per year.

a. How much interest will she earn each year if she chooses Option A?

b. Establish an expression for the value of Stephanie's investment at the beginning of the *n*th year under Option A.

c. Find the value of Stephanie's investment at the beginning of the first three years under Option B and establish an expression for the value of her investment at the beginning of the *n*th year using this option.

d. Graph the two options on the same set of axes for $1 \leq n \leq 10$

e. After how many years is the investment using Option B worth more than the investment using Option A?

A. a. 10% of the initial investment: 10/100 × 1 000 = 100. She will earn K100 each year under Option A.

b. If V_n is the value at the beginning of the *n*th year then $V_1 = 1\ 000$, $V_2 = 1\ 100$, $V_3 = 1\ 200$ etc: the terms of an arithmetic sequence with $a = 1\ 000$, and $d = 100$.

$V_n = 1\ 000 + (n - 1)100$

$= 900 + 100n$

c. If C_n is the value of the investment at the beginning of the *n*th year, then:

$C_1 = 1\ 000$

$C_2 = 1\ 000 + 8/100 \times 1\ 000 = 1\ 080$

$C_3 = 1\ 080 + 10/100 \times 1\ 080 = 1\ 166.4$ etc: the terms of a geometric sequence with $a = 1\ 000$ and $r = 1.08$

$C_n = 1\ 000(1.08)^{n-1}$

d. Using the calculator, set up a table of values for both sequences.

The **graph** of the two sequences is below:

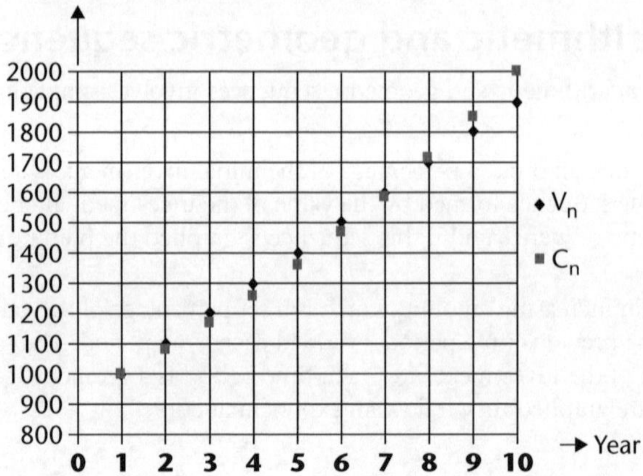

The graph of V_n is a straight line, whereas the graph of C_n is an exponential curve.

e. From the graph (and your table) it can be seen that the value of the investment is more under Option B at the *beginning* of the 8th year. Hence the investment using Option B is more *after* 7 years.

Unit 12.1 Activity 8B: Contrasting arithmetic and geometric sequences

1. a. On the same set of axes, graph the first 8 terms of:

i. A_n: an arithmetic sequence with $a = 1$ and $d = 5$

ii. G_n: a geometric sequence with $a = 1$ and $r = 2$

b. Find the least value of n, such that $G_n > A_n$

2. Match the graphs below to the sequences:

a. Geometric with $-1 < r < 0$

b. Geometric with $r > 1$

c. Arithmetic with $d > 0$

d. Geometric with $r < -1$

e. Arithmetic with $d < 0$

f. Geometric with $0 < r < 1$

A.

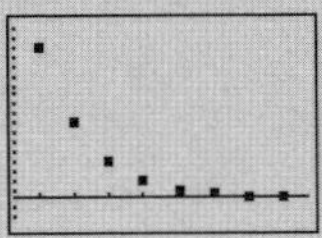

B.

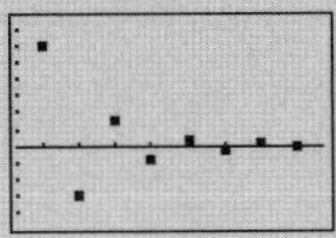

C.

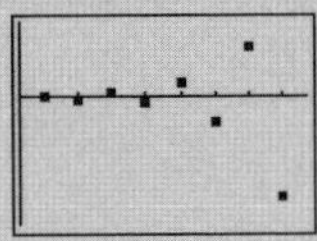

E.

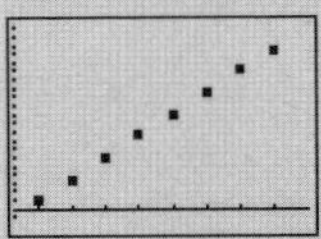

F.

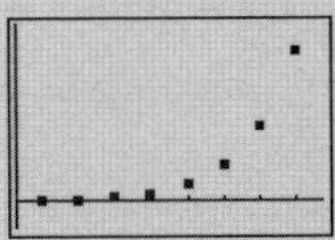

F.

3. Kila wants to invest her K5 000 savings for at least 5 years and she has the choice of an account paying 9% simple interest or an account paying 8% compound interest. Which account should she choose? How much more will she have after 5 years if she chooses this account?

4. On his 10th birthday Moses is offered the following options for his weekly pocket money:

Option A: K5 plus K2 for each year over age 10.

Option B: K5 plus 20% increase each year after the age of 10.

a. Establish a general expression for each option that gives the amount of weekly pocket money paid in the nth year after his 10th birthday.

b. For both options, use your calculator to generate a table of weekly pocket money values from age 10 to 21.

c. Plot these values on the same set of axes.

d. At what age does Option B pay more than Option A?

e. How much more is paid per week at age 21 with Option B?

f. For each option calculate the total amount paid to Moses from age 10 up to his 21st birthday.

5. A comparison of the cost of phone calls to a particular city from two service providers has revealed the following cost structures:

A: K1 for the first minute, reducing by 5 toea for each subsequent minute until the cost per minute is zero; thereafter the cost is zero.

B: K1 for the first minute, reducing by 5% for each minute thereafter.

Giving your answers to the nearest cent:

a. Find the cost of a phone call in the fifth minute for each cost structure.

b. Find the cost of a 3-minute phone call for each cost structure.

c. Establish a general expression for the cost of a phone call lasting n minutes for each cost structure.

d. After how many minutes is the total cost of a phone call cheaper using cost structure B?

e. After how many minutes is the cost per minute zero with cost structure A?

f. Find the maximum cost for a phone call when using each cost structure.

6. The five participants in a golf competition are to receive a share of the total prize pool of K10 000.

Two proposals are produced:

A: The prizes are to form an arithmetic sequence with 5th prize being K500 (1st prize naturally receiving the most money).

B: The prizes are to form a geometric sequence with the fifth prize being K500 (1st prize gets the most).

Find the prizes allotted in each case.

Unit 12.1 Patterns and Algebra

Topic 9: Binomial Theorem

Continuing the focus on different methods of solving real-life problems, Topic 9 looks at the Binomial Theorem. In particular, and in line with the Syllabus (p. 18), this Topic is about:

- Discussing and applying Pascal's Triangle in expanding binomials.
- Discussing and applying Binomial Theorem and its properties.

Binomial expansions

A **binomial** has two terms. Expansion of binomials raised to powers, of the form $(a + b)^n$ where $n \in$ **whole numbers**, gives the following pattern:

$n = 0, (a + b)^0 = 1$

$n = 1, (a + b)^1 = a + b$

$n = 2, (a + b)^2 = a^2 + 2ab + b^2$

$n = 3, (a + b)^3 = a^3 + 3a^2b + 3ab^2 + b^3$

$n = 4, (a + b)^4 = a^4 + 4a^3b + 6a^2b^2 + 4ab^3 + b^4$

It is clear from the pattern above that the powers of a and b ascend or descend by 1 from n, and that their sum is n for each term in the expansion. For example, when $n = 4$ the first term has a^4 (and b^0), the second term has a^3 (and b^1), the third term has a^2 (and b^2), and so on.

A method of evaluating the coefficients of the terms was established by Blaise Pascal (1623–1662), who arranged the coefficients in the shape of a triangle.

Pascal's Triangle

1

1 1

1 2 1

1 3 3 1

1 4 6 4 1

1 **5** **10** 10 5 1

1 6 **15** 20 15 6 1

1 7 21 35 35 21 7 1

and so on.

Each number in this pattern, other than 1, is the sum of the two numbers standing above it to the left and to the right, eg 15 = 10 + 5.

For example, using the last line written above in Pascal's Triangle allows expansion of $(a + b)^7$.

$(a + b)^7 = a^7 + 7a^6b + 21a^5b^2 + 35a^4b^3 + 35a^3b^4 + 21a^2b^5 + 7ab^6 + b^7$

Using this array of coefficients allows expansion of $(a + b)^n$, where n is a whole number.

Example A

Q. Expand $(x - 2y)^6$

A. First expand $(a + b)^6$ and identify that a is x and b is $-2y$.

$$(a + b)^6 = a^6 + 6a^5b + 15a^4b^2 + 20a^3b^3 + 15a^2b^4 + 6ab^5 + b^6$$

$$(x - 2y)^6 = (x)^6 + 6(x)^5(-2y) + 15(x)^4(-2y)^2 + 20(x)^3(-2y)^3 + 15(x)^2(-2y)^4 + 6(x)(-2y)^5 + (-2y)^6 \quad \text{[substituting } a = x, b = -2y\text{]}$$

$$(x - 2y)^6 = x^6 - 12x^5y + 60x^4y^2 - 160x^3y^3 + 240x^2y^4 - 192xy^5 + 64y^6 \quad \text{[simplifying]}$$

Note: The use of brackets is advisable as a and b are replaced by (x) and $(-2y)$, otherwise errors in computation may arise.

The expansions are much easier if the first term in the brackets is 1.

Example B

Q. Expand $(1 - 2x)^5$.

A. $(a + b)^5 = a^5 + 5a^4b + 10a^3b^2 + 10a^2b^3 + 5ab^4 + b^5$

$$(1 - 2x)^5 = 1^5 + 5.1^4(-2x) + 10.1^3(-2x)^2 + 10.1^2(-2x)^3 + 5.1.(-2x)^4 + (-2x)^5$$

[substitute $a = 1$, $b = -2x$]

Since $1^n = 1$, powers of 1 can be ignored after the 1st term

$$(1 - 2x)^5 = 1 - 10x + 40x^2 - 80x^3 + 80x^4 - 32x^5$$

Investigations

1. Starting at X each time and moving *down* the page only, find how many routes there are to A, B, C, …

See if you can extend the pattern.

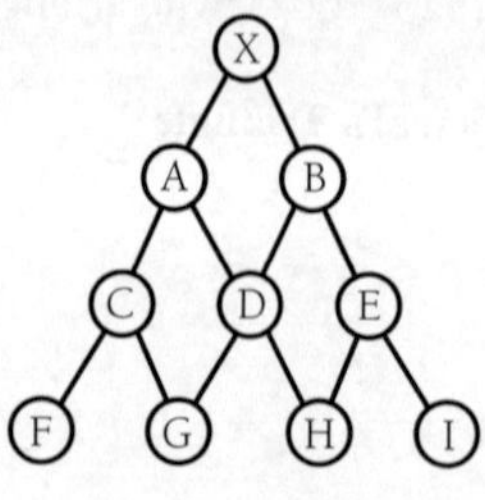

2. Pascal's Triangle was known to be in existence 600 years before Pascal developed various properties of the numbers forming the triangle.

1	1	1	1	1	1	1
1	2	3	4	5	6	
1	3	6	10	15		
1	4	10	20			
1	5	15				
1	6					
1						

a. Complete the rest of the numbers in the box.

b. Find the sum of the numbers highlighted in the first column. What do you notice?

c. Find the sum of the second column of numbers, and look for patterns.

d. Find the sum of the numbers along the diagonals, and look for patterns.

Unit 12.1 Activity 9A: Expanding

Using Pascal's Triangle, answer the following questions:

1. Expand the following:

a. $(1 + x)^5$ **b.** $(1 - a)^8$ **c.** $(x + y)^5$ **d.** $(x - y)^6$

e. $(1 - 3y)^6$ **f.** $(2 + 3x)^5$ **g.** $(4 - x)^4$ **h.** $(1 - \frac{1}{2}x)^5$

2. Find the first four terms in the expansion of:

a. $(1 - 2x)^9$ **b.** $(x + 2y)^7$ **c.** $(x - 2y)^{10}$

3. a. Expand $(a + b)^4$

b. Use your expansion to verify that:

i. $2^4 = 16$ **ii.** $3^4 = 81$

4. a. Expand $(1 - 2x)^4$ **b.** Now expand $(1 + x)(1 - 2x)^4$

5. Find the first three terms in ascending powers of x in the expansion of $(1 + x)^4(1 - x)^5$

6. Using algebraic long division, find the first four terms in the series for $\frac{1}{1 + x}$. Now find the first three terms in the expansion of $\frac{(1 + x)^4}{1 + x}$

7. Write each of the following in the form $(a + b)^n$

a. $1 + 5x + 10x^2 + 10x^3 + 5x^4 + x^5$

b. $1 - 12x + 60x^2 - 160x^3 + 240x^4 - 192x^5 + 64x^6$

c. $16 + 96x + 216x^2 + 216x^3 + 81x^4$

It is sometimes necessary to identify and extract one particular term in the expansion of $(a + b)^n$. In order to do this a theorem must be established for the expansion of $(a + b)^n$ which is not tied to Pascal's Triangle. This requires two basic 'building blocks'. One is **factorials** and the other is **combinations** (both concepts are used in **probability**).

Factorials

The product of the first n natural numbers is called ***n* factorial** and is written $n!$

$$n! = n(n - 1)(n - 2) \ldots 3.2.1$$

Example C

1! = 1

4! = 4 x 3 x 2 x 1 = 24

6! = 6 x 5 x 4 x 3 x 2 x 1 = 720

Simplification of factorials is useful.

Example D

$$\frac{8!}{6!} = \frac{8 \times 7 \times 6 \times 5 \times 4 \times 3 \times 2 \times 1}{6!}$$

$$= \frac{8 \times 7 \times 6!}{6!}$$

$$= 56 \qquad \text{[cancelling 6!]}$$

Most scientific calculators have a factorial button to simplify calculations.

Selections

Using Pascal's Triangle, $(a + b)^5 = a^5 + 5a^4b + 10a^3b^2 + 10a^2b^3 + 5ab^4 + b^5$.

When **expanding** brackets such as $(a + b)^5$, which is $(a + b)(a + b)(a + b)(a + b)(a + b)$, terms in the expansion are formed by multiplying together one term from each of the five brackets. For example, if b is selected from each bracket, the term b^5 results; if b is selected from three of the brackets and a is selected from the other two brackets, then the term a^2b^3 results. (Since there are two choices of term from each bracket there are $2^5 = 32$ different selections possible and thus 32 terms in the expansion.)

While there is only one way of selecting b^5, there are in fact ten different ways of creating the term a^2b^3, thus in the expansion of $(a + b)^5$ the term in a^5 has coefficient 1 and the term in a^3b^2 has coefficient 10. (Note that the sum of the coefficients in the expansion of $(a + b)^5$ is $1 + 5 + 10 + 10 + 5 + 1 = 32$, the total number of different selections possible.)

Fortunately there is a formula that gives us the number of ways of choosing r elements from a group of n elements, without regard for order. This is written as $\binom{n}{r}$ and the number of ways is given by:

$$\boxed{\binom{n}{r} = \frac{n!}{(n-r)!\,r!}}$$

The number of ways of choosing three b's from 5 different brackets is thus:

$$\binom{5}{3} = \frac{5!}{(5-3)!\,3!} \qquad \text{[substituting into the formula]}$$

$$= \frac{5!}{2!\,3!}$$

$$= \frac{5 \times 4}{2 \times 1} \qquad \text{[cancelling 3!]}$$

$= 10$ ways.

These selections are often called combinations.

Note: Choosing b from three of the brackets means choosing a from the other two brackets.

(*Notation:* An alternative notation for $\binom{n}{r}$ is ${}^{n}C_{r}$ and this is also in common use. Many scientific calculators have a ${}^{n}C_{r}$ button.)

Binomial Theorem

Using combinations the expansion of $(a + b)^5$ can now be restated as:

$$(a + b)^5 = \binom{5}{0}a^5 + \binom{5}{1}a^4b + \binom{5}{2}a^3b^2 + \binom{5}{3}a^2b^3 + \binom{5}{4}ab^4 + \binom{5}{5}b^5.$$

Usually $\binom{5}{0}$ and $\binom{5}{5}$ are omitted because they both equal one.

This result holds true generally, and the **Binomial Theorem** is now stated for the expansion of $(a + b)n$ where $n \in$ W.

$$(a + b)^n = a^n + \binom{n}{1}a^{n-1}b + \binom{n}{2}a^{n-2}b^2 + \binom{n}{3}a^{n-3}b^3 + \ldots + \binom{n}{r}a^{n-r}b^r + \ldots + b^n$$

There are $(n + 1)$ terms in the expansion of $(a + b)^n$

General term:

The $(r + 1)th$ term of the binomial expansion $(a + b)^n$ is

$$\binom{n}{r} a^{n-r} b^r$$

This theorem provides the method for identifying a particular term in an expansion.

Example E

Q. Find the fifth term in the expansion of $(x - 2y)^{11}$

A. The fifth term is the general term, ie the $(r + 1)$th term, where $r = 4$

General term is $\binom{n}{r} a^{n-r} b^r$ where $r = 4$, $a = x$, $b = -2y$, $n = 11$

5th term is $\binom{11}{4}(x)^7(-2y)^4 = 330\, x^7 2^4 y^4$ $[n - r = 11 - 4 = 7]$

$= 5\,280\, x^7 y^4$

Example F

Q. Find the coefficient of x^5 in the expansion of $(1 - 2x)^8$

A. The general term $\binom{n}{r} a^{n-r} b^r$ is used to represent the term containing x^5. When b is replaced by $-2x$, it can be seen that x^5 occurs when $r = 5$

The required term is $\binom{8}{5}(1)^{8-5}(-2x)^5 = 56 \times (-32)x^5$

The coefficient of x^5 is $-1\,792$

The term independent of *x*

In the expansion of $\left(x^2 + \frac{1}{x^2}\right)^4$, which is set out below, the third term is independent of x, ie is a constant term.

$$\left(x^2 + \frac{1}{x^2}\right)^4 = x^8 + 4x^4 + 6 + \frac{4}{x^4} + \frac{1}{x^8}$$

Using the formula for the $(r + 1)$th term, the constant term can be identified without writing out the expansion in full.

Example G

Q. Find the term independent of x in the expansion of $\left(x^2 - \frac{3}{x}\right)^6$

A. The general $(r + 1)$th term is

$\binom{6}{r}(x^2)^{6-r}\left(\frac{-3}{x}\right)^r$ [using $\binom{n}{r}a^{n-r}b^r$ where $n = 6$, $a = x^2$, $b = \frac{-3}{x}$]

$= \binom{6}{r}x^{12-2r}\frac{(-3)^r}{x^r}$ [expanding using index laws]

$= \binom{6}{r}(-3)^r x^{12-3r}$ [division law for indices]

For a constant term, the power of x is zero

$12 - 3r = 0$

$r = 4$ [solving]

Substituting $r = 4$ in general term gives

Constant term $= \binom{6}{4}(-3)^4 x^0$

$= 1\ 215$

Note: An interesting result about finding the sum of the rows of numbers in Pascal's Triangle was explored in an investigation earlier in this Topic. It is now stated using combinations and the Binomial Theorem. Expanding:

$$(a + b)^6 = \binom{6}{0}a^6 + \binom{6}{1}a^5b + \binom{6}{2}a^4b^2 + \binom{6}{3}a^3b^3 + \binom{6}{4}a^2b^4 + \binom{6}{5}ab^5 + \binom{6}{6}b^6$$

Replacing a and b by one, gives:

$$2^6 = \binom{6}{0} + \binom{6}{1} + \binom{6}{2} + \binom{6}{3} + \binom{6}{4} + \binom{6}{5} + \binom{6}{6}$$

Unit 12.1 Activity 9B: Binomial series

1. Find:

a. The 4th term in the expansion of $(1 + 3x)^8$
b. The 5th term in the expansion of $(1 - 2x)^{10}$
c. The 6th term in the expansion of $(x + 2y)^7$
d. The 5th term in the expansion of $(x - 2y)^8$
e. The middle term in the expansion of $(2 - x)^8$
f. The 3rd term in the expansion of $(1 - x^2)^6$

2. Find the coefficient of x^4 in the expansion of:

a. $(1 + x)^9$ **b.** $(1 - 2x)^8$ **c.** $(1 + \frac{1}{2}x)^7$ **d.** $(2 - 3x)^6$
e. $(a - x)^n$ **f.** $(2x - 1)^7$

3. Find the $(r + 1)$th term in the expansion of:
a. $(2 + x)^8$ **b.** $(1 - 2x)^8$ **c.** $(x + \frac{1}{x})^9$

4. Find the term independent of x in the expansion of:
a. $\left(2x + \frac{3}{x^2}\right)^6$ **b.** $\left(x - \frac{1}{x}\right)^6$ **c.** $\left(2x^2 - \frac{1}{x}\right)^{12}$

5. a. Find the coefficient of x^{-4} in the expansion of $\left(x^2 - \frac{1}{x}\right)^{10}$

b. Simplify the 4th term in the expansion of $(\sqrt{x} + x)^6$

6. The coefficient of x^3 in the expansion of $(1 + x)^n$ is four times the coefficient of x^2. Find the value of n.

7. Use the expansion of $(1 + x)^3$ to evaluate $(1.03)^3$ correct to 2 dp.

8. If in the expansion of $(1 + kx)^8$ the coefficient of x^3 is $-1\ 512$, find k.

9. Using a binomial expansion, prove that $(1 + \sqrt{2})^6 = 99 + 70\sqrt{2}$.

10. Prove that $(1 + \sqrt{2})^5$ is an irrational number.

11. Show that $(1 + \sqrt{3})^5 + (1 - \sqrt{3})^5$ is a **rational** number.

12. Using $(a + b)^7$ evaluate: $\binom{7}{0} + \binom{7}{1} + \binom{7}{2} + \binom{7}{3} + \binom{7}{4} + \binom{7}{5} + \binom{7}{6} + \binom{7}{7}$

Unit 12.1 Activity 9C: Revision questions

1. Write $x^2 + 3x - 1$ in the form $(x + a)^2 + b$

2. Expand $(x - 3)(x^2 - 1)(5 - x - x^2)$

3. Find a and b so that $(x^4 + 1) = (x^2 + ax + 1)(x^2 + bx + 1)$

4. Write as a single fraction $1 + \frac{a}{1 + a} + \frac{1 - a}{a}$

5. Solve $\frac{7 - x}{x + 1} - \frac{3x - 1}{x - 1} = -3$

6. Find all values of x for which $y = 0$ where $y = \frac{(2x - 1)(3x + 4) - (5x^2 + 2x + 6)}{(5x^2 + 2x + 1)}$

7. Find a and b such that $(x^2 + ax + b)(x^2 + 2x - 1) = x^4 - 3x^3 - 8x^2 + 11x - 3$

8. Express x in terms of y and z where $\frac{1}{x} + \frac{1}{y} = \frac{1}{z}$

9. Solve $(x + 1)^2 = 3$

10. Simplify each of the following where possible:

a. $\dfrac{x + x^2}{x + 1}$ **b.** $\dfrac{1 + x^2}{1 + x}$ **c.** $\left(\dfrac{2^{n+1}}{(2^n)^{n-1}}\right) \div \left(\dfrac{4^{n+1}}{(2^n)^{n-1}}\right)$ **d.** $\dfrac{x(x - 5) - 2(3 - 2x)}{(x + 1)^2 + (x + 1)}$

11. Solve $\dfrac{x + 5}{x + 1} + \dfrac{2}{3} = -\dfrac{3}{x}$

12. Find all values of x for which $y = 0$ if $y = \dfrac{x^2 + 2x + 5 - (x + 1)(2x + 2)}{(x^2 + 2x + 5)^2}$

13. Find the real solutions of the equation $x^3 - 2x^2 - 5x + 6 = 0$

14. Complete the following calculations giving exact solutions in the form $a + b\sqrt{3}$

a. $(2 + \sqrt{3})(4 + \sqrt{3})$ **b.** $\dfrac{14}{2 + \sqrt{3}}$

15. a. Use the **Remainder Theorem**, or some other method, to find the remainder when the cubic function $f(x) = x^3 - 5x^2 + 4x + 6$ is divided by $(x - 2)$.

b. The cubic equation $x^3 - 5x^2 + 4x + 6 = 0$ has one rational and two irrational solutions. Solve the cubic, leaving your irrational solution in **surd** form (ie do not give decimal approximations).

16. a. Solve for x:

i. $(x - 2)^2 = 9$ **ii.** $x + \dfrac{8}{x} = 6$ **iii.** $\sqrt{x} = x - 2$

b. Write $2 \log x + \log 6x - \log 2x$ in the form $\log a$

c. Write $(2 + \sqrt{3})(3 - 2\sqrt{3})$ in the form $a + b\sqrt{3}$

17. Solve the equations for x:

a. $\log_2 x = 3$ **b.** $\ln (x + 3) = \ln x + \ln 3$ **c.** $e^x = 5$

18. Solve for x, $2^{\ln x} = 3$. Give your answer to 3 decimal places.

19. Solve for s the equation $\dfrac{s - 1}{t - 1} - \dfrac{s + 1}{2s} + \dfrac{1}{s} = 0$

(*Hint:* there are two solutions, one of which depends on t.)

20. Find all x, y, z satisfying the equations: $x = yz$, $y = xz$, $z = xy$

21. If $\dfrac{b}{2 - a} + 1 = \dfrac{1}{b}$, find an expression for a in terms of b.

22. If $\dfrac{1 - x^4}{1 - x} = Ax^3 + Bx^2 + Cx + D$, find A, B, C and D.

23. Express $\dfrac{x^3 + x^2 + x + 1}{x^2 - 2x + 4}$ in the form $ax + b + \dfrac{cx + d}{x^2 - 2x + 4}$ and find the values of a, b, c and d.

24. Given $x = \dfrac{1-a}{1+a}$ and $y = \dfrac{1-a^2}{1+a^2}$, find an expression for a in terms of x and an expression for y in terms of x.

25. Solve for all $x > 0$: $\sqrt{x} - \dfrac{2}{\sqrt{x}} = 1$

26. Verify that $\dfrac{x+\sqrt{x}}{1+x+\sqrt{x}} = \dfrac{x^2+\sqrt{x}}{1+x+x^2}$ for all $x \geq 0$

27. Find the value of the real number p so that the real **roots of the equation** $2x^2 - 12x + p + 2 = 0$ are of the form a, $a + 2$

28. By **completing the square**, obtain the formula for solving the **quadratic equation** $ax^2 + bx + c = 0$

29. Determine the value of the term that is independent of x in the expansion of $(x^2 + \dfrac{2}{x})^9$

30. **a.** Use the Binomial Theorem to determine a and b so that $(1 + x)^n \approx 1 + ax + bx^2$ for $n \in \mathbb{N}$.

b. Use this quadratic approximation to determine the value of $(1.03)^{20}$. Record your result to three decimal places.

Unit 12.1 Patterns and Algebra
Topic 10: Determinants – simultaneous linear equations

Another method of solving real-life problems is the use of determinants and Topic 10 looks at systems of linear equations:

- Solving systems of three linear equations in three variables, where there is a unique solution. This may involve rearrangement of equations and/or interpreting solutions.
- Forming and solving a 3 × 3 system of linear equations.
- Discussing the consistency or non-independence of 3 × 3 systems of equations, including geometric representations.

Solving linear equations

One-variable **linear** equations of the following types should be very familiar:

$$3x + 1 = 7$$

$$4x - 1 = 4 - 3x$$

$$\frac{3}{4}(x + 1) - \frac{2}{3}(x - 1) = 3$$

The method of finding values for the **variables** that **satisfy** the equation is called **solving** the equation.

Example A

1. $3x + 1 = 7$

$\therefore 3x = 6$ [subtracting 1]

$\therefore x = 2$ [dividing by 3]

2. $4x - 1 = 4 - 3x$

$\therefore 7x - 1 = 4$ [adding $3x$]

$\therefore 7x = 5$ [adding 1]

$\therefore x = \frac{5}{7}$ [dividing by 7]

3. $\frac{3}{4}(x + 1) - \frac{2}{3}(x - 1) = 3$

$\therefore 9(x + 1) - 8(x - 1) = 36$ [multiplying by 12]

$\therefore 9x + 9 - 8x + 8 = 36$ [expanding brackets]

$\therefore x + 17 = 36$ [simplifying]

$\therefore x = 19$ [simplifying]

Word problems

Example B

Q. Write an equation that **models** the following problem, and solve it.

Tracy bought some meat for a barbecue which cost K6.70 per kg. In addition, she bought K6.85 worth of vegetarian sausages. Her total bill was K23.60.

Find how much meat she bought.

A. Let the weight of the meat she bought be W.

$\therefore W \times$ K6.70 + K6.85 is the total amount of money she spent.

$\therefore 6.7\,W + 6.85 = 23.60$ [total amount spent was K23.60]

$\therefore 6.7\,W = 16.75$ [subtracting 6.85]

$\therefore W = 2.5$ [dividing by 6.7]

Hence Tracy purchased 2.5 kg of meat.

Unit 12.1 Activity 10A: Linear equations

1. Solve these equations:

a. $4x - 1 = 9$ **b.** $\frac{3x}{4} + 3 = 6$ **c.** $\frac{2}{3}(x - 1) = 7$

d. $\frac{3}{4}(x + 1) - \frac{2}{7}(x - 2) = 2\frac{5}{7}$ **e.** $\frac{x + 1}{3} = 7 - x$ **f.** $\frac{x + 3}{2} - \frac{x - 1}{3} = \frac{5}{3}$

g. $\frac{2}{x + 1} = \frac{3}{x - 1}$ **h.** $\frac{2x - 3}{x + 7} = 4$ **i.** $\frac{2}{3} = \frac{4}{2 - x}$

j. $2\frac{1}{3} - \frac{2}{3x - 4} = 1\frac{1}{6}$

2. Cathy weighs 4.2 kg more than Rachael. Rachael weighs 2.7 kg more than Lucy. The three girls together weigh 177.9 kg. What does Rachael weigh?

3. This week I worked for 8 hours. Last week I worked for $6\frac{1}{2}$ hours. I got paid an hourly rate and also bonuses. During the last two weeks my bonuses added up to K97 and my total earnings were K224.60. What is my hourly rate?

4. Wamsie, Hosea and David went shopping. They spent a total of K86.50. Wamsie spent twice as much as Hosea and K36 more than David. How much did each spend?

5. On Saturday Benedict cycled a certain distance at an average speed of 36 km h^{-1}. The next day Benedict cycled 16 km further than he did on Saturday, and his average speed was 35 km h^{-1}. If he cycled a total of 3.5 hours, how far did he cycle altogether over the two days?

6. A village has two water tanks. The capacity of the smaller tank is 250 L less than the capacity of the larger tank. Currently the larger tank is three-quarters full and the smaller tank is nine-tenths full. If the larger tank has 100 L more water in it than the smaller tank has, what is the capacity of the smaller tank?

Simultaneous equations

Two or more equations involving more than one unknown which hold true at the same time are called **simultaneous equations**. For example:

Two simultaneous equations in two unknowns or variables
$$\begin{cases} 2x - 2y = 1 \\ x + 3y = 2 \end{cases}$$

Three simultaneous equations in three unknowns or variables
$$\begin{cases} 3x - y + 2z = 7 \\ x + y + z = 4 \\ x - y + z = 2 \end{cases}$$

Solution of simultaneous equations in two unknowns

One method of solution is called **elimination**. In this process, the equations are multiplied by a constant (if necessary) and then added or subtracted together, so that one variable is *eliminated*. This allows the equation to be solved for the other variable. The process can be repeated to find the other variable.

Example C

Q. Solve the simultaneous equations: $2x - 3y = 1$
$x + 4y = 2$

A. Number the equations as shown:

$2x - 3y = 1$ ... (1)

$x + 4y = 2$ ... (2)

Multiply equation (2) by 2 so that the coefficients of x are the same in both equations. This gives:

$2x - 3y = 1$ ... (1)

$2x + 8y = 4$ ... (3)

$-11y = -3$ [subtracting (3) from (1)]

$\therefore y = \frac{3}{11}$ [dividing by –11]

To find the value of x, y is now eliminated from the original set of equations.

Multiply (1) by 4 and (2) by 3 to give:

$8x - 12y = 4$ ... (4)

$3x + 12y = 6$ ... (5)

$11x = 10$ [adding (4) and (5)]

$\therefore x = \frac{10}{11}$ [dividing by 11]

Note how x was first eliminated to find y, and then y eliminated to find x.

It is possible, having found one variable, to substitute into one of the original equations in order to find the other variable, rather than do a second elimination. In cases where a solution is a small **integer** this is usually a quicker method.

Example D

Q. A student buys some soft drinks which come in large and small sizes. The large costs K1.10 a bottle, the small K0.85. He buys a total of 18 bottles and spends a total of K18.05. How many of each type did he buy?

A. Let L be the number of large bottles and S the number of small bottles.

$L + S = 18$ ... (1) [as there are 18 bottles]

$1.1\,L + 0.85\,S = 18.05$ ... (2) [as each large bottle costs K1.10 and each small bottle costs K0.85 and the total cost is K18.05]

Multiply (1) by 1.1 to give the same coefficient L in both equations:

$$
\begin{aligned}
1.1\,L + 1.1\,S &= 19.8 && \text{... (3)}\\
1.1\,L + 0.85\,S &= 18.05 && \text{... (2)}\\
0.25\,S &= 1.75 && \text{[subtracting (2) from (3)]}\\
\therefore S &= 7 && \text{[dividing by 0.25]}\\
\therefore L + 7 &= 18 && \text{[substituting } S = 7 \text{ in (1)]}\\
\therefore L &= 11
\end{aligned}
$$

Hence the student purchased 11 large bottles and 7 small bottles.

Interpretation of solutions

The solutions to the previous two examples are the values of the variables for which both equations hold true. If the two equations are considered as lines on a **Cartesian plane**, then the solution (x, y) gives the point on the Cartesian plane where the two lines intersect.

In Example C, $(\frac{10}{11}, \frac{3}{11})$ is the point of **intersection** of the straight lines $2x - 3y = 1$ and $x + 4y = 2$.

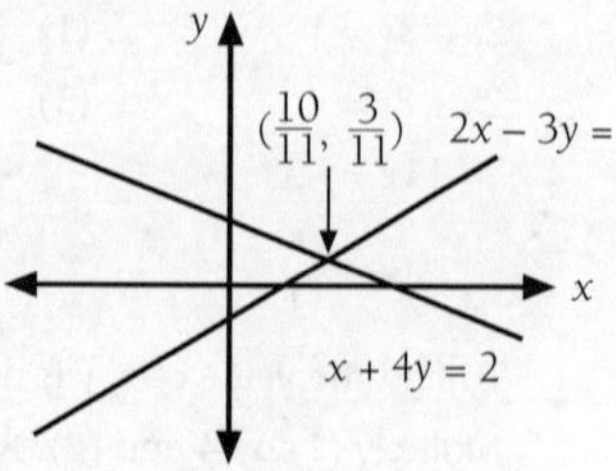

It is also possible to have two equations with two unknowns for which there are *no* solutions, as well as cases where there are *infinitely* many solutions.

- No solutions occur when the equations represent **parallel lines** which have no point of intersection.

  Lines are parallel when their equations can be written in the form $ax + by = c$ and $ax + by = d$ where $c \neq d$. For example, the lines $2x + 3y = 7$ and $2x + 3y = 8$ are parallel.

- Infinitely many solutions are obtained when the two equations represent the same set of points, ie are the same line. (Thus, every point on the first line also lies on the second line and therefore is a solution.) Lines of the form $ax + by = c$ and $kax + kby = kc$ are the same line.

Example E

Q. Solve the systems of equations: **1.** $x - 4y = 8$, $x - 4y = 7$ **2.** $2x - 8y = 16$, $x - 4y = 8$

A. **1.** Subtracting one equation from the other yields $0 = 1$, which is impossible. Graphically the situation is a pair of parallel lines (with no points in common).

2. Doubling the second equation yields the first equation. Thus any pair of values satisfying one equation will satisfy both, ie both equations represent the same line. Doubling the second equation and subtracting from the first equation gives:

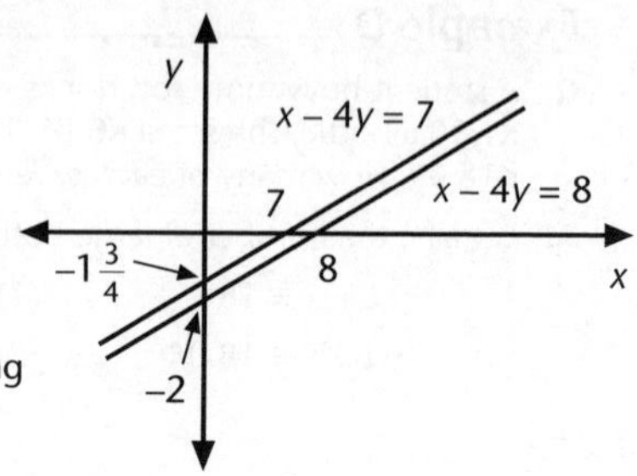

$$2x - 8y = 16 \quad \dots (1)$$
$$2x - 8y = 16 \quad \dots (3)$$
$$0 = 0 \quad \text{[subtracting]}$$

If such a solution, ie $0 = 0$, is obtained, it means that there is an infinite set of solutions.

For example, $x - 4y = 8$ and $2x - 8y = 16$ are both satisfied by:

$x = 0, y = -2, \quad x = 1, y = -1\frac{3}{4}, \quad x = 2, y = -1\frac{1}{2},$ etc.

Consistency of simultaneous equations in two unknowns

If a system of two linear equations has either a **unique solution** or an infinite set of solutions, then the equations are said to be **consistent**.

If they have no solution, they are said to be **inconsistent**.

In Example E the equations in **1**. are inconsistent because there is no solution while those in **2**. are consistent, even though there is an infinite number of solutions.

Example F

Q. In the system of equations:

$$8x + ay = k$$
$$2x + y = 1$$

find a value of a and k that makes the equations:

1. Inconsistent. **2.** Consistent with infinitely many solutions.

A. 1. Number the equations

$$8x + ay = k \quad \dots (1)$$
$$2x + y = 1 \quad \dots (2)$$

Equations are inconsistent when lines are parallel, ie coefficients of x are equal and coefficients of y are equal but the constants are not equal.

$$8x + ay = k \quad \dots (1)$$

Multiplying (2) by 4 gives $\quad 8x + 4y = 4 \quad \dots (3)$

Choosing $a = 4$ and k equal to any value except 4 will provide a system of inconsistent equations.

For example, if $k = 5$ the equations become:

$$8x + 4y = 5 \quad \dots (1)$$
$$8x + 4y = 4 \quad \dots (3)$$

and are inconsistent. So $a = 4$ and $k \neq 4$.

2. Equations are consistent with infinitely many solutions when they are the same line. Multiplying (2) by 4 gives:

$$8x + ay = k \quad \dots (1)$$
$$8x + 4y = 4 \quad \dots (3)$$

Choosing $a = 4$ and $k = 4$ makes equations (1) and (3) the same, so the system has infinitely many solutions.

Cramer's rule

Cramer's rule uses determinants to solve simultaneous linear equations. Cramer's rule provides a systematic approach to solving the equations but requires care and practice if it is to be used often.

Consider the following set of linear equations in two unknowns, x and y:

$$ax + by = p$$
$$cx + dy = q$$

For this set of equations the determinant, D, is written as:

$$D = \begin{vmatrix} a & b \\ c & d \end{vmatrix}$$

The first column contains the coefficients of x in the equations.

The second column contains the coefficients of y in the equations.

To evaluate D we multiply the values in the leading diagonal and then subtract the product of the values in the other diagonal:

$$D = \begin{vmatrix} a & b \\ c & d \end{vmatrix} = ad - bc$$

Note: If $D = 0$ then there is no unique solution to the set of equations.

Cramer's rule tells us that the solutions to the equations is given by:

$$x = \frac{D_x}{D} \text{ and } y = \frac{D_y}{D} \text{ where } D_x = \begin{vmatrix} p & b \\ q & d \end{vmatrix} \text{ and } D_y = \begin{vmatrix} a & p \\ c & q \end{vmatrix}$$

The 'x' column is replaced by the values in the right-hand side of the equations.

The 'y'column is replaced by the values in the right-hand side of the equations.

Example G

Q. Use Cramer's rule to solve the following simultaneous equations:

$2x + 3y = 7$

$5x - 7y = 3$

A. The determinant $D = \begin{vmatrix} 2 & 3 \\ 5 & -7 \end{vmatrix} = 2 \times -7 - 5 \times 3 = -14 - 15 = -29$

$D_x = \begin{vmatrix} 7 & 3 \\ 3 & -7 \end{vmatrix} = 7 \times -7 - 3 \times 3 = -49 - 9 = -58$

$x = \frac{D_x}{D} = \frac{-58}{-29} = 2$

$D_y = \begin{vmatrix} 2 & 7 \\ 5 & 3 \end{vmatrix} = 2 \times 3 - 5 \times 7 = 6 - 35 = -29$

$$y = \frac{D_y}{D} = \frac{-29}{-29} = 1$$

The solution is $x = 2$ and $y = 1$(or the point (2, 1))

Unit 12.1 Activity 10B: Simultaneous linear equations in two unknowns

1. Solve these pairs of simultaneous equations:

a. $x + y = 1$
$2y - x = 5$

b. $3x - 2y = 14$
$2x + 5y = 3$

c. $3x = 4y - 3$
$2x - 3y = y - x + 2$

d. $3x - y = 4$
$2x - 2y - 2 = 3y - x + 2$

e. $2x + y = 3x - y$
$\frac{1}{3}(x + y) = 2x - y - 2$

f. $\frac{1}{2}x - \frac{1}{3}y = \frac{1}{3}y + 5$
$3x - 4y = 30$

2. Use Cramer's rule to solve the following pairs of simultaneous equations:

a. $2x - 7y = -4$
$x - 5y = 7$

b. $5x - 7y = -1$
$x - 5y = 7$

c. $6x + 5y = 1$
$3y - 8y = 18$

d. $8x + 9y = 2$
$14x + 10y = 15$

3. James gets K1.50 more per hour than Gandhi. James works for 8 hours, Gandhi for 9 hours and together they earn K182. What is the hourly rate for each?

4. Towns A, B and C all lie along the same road in that order. Town C is 52 km from town B. Gevena drives from B to A then to C then back to A. In doing so, she travels a total distance of 266 km. How far is A from B?

5. A total of 25 toy vehicles are ordered by a store. Some are tricycles and some are bicycles. The vehicles have a total of 64 wheels. How many are tricycles?

6. In an election for a district there were two candidates: Menai and Bais. A total of 25 500 people voted and Bais got a majority of 2 500 votes. How many people voted for Menai?

7. A group of adults and children went to a cinema. Tickets for adults were K12 each and tickets for children were K9 each. Altogether there were 20 people in the group and tickets cost a total of K204. How many children went to the cinema?

8. The graph of $3x - y = 2$ is shown on the axes alongside.

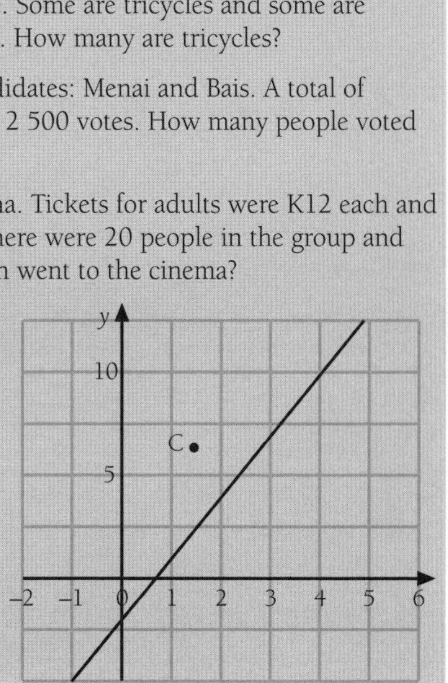

a. Copy this diagram. Sketch the graph of another linear equation that passes through the point labelled C, so that the pair of equations is inconsistent.

b. For the pair of equations
$3x - y = 2$
$9x + ay = b$
state a value for a and b so that the pair of equations has more than one solution.

9. Consider the following system of two linear equations where k is a constant:

$y = kx + 2$
$y = 1$

Give a value for the constant k for which the system is *inconsistent* and interpret this inconsistent system geometrically.

10. Consider the following system of two linear equations, where c is a constant:

$2x + 5y = 16$
$4x + cy = 25$

a. Solve the equation when $c = 3$.

b. Give a value of the constant c for which the system is inconsistent.

c. If c is chosen so that the system is consistent, explain in geometric terms why there is a unique solution.

11. The system of equations below is consistent with infinitely many solutions, where a and b are constants:

$y = 3x + 4$
$ax + by + 6 = 0$

Find the values of a and b.

Unit 12.1 Patterns and Algebra

Topic 11: Determinants – simultaneous linear equations in three unknowns

Topic 11 builds on the previous Topic and extends the treatment of linear equations as a method of solving real-life problems by looking at linear equations in three unknowns.

The method of elimination can be extended to solve three-variable simultaneous equations. One variable is chosen to be eliminated from two of the equations, leaving a single two-variable equation. This is repeated with a different pair of equations to give a second two-variable equation. These new equations can then be solved using the two-variable method.

Equations with a unique solution

An example of a consistent set of equations with a unique solution is shown below.

Example A

Q. Solve the system of simultaneous equations

$3x + y - 2z = 9$ ... (1)

$2x + 3y + z = -1$ ... (2)

$x + 2y + 3z = -6$ ... (3)

A. Begin by eliminating z from equations (1) and (2)

$3x + y - 2z = 9$ ... (1)

$2x + 3y + z = -1$ ... (2)

Multiply (2) by 2 (to get opposite coefficients of z) $4x + 6y + 2z = -2$ ... (4)

Add (1) and (4) $7x + 7y = 7$ ... (5)

Now eliminate z from equations (2) and (3)

$2x + 3y + z = -1$ ... (2)

$x + 2y + 3z = -6$ ... (3)

Multiply (2) by 3 (to get equal coefficients of z) $6x + 9y + 3z = -3$ ... (6)

Subtract (3) from (6) $5x + 7y = 3$ ... (7)

[note $-3 - -6 = 3$]

Now solve (5) and (7) simultaneously. Subtracting (7) from (5) gives

$2x = 4$

$x = 2$

Another elimination could now be used to find y. However, because x is an easy number to calculate with, substitute $x = 2$ into one of (5) or (7).

Substituting into (5) gives: $14 + 7y = 7$

$\therefore y = -1$ [solving]

Now that two unknowns have been solved, the remaining unknown, z, can be calculated by substituting the values found for x and y into any one of the original equations, (1), (2), or (3).

Choosing (2) gives: $4 - 3 + z = -1$

$\therefore z = -2$ [solving]

Thus the solution is $(x, y, z) = (2, -1, -2)$.

Geometric interpretation of solution

Each of the three-variable equations can be represented by a plane. In Example A the unique solution (2, –1, –2) is the point of intersection of the three planes. It is helpful to consider a diagram to see that the planes have a unique point of intersection.

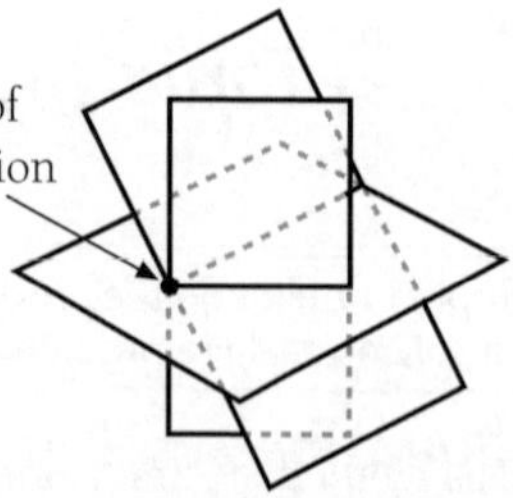

Equations with infinitely many solutions

As with two-variable equations, systems of equations in three unknowns can be consistent, but with *infinitely* many solutions.

Example B

Q. Solve the simultaneous equations

$$2x + y + 3z = 3 \quad \ldots (1)$$
$$x - y + z = 1 \quad \ldots (2)$$
$$x + 2y + 2z = 2 \quad \ldots (3)$$

A. Eliminate z from equations (1) and (2)

$$2x + y + 3z = 3 \quad \ldots (1)$$
$$x - y + z = 1 \quad \ldots (2)$$

Multiply (2) by 3 $\quad 3x - 3y + 3z = 3 \quad \ldots (4)$

Subtract (4) from (1) $\quad -x + 4y = 0 \quad \ldots (5)$

Eliminate z from equations (2) and (3)

$$x - y + z = 1 \quad \ldots (2)$$
$$x + 2y + 2z = 2 \quad \ldots (3)$$

Multiply (2) by 2 $\quad 2x - 2y + 2z = 2 \quad \ldots (6)$

Subtract (6) from (3) $\quad -x + 4y = 0 \quad \ldots (7)$

Equations (5) and (7) are the same, thus $\quad y = \frac{1}{4}x \quad$ [rearranging]

From (2) $\quad z = 1 - x + y \quad$ [rearranging]

$$= 1 - x + \frac{1}{4}x$$

[substituting $y = \frac{1}{4}x$]

$$= 1 - \frac{3}{4}x$$

Thus any (x, y, z) of the form $(x, \frac{1}{4}x, 1 - \frac{3}{4}x)$ will satisfy the equations.

For example, if $x = 4$

then $y = 1$

and $z = -2$

and $(x, y, z) = (4, 1, -2)$ is a solution.

Geometric interpretation

In Example B, when equation (2) is subtracted from equation (1) the result is the same as equation (3). Thus the intersection of equations (1) and (2) lies on the third plane. Diagrammatically this means that the three planes have a common line of intersection.

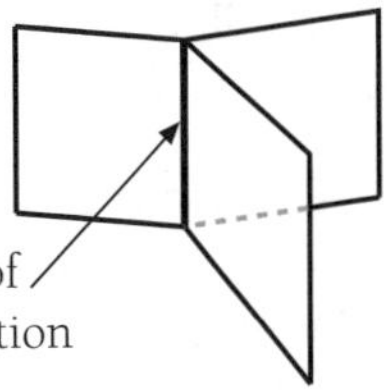

Equations with no solutions

A set of three equations in three unknowns may be inconsistent and have no solution, as in the following example:

Example C

Q. Solve the simultaneous equations

$$2x - 3y + 5z = 7 \quad \dots (1)$$
$$x - 2y + 3z = 4 \quad \dots (2)$$
$$x - y + 2z = 5 \quad \dots (3)$$

A. Try eliminating x

Doubling (2) and subtracting from (1) gives

$$2x - 3y + 5z = 7 \quad \dots (1)$$
$$2x - 4y + 6z = 8 \quad \dots (4) \text{ [double (2)]}$$
$$y - z = -1 \quad \dots (5)$$

Subtract (2) from (3):

$$x - y + 2z = 5 \quad \dots (3)$$
$$x - 2y + 3z = 4 \quad \dots (2)$$
$$y - z = 1 \quad \dots (6)$$

Equations (5) and (6) are inconsistent (have no solutions)

$$y - z = -1 \quad \dots (5)$$
$$y - z = 1 \quad \dots (6)$$

Subtracting (6) from (5)

$$0 = -2 \quad \dots (7)$$

$\therefore$ there are no solutions to the three simultaneous equations.

Geometric interpretation

If simultaneous equations in two variables are inconsistent then the lines they represent are separate and parallel. In the three-variable situation there are three separate cases.

Three parallel planes

In the following example the equations represent three parallel planes (with no point of intersection):

Example D

When planes are parallel, the coefficients of the variables are the same or a constant multiple of each other, while the numbers on the right are different.

An example is:

$x + y + z = 2$

$x + y + z = 3$

$2x + 2y + 2z = 7$

Here is how this might look:

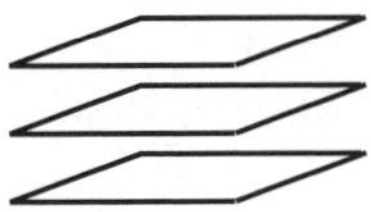

Two parallel planes

In the following example two of the planes are parallel, while the third one is not, and thus cuts through the other two planes.

Example E

In this case the coefficients of the variables in two of the equations are the same or a constant multiple of each other, while the numbers on the right are different.

An example is:

$x - y + 3z = 6$

$x - y + 3z = 5$

$x + y + z = 1$

Here is how this might look:

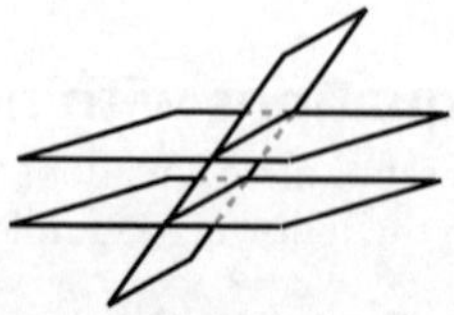

Parallel lines of intersection

In this case, elimination of the same variable from different pairs of equations results in lines which are parallel (and have no point of intersection).

Example F

For this case, the coefficients of the variables are neither the same nor multiples of each other.

In this situation it is often found that one pair of equations can be combined to make an equation which has the same expression on the left-hand side as the third equation, but a different constant on the right-hand side.

As an example, the following system of equations has no solution:

$3x + 4y + 2z = 11$ ... (1)

$2x + y + z = 3$ ... (2)

$x + 3y + z = 5$ ... (3)

By subtracting (2) from (1) the equation $x + 3y + z = 8$ is formed. Comparing this with the 3rd equation $x + 3y + z = 5$ it can easily be seen that there is no point of intersection.

Here is how this might look:

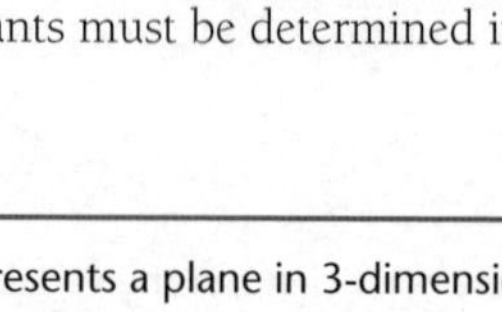

In some problems the values of constants must be determined in order to achieve solutions (or no solutions) of a certain type.

Example G

Q. Each of the equations below represents a plane in 3-dimensional space:

$4x - 2y - z = 1$ ... (1)

$x + 3y + z = 5$ ... (2)

$3x + Ay + Bz = C$ ... (3)

1. Select values for A, B and C so that the three planes meet on a line.
2. Select values for A, B and C so that the three equations are inconsistent and their geometrical description is that two planes are parallel.

A. **1.** Subtracting the second equation from the first gives $3x - 5y - 2z = -4$
which is similar to the third equation $3x + Ay + Bz = C$

By selecting $A = -5$, $B = -2$ and $C = -4$, the third equation becomes the same as the sum of the first two equations and thus the three planes meet on a line and there is an infinite number of solutions.

2. Considering equations (2) and (3)

$$x + 3y + z = 5 \quad \ldots (2)$$
$$3x + Ay + Bz = C \quad \ldots (3)$$

Multiplying (2) by 3 gives

$$3x + 9y + 3z = 15 \quad \ldots (4)$$

Choosing $A = 9$, $B = 3$ and $C \neq 15$ will make (3) and (4) a pair of parallel planes (since the coefficients of x, y and z are equal but the constants on the right-hand side are unequal).

Cramer's rule for the solution of simultaneous equations in three unknowns

Although Cramer's rule provides a systematic approach to solving simultaneous equations in three unknowns, it is more complex than the rule for two unknowns. We will use an example to demonstrate the process:

Solve the following simultaneous equations in three unknowns (previous Example A):

$$3x + y - 2z = 9$$
$$2x + 3y + z = -1$$
$$x + 2y + 3z = -6$$

The determinant for the coefficients of x, y and z is given below:

$$D = \begin{vmatrix} 3 & 1 & -2 \\ 2 & 3 & 1 \\ 1 & 2 & 3 \end{vmatrix}$$

To evaluate D:

Using the values from D:

$$D = \begin{vmatrix} 3 & & \\ & 3 & 1 \\ & 2 & 3 \end{vmatrix} - \begin{vmatrix} & 1 & \\ 2 & & 1 \\ 1 & & 3 \end{vmatrix} + \begin{vmatrix} & & -2 \\ 2 & 3 & \\ 1 & 2 & \end{vmatrix}$$

$$D = 3 \times \begin{vmatrix} 3 & 1 \\ 2 & 3 \end{vmatrix} - 1 \times \begin{vmatrix} 2 & 1 \\ 1 & 3 \end{vmatrix} + (-2) \times \begin{vmatrix} 2 & 3 \\ 1 & 2 \end{vmatrix}$$

$$\begin{aligned} D &= 3 \times (3 \times 3 - 2 \times 1) - 1 \times (2 \times 3 - 1 \times 1) + -2(2 \times 2 - 1 \times 3) \\ &= 3 \times (9 - 2) - (6 - 1) - 2\,(4 - 3) \\ &= 3 \times 7 - 5 - 2 \times 1 \\ &= 21 - 5 - 2 \\ &= 14 \end{aligned}$$

Cramer's rule states that:

$$x = \frac{D_x}{D};\ y = \frac{D_y}{D} \text{ and } z = \frac{D_z}{D}$$

where $D_x = \begin{vmatrix} 9 & 1 & -2 \\ -1 & 3 & 1 \\ -6 & 2 & 3 \end{vmatrix}$; replacing the '$x$' column by the values in the right-hand side of the equation.

Similarly:

$$D_y = \begin{vmatrix} 3 & 9 & -2 \\ 2 & -1 & 1 \\ 1 & -6 & 3 \end{vmatrix} \text{ and } D_z = \begin{vmatrix} 3 & 1 & 9 \\ 2 & 3 & -1 \\ 1 & 2 & -6 \end{vmatrix}$$

$$D_x = 9\begin{vmatrix} 3 & 1 \\ 2 & 3 \end{vmatrix} - 1\begin{vmatrix} -1 & 1 \\ -6 & 3 \end{vmatrix} + (-2)\begin{vmatrix} -1 & 3 \\ -6 & 2 \end{vmatrix}$$

$= 9(9 - 2) -1(-3 + 6) -2(-2 + 18)$

$= 63 - 3 - 32$

$= 28$

$$D_y = 3\begin{vmatrix} -1 & 1 \\ -6 & 3 \end{vmatrix} - 9\begin{vmatrix} 2 & 1 \\ 1 & 3 \end{vmatrix} + (-2)\begin{vmatrix} 2 & -1 \\ 1 & -6 \end{vmatrix}$$

$= 3(-3 + 6) - 9(6-1) -2(-12 + 1)$

$= 9 - 45 + 22$

$= -14$

$$D_z = 3\begin{vmatrix} 3 & -1 \\ 3 & -6 \end{vmatrix} - 1\begin{vmatrix} 2 & -1 \\ 1 & -6 \end{vmatrix} + 9\begin{vmatrix} 2 & 3 \\ 1 & 2 \end{vmatrix}$$

$= 3(-18 + 2) - 1(-12+1) + 9(4-3)$

$= -48 + 11 + 9$

$= -28$

Using $x = \frac{D_x}{D}$; $y = \frac{D_y}{D}$ and $z = \frac{D_z}{D}$

we get the solutions

$$x = \frac{D_x}{D} = \frac{28}{14} = 2;\ y = \frac{D_y}{D} = \frac{-14}{14} = -1 \text{ and } z = \frac{D_z}{D} = \frac{-28}{14} = -2$$

Unit 12.1 Activity 11A: Simultaneous linear equations in three unknowns

1. Solve the following simultaneous equations:

a. $x + y + 2z = 7$
$2x - y + 3z = 14$
$3x + 2y - 4z = -8$

b. $4x - 2y + z = 4$
$2x + y + 3z = 10$
$x + \frac{y}{2} = z$

c. $2x + 3y = z + 4$
$x + z = 6 - 4y$
$x - y = 2z - 2$

2. Use Cramer's rule to solve the following simultaneous equations:

a. $x + 2y - z = 7$
$2x - y + 3z = 1$
$4x + 4y - z = 20$

b. $3x + 2y + z = 1$
$6x + 4y + z = 3$
$3x - 2y - 2z = 2$

c. $x - 2y + z = 2$
$x + 3y + 2z = 3$
$3x - y + 4z = 8$

d. $3x + y + z = 1$
$z - x - y = 0.2$
$2x + 2y + 3 = 2.1$

3. The number of children in the Revaki family can be found by solving these three equations given below (one of the variables stands for the number of children in the Revaki family).

$x + y = 2z - 3$
$y - x + 3z = 16$
$5(y - x) = z$

If it is also known that the number of children in the Revaki family cannot exceed four, find how many children there are in the Revaki family.

4. Jason is told that the number of jobs he will have can be found by solving these equations for x:

$x + 2y - z = 2$
$4x - y + z = 5$
$2x = 1 + 5y - 3z$

Determine if this is true, giving reasons.

5. A club serves a meal consisting of three kinds of food D, E and F. The number of units of each of these foods, d, e and f (respectively), used in this meal can be found by solving the following system of linear equations:

$6d + e + 14f = 70$
$d + e + 2f = 13$
$2d + e + 4f = 24$

Find d, e and f and interpret the solution.

6. Consider the following system of equations in x, y and z:

$2x + 2y + 2z = 9$
$x + 3y + 4z = 5$
$Ax + 8y + 10z = B$

Give possible values of A and B in the third equation which make this system:

a. Inconsistent.

b. Consistent but with an infinite number of solutions.

7. A problem results in the following system of equations:

$2x + 2y - 4z = 4$ ... (1)
$x - 2y + 3z = -4$ ... (2)
$-7x + 2y - z = 2$ ... (3)

a. Eliminate y using equations (1) and (2), and hence form a single equation.

b. Eliminate y using equations (2) and (3), and hence form a single equation.

c. Hence or otherwise show that the resulting pair of equations has no solution.

d. Each of the three equations represents a plane in three-dimensional space. The three planes do not meet at a unique point. Explain geometrically how this occurs in this situation.

8. Each equation below represents a plane in three-dimensional space:

$-3x + 2y + z = -6$

$x - 3y + 2z = -5$

$2x + Ay + Bz = C$

a. If $A = 2$, $B = -1$ and $C = 5$, solve the system of equations for x, y and z.

b. Select a set of values for A, B and C so that the three planes meet on a line.

c. Select a set of values for A, B and C so that the three equations are inconsistent and their geometrical description is that two planes are parallel.

9. Three linear equations of the form $ax + by + cz = d$ represent three planes in three-dimensional space. The three planes all pass through the origin. Describe the **nature** of the solutions to these equations.

10. Consider this set of simultaneous equations:

$x + 2y - 3z = 2$ ... (1)

$2x - y - z = 4$ ... (2)

$3x + 5y + pz = q$ ... (3)

By eliminating x and y from the equations, find the values of p and q that:

a. Make the set consistent but not unique.

b. Make the set inconsistent.

c. Make the set have a unique solution.

Solution of simultaneous equations using calculators

There are many calculators on the market which have modes that enable the user to solve two- and three-variable simultaneous equations.

Equations are first written in the form $ax + by = c$ for two-variable equations, and in the form $ax + by + cz = d$ for three-variable equations.

Coefficients of the variables and the constant are then entered into the calculator.

For example, for the equations $2x + 4y = 7$

$$x - 3y = 5$$

the values 2, 4, 7 and 1, –3, 5 are entered, then SOLV is pressed, resulting in a solution of $x = 4.1$, $y = -0.3$.

Three-variable equations are entered and solved in a similar way.

Simultaneous equations whose solution is not unique

If simultaneous equations do not have a unique solution then the calculator simply returns an error message. This occurs whether the equations have either infinitely many solutions or inconsistent solutions. Further investigation of the type suggested in this chapter is required in order to determine the **nature** of the equations.

Word problems

Many practical problems will be expressed in words, and the information given needs to be written as an equation (or set of equations). Assign a letter to each unknown and form equations using these variables. The equations are then solved simultaneously to give the values of the unknowns.

Example H

Q. A retailer orders 100 rugby jumpers. The large size costs her K56 each, the medium size K50 each, and the small size K44 each. She spends a total of K4 970. She makes a profit of K8 on the large and medium jumpers, and K6 on the small. Her total profit is K740. Find how many of each type of jumper she purchased.

A. Let L stand for the number of large jumpers, M for the number of medium jumpers, and S for the number of small jumpers. As she orders 100 jumpers in total, the resulting equation is:

$L + M + S = 100$

Using the cost information gives the equation:

$56L + 50M + 44S = 4\,970$

and the profit information gives the equation:

$8L + 8M + 6S = 740$

Solving these three equations simultaneously gives $L = 25$, $M = 45$, $S = 30$.

$\therefore$ she purchased 25 large jumpers, 45 medium jumpers and 30 small jumpers.

Unit 12.1 Activity 11B: Simultaneous linear equations and word problems

1. There are three numbers. The largest number is two and a half times bigger than the smallest number. The sum of all three numbers is 24.5. Double the largest number plus the other two numbers is 37. Find the numbers.

2. A garage owner hires three people. Their combined hourly wage is K43. The difference between the highest wage and the lowest wage is K2 per hour. The highest- and lowest-paid workers work for 2 hours; the other works for 3 hours. The total wage bill is K100. What are the hourly rates for each of the workers?

3. A fisherman catches a total of 120 fish. The fish are of three types: A, B and C. A has average weight 3 kg, B has average weight 5 kg, and C has average weight 6.5 kg. A sells on average for K3 per fish, B for K8 and C for K10.50. The total weight of fish caught is 513 kg; the total value of fish caught is K715. Find the number of each type of fish caught.

4. A supermarket stocks three types of flour on the shelves. The types are distinguished by the bags they come in. Two bags of A have the same weight as 3 bags of B, and the same weight as 5 bags of C. There is a total of 50 bags of flour in the shop. There are 12 bags of A and 18 bags of B. The total weight of flour in the shop is 480 kg. What are the weights of individual bags of A, B and C?

5. Technicians at a company get K52 per week more than labourers. Accountants get K36 per week more than technicians. Each week this company pays a total of K12 814 in wages to its workers coming into these three categories. There are 5 accountants, 12 technicians, 30 labourers. Find out how much each type of worker gets paid.

6. An alloy consists of a mixture of three different types of metals. Type A has a weight of 3 kg per unit, B has weight of 2.5 kg per unit and C has a weight of 1.5 kg per unit. A quantity of 77 kg of the alloy is made which requires a combined total of 30 units of the three metal types. There are twice as many units of A as of B. How many units of each type of metal were required to make this 77 kg quantity of alloy?

7. A company uses three types of car. To run type A costs K0.30 per km, type B costs K0.80 per km and type C costs K0.40 per km. The total number of cars in the fleet is 36. On average, cars of type A are driven 100 km per day, cars of type B are driven 30 km per day, and cars of type C are driven 55 km per day. The average distance driven per day for the entire fleet is 2 870 km. The average cost per day of running the fleet is K976. Find out how many cars of each type the company runs.

8. Another company has 26 cars of type D, 15 of type E and 7 of type F. Each car in the fleet averages 120 km per day. The cost for running one car of each type for 1 km simultaneously is K1.65. The cost of running the D and E cars is K2 712 per day; the cost of running the E and F cars is K1 956 per day. Find the cost per kilometre of running each type of car.

9. A secondhand bookseller specialises in the sale of textbooks. Maths books sell for K18 each, science books for K22 each and commerce books for K14 each. On Monday he sells a total of K240 of these kinds of books; on Tuesday he sells twice as many maths books but only half the number of commerce books he did on the previous day; sales of science books remain constant, yielding a total sale of K288. On Wednesday sales of commerce books return to Monday's volume, maths books are three up on the previous day, and sales of science books twice the level of Tuesday. Total sales on Wednesday amount to K450. Find out how many books in each subject area were sold on Monday.

10. A quadratic function has the formula: $f(x) = Ax^2 + Bx + C$. $f(1) = 21, f(2) = 35$ and $f(-1) = 11$. Find A, B and C.

11. A pupil sits three exams: English, history and geography. Past experience has shown that this pupil's English mark is usually higher by 10 than the history mark, and the geography mark is higher by 5 than the English mark. In an exam the combined total for the pupil's marks in these three subjects was 220. What were the pupil's marks in each subject?

12. A plane has the points (1, 1, 1), (2, 2, 0) and $(\frac{1}{3}, 1, 2)$ on it. Find A, B, C if the equation of the plane is $Ax + By + Cz = 4$.

13. A hospital serves a special meal consisting of three kinds of food, *A*, *B* and *C*.

Each unit of food *A* has 20 g of carbohydrate, 2 g of fat and 4 g of protein.

Each unit of food *B* has 5 g of carbohydrate, 1 g of fat and 2 g of protein.

Each unit of food *C* has 80 g of carbohydrate, 3 g of fat and 8 g of protein.

The dietician designs a special meal so that it contains 140 g of carbohydrate, 11 g of fat and 24 g of protein. Let a, b and c be the number of units of food *A*, *B* and *C* (respectively) used in the special meal.

a. Set up a system of three simultaneous equations relating a, b and c.

b. Solve the equations.

14. When newsprint is supplied to a newspaper company it is rewound from large reels to smaller reels. During this rewinding process, the tension on the paper should not change. This is achieved by altering the load on the smaller reel as its diameter increases.

The relationship between the load, L kiloNewtons per metre (L kN m^{-1}), and the reel diameter, x metres, is modelled by a graph consisting of two parabolic arcs, AB and BC, as shown.

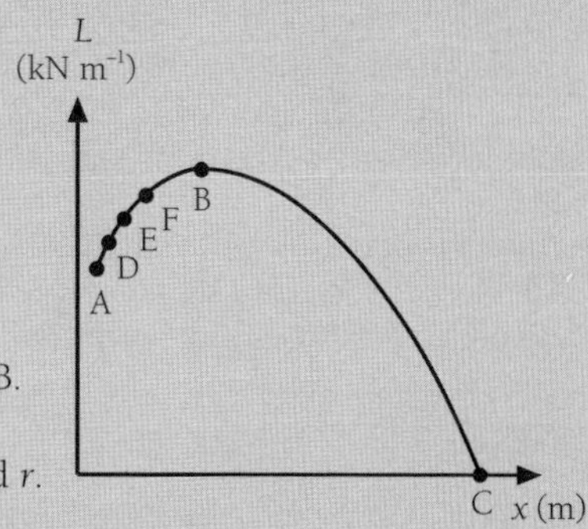

a. Arc AB is part of the parabola $L = px^2 + qx + r$. Points D (0.1, 2.025), E (0.2, 2.9) and F (0.3, 3.425) lie on arc AB.

i. Set up a system of three simultaneous equations relating p, q and r.

ii. Solve the equations.

b. Arc BC is part of the parabola $L = sx^2 + tx + u$. The coefficients s, t and u can be found by solving the following system of linear equations:

$0.25s + 0.5t + u = 3.5$

$0.36s + 0.6t + u = 3.2$

$0.64s + 0.8t + u = 2.0$

Find s, t and u.

15. There are three companies selling a product in a town. The total value of the product in June is K200 000. During June company A sells K3 200 more than company B and company C combined, and company B sells twice as much as company C. Set up and solve a system of equations to find the value of the product that each company sold in June.

16. A takeaway bar has three types of burger. Information about these is given in the table below:

Burger	A	B	C
Type of filling	Pork	Beef	Vegetarian
Cost to purchase	K3	K4.50	K1.90
Weight of filling	180 g	200 g	200 g

In a week of trading a total of 460 of these burgers were sold. The total weight of filling was 88.5 kg and the total income from the burgers was K1 547.50. Set up and solve a system of equations in order to determine the number of each type of burger sold during that week.

17. In a sale some books are sold at K10, K15 and K20. A customer purchased 20 of the books. He paid K285 and purchased as many of the cheapest books as he did of the other two types combined. Set up a system of the three equations and solve it to find out how many of each type was purchased.

18. Tracey is paid K849 a week. She pays board which covers all her living expenses, she invests some and spends the rest. Her board is K105 more than the combined amounts she invests and spends. She invests K60 more than she spends. Set up a system of equations and determine how much Tracey spends on board, invests and spends otherwise.

19. A garden centre makes three types of compost: azalea, bulb and all-purpose, to suit customers' requirements. To make these types of compost, nitrogen, phosphorus and potassium are added in different proportions. The ratio of nitrogen: phosphorus: potassium for the three types of compost is as follows: azalea 1:1:1, bulb 4:5:6 and all-purpose 5:6:7. For a particular week's production there are 5.3 tonnes of nitrogen, 8.4 tonnes of phosphorus and 11.5 tonnes of potassium available.

Using this information, a system of equations can be formed to calculate the amount of each type of compost able to be manufactured, using all the nitrogen, phosphorus and potassium available. If a, b and c represent respectively, in tonnes, the amount of azalea, bulb and all-purpose compost manufactured during this week, then two of these equations are:

$$30a + 24b + 25c = 477$$
$$30a + 30b + 30c = 756$$

a. Find the third equation.
b. Find the general solution for this system of equations in terms of c.
c. Comment on the practical consequences of the solution in part **b**.

Unit 12.2 Trigonometry and Vectors

Topic 1: Basic trigonometry, circular functions and radian measure

Trigonometry is used in a wide range of activities in modern society: mapping, planning, designing, surveying and navigation. It emphasises the importance of modelling practical situations. Vectors enforce the geometrical applications.

Sine, cosine and tangent

Consider the angle x in the right-angled triangle shown. If the size of angle x stays the same while the size of the triangle changes, the values of the ratios $\frac{o}{h}$, $\frac{a}{h}$ and $\frac{o}{a}$ remain the same. These values are called the **sine**, **cosine** and **tangent** of the angle x.

$\frac{o}{h} = \text{sine } x$

$\frac{a}{h} = \text{cosine } x$

$\frac{o}{a} = \text{tangent } x$

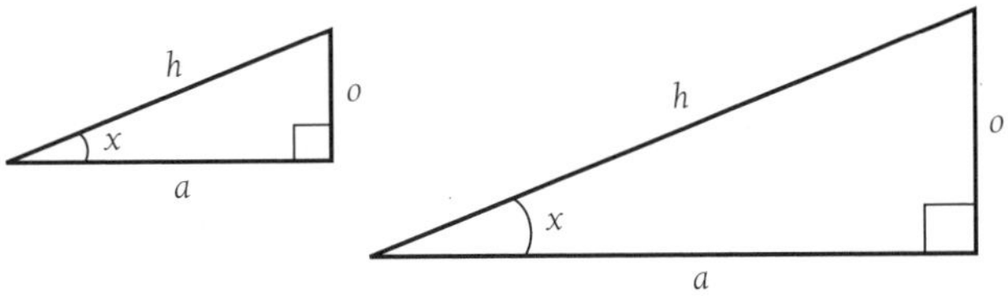

Note: o refers to the side *opposite* angle x.
a refers to the side *adjacent* to angle x.
h refers to the **hypotenuse**, the side opposite to the right angle.
The hypotenuse is always the longest side of a right-angled triangle.

The trigonometrical ratios sine x, cosine x and tangent x are commonly abbreviated to $\sin x$, $\cos x$ and $\tan x$.

Example A

Q. Prove that $\sin 45° = \cos 45° = \frac{1}{\sqrt{2}}$ and $\tan 45° = 1$.

A. In an **isosceles** right-angled triangle with equal sides of length 1, the hypotenuse is of length $\sqrt{2}$.

The two equal angles are each 45°.

From the diagram,

$\sin 45° = \frac{o}{h} = \frac{1}{\sqrt{2}}$

$\cos 45° = \frac{a}{h} = \frac{1}{\sqrt{2}}$

$\tan 45° = \frac{o}{a} = \frac{1}{1} = 1$

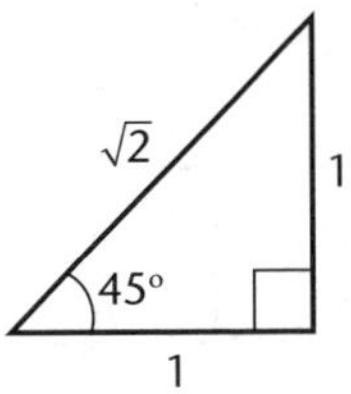

Example B

Q. Prove that $\sin 30° = \frac{1}{2}$, $\cos 30° = \frac{\sqrt{3}}{2}$ and $\tan 30° = \frac{1}{\sqrt{3}}$.

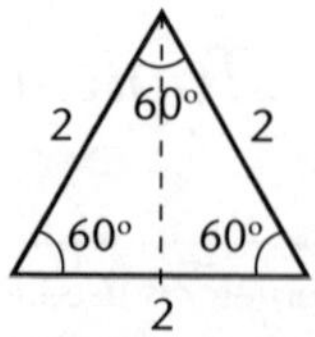

A. In an equilateral triangle of side length 2, all angles are 60°.

This triangle is bisected (by the dotted **axis of symmetry** shown) to create two **congruent** right-angled triangles as shown.

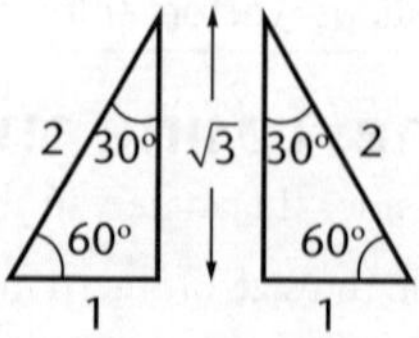

Using the theorem of **Pythagoras**, the vertical side length is $\sqrt{3}$.

From either right-angled triangle the following trigonometrical ratios can be read off:

$$\sin 30° = \frac{o}{h} = \frac{1}{2}$$

$$\cos 30° = \frac{a}{h} = \frac{\sqrt{3}}{2}$$

$$\tan 30° = \frac{o}{a} = \frac{1}{\sqrt{3}}$$

Note: The values of sin 60°, cos 60° and tan 60° can also be read off easily from these right-angled triangles.

Table of exact trigonometric values

The following table summarises some important exact trigonometric values frequently used in mathematics.

It is useful to learn these values.

Angle (in **degrees**)	Sin	Cos	Tan
0	0	1	0
30	$\frac{1}{2}$	$\frac{\sqrt{3}}{2}$	$\frac{1}{\sqrt{3}}$
45	$\frac{1}{\sqrt{2}}$	$\frac{1}{\sqrt{2}}$	1
60	$\frac{\sqrt{3}}{2}$	$\frac{1}{2}$	$\sqrt{3}$
90	1	0	Undefined

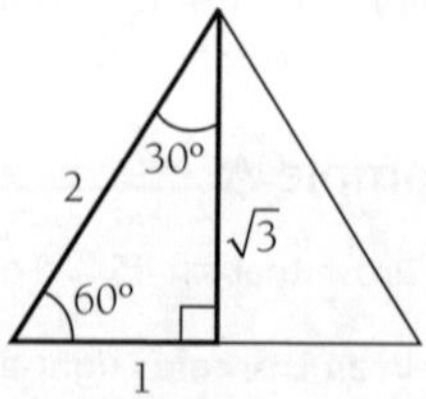

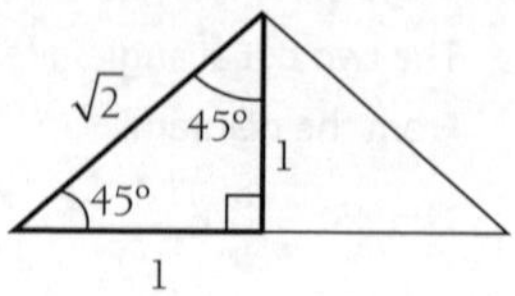

Note: Values (such as $\frac{\sqrt{3}}{2}$ or $\frac{1}{\sqrt{2}}$) with $\sqrt{\ }$ signs in them are called **surds**.

Circular functions: introduction and definitions

A **circle** of radius 1 with its centre at the origin is shown with P in the first **quadrant** and the angle x **acute**. Point P at the end of the radius has $\cos x$ as its x-coordinate, and $\sin x$ as its y-coordinate, that is, P is the point $(\cos x, \sin x)$.

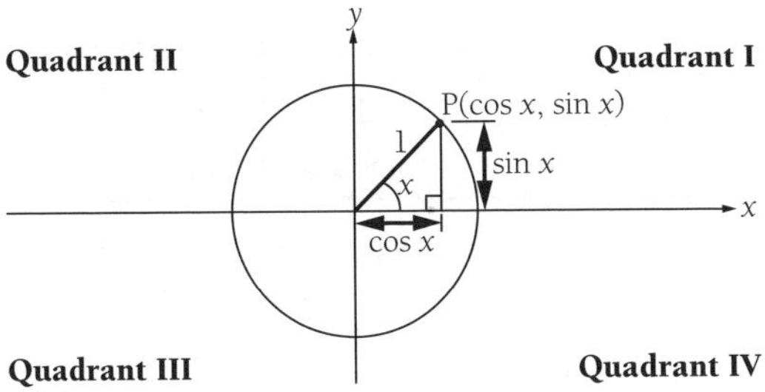

The observations above can be extended to *all angles in any quadrant* using the following definitions. If P is the point at the end of the radius of length 1, which makes an angle x with the positive horizontal axis, then:

$$\cos x = \text{horizontal coordinate of P}$$
$$\sin x = \text{vertical coordinate of P}$$
$$\tan x = \frac{\sin x}{\cos x}$$

The following diagram shows the angle x between 90° and 180° (ie in the 2nd quadrant) and illustrates an important relationship between x and $(180 - x)$.

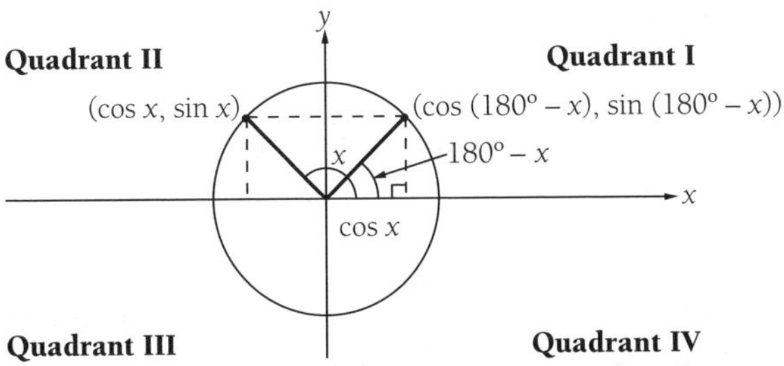

In the 2nd quadrant the radius (making an angle x with the positive horizontal axis) is the **reflection in the y-axis** of the radius which forms the **acute angle** $(180° - x)$.

It can be seen from the previous diagram and definitions that:

$\cos x = -\cos (180° - x)$ [since the x-coordinate of the two points on the circumference corresponding to angles x and $(180° - x)$ are equal and opposite]

$\sin x = \sin (180° - x)$ [since the y-coordinates of the two points are the same]

$\tan x = \dfrac{\sin x}{\cos x}$ [by definition]

$= \dfrac{\sin (180° - x)}{-\cos (180° - x)}$ [substituting $\sin x = \sin (180° - x)$ and $\cos x = -\cos (180° - x)$]

$= -\tan (180° - x)$ [simplifying]

The table below shows how the **trigonometric function** of an angle greater than 90° can be expressed as the trigonometric function of an angle in the first quadrant.

Angle	Quadrant	cos x	sin x	tan x
$0° < x \leq 90°$	I	$\cos x$	$\sin x$	$\tan x$
$90° < x \leq 180°$	II	$-\cos (180° - x)$	$\sin (180° - x)$	$-\tan (180° - x)$
$180° < x \leq 270°$	III	$-\cos (x - 180°)$	$-\sin (x - 180°)$	$\tan (x - 180°)$
$270° < x \leq 360°$	IV	$\cos (360° - x)$	$-\sin (360° - x)$	$-\tan (360° - x)$

Note: The trigonometrical functions are frequently called the **circular functions** because they are defined by the co-ordinates of the points on a circle.

Example C

1. $\sin 130° = \sin (180° - 130°)$ [130° in quad II, $\sin x = \sin (180° - x)$]

$= \sin 50°$

2. $\cos 220° = -\cos (220° - 180°)$ [220° in quad III, $\cos x = -\cos (x - 180°)$]

$= -\cos 40°$

The mnemonic **All Science Teachers Count** is applied from quadrant I, going **anticlockwise** to decide which quadrants have positive values for each of the trigonometrical functions:

A refers to quadrant I and stands for *all* are positive.

S refers to quadrant II and stands for *sine only* is positive.

T refers to quadrant III and stands for *tangent only* is positive.

C refers to quadrant IV and stands for *cosine only* is positive.

The following diagram summarises the **data** from the table above and combines it with the mnemonic.

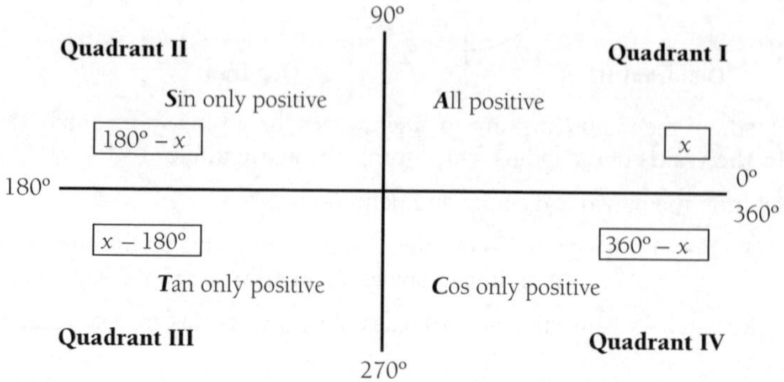

The sine, cosine and tangent of angles between 90° and 360° can be evaluated using this summary of results.

Example D

1. $\sin 135° = \sin(180° - 135°)$ [since 135° is in quadrant II]

$= \sin 45°$

$= \frac{1}{\sqrt{2}}$

2. $\tan 330° = -\tan(360° - 330°)$ [since 330° is in quadrant IV]

$= -\tan 30°$

$= \frac{-1}{\sqrt{3}}$

Angles greater than 360°

The trigonometrical functions of angles greater than 360° can be expressed in terms of an acute angle in the following way:

Example E

$750° = 2 \times 360° + 30°$ [2 revolutions plus 30°]

$\therefore \cos 750° = \cos 30°$

The following diagrams illustrate this:

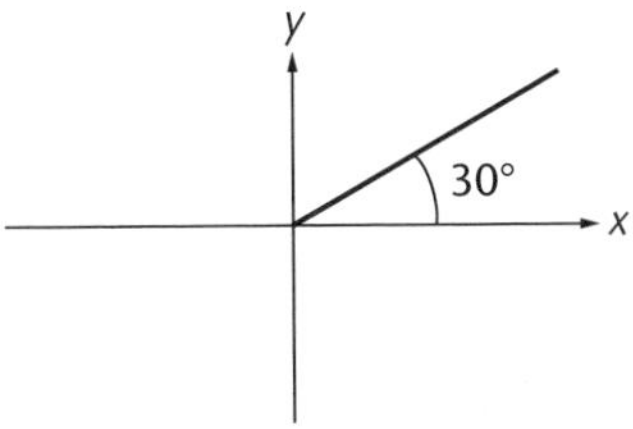

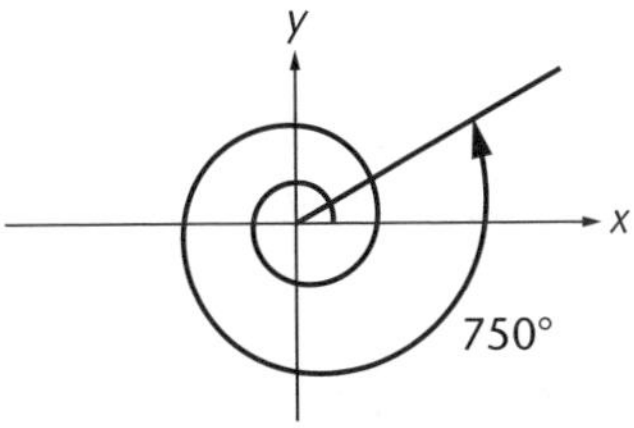

Unit 12.2 Activity 1A: Finding sines, cosines and tangents

1. Write each of the following as positive or negative sine, cosine or tangent of an acute angle: [example: $\sin 312° = -\sin 48°$]

a. cos 237°	**b.** sin 136°	**c.** tan 325°	**d.** tan 173°
e. cos 310°	**f.** sin 256°	**g.** tan 214°	**h.** sin 342°
i. cos 162°	**j.** sin 500°	**k.** cos 392°	**l.** tan 1 060°
m. cos 1 126°	**n.** sin 3 647°	**o.** cos 53°	

2. Find the **exact values** of the following, leaving answers in surd, fraction or whole-number form:

a. cos 45°	**b.** sin 60°	**c.** tan 30°	**d.** cos 150°
e. sin 210°	**f.** tan 240°	**g.** tan 330°	**h.** cos 540°
i. sin 420°	**j.** cos 675°	**k.** cos 225°	**l.** sin 225°
m. sin 315°	**n.** cos 330°	**o.** sin 300°	

Introduction to radian measure

The formulae for the circumference and **area of a circle** are $C = 2\pi r$ and $A = \pi r^2$ respectively, where r is the radius. These formulae involve π (pronounced 'pi'), an **irrational** number

which has approximate values of $\frac{22}{7}$, 3, 3.1, 3.14, 3.142, …, depending on the degree of accuracy required.

By associating the size of an angle with the length of the **arc** it **subtends**, an alternative way of measuring the sizes of angles is defined, in which angles are measured in **radians** rather than degrees. The circumference of a circle of radius 1 is 2π, so a 360° angle is called 2π radians. An important conversion resulting from this is:

$$\boxed{\pi \text{ radians} = 180^\circ}$$

Thus 3 radians are approximately 180°, or, more accurately, 3.142 radians are approximately 180°. One radian is approximately 57.3°.

Conversion between radians and degrees

An angle measured in degrees can be changed to radians and vice versa using the relationship, π radians = 180°.

Example F

Q. Change: **1.** 48° to radians. **2.** $\frac{2\pi}{15}$ radians to degrees.

A. 1. $180^\circ = \pi$ radians

$\therefore 1^\circ = \frac{\pi}{180}$ [dividing by 180]

$\therefore 48^\circ = \frac{48 \times \pi}{180}$ [multiplying by 48]

= 0.838 radians (3 dp) [by calculator]

2. π radians = 180°

$\therefore \frac{2\pi}{15}$ radians $= \frac{2 \times 180^\circ}{15}$ [dividing by 15 and multiplying by 180]

= 24°

Example G

Q. A wheel of radius 1.5 m does three revolutions in 1 minute.

1. How many radians does the wheel turn through in 45 seconds?

2. How fast does a point on the circumference travel in metres per second?

A. 1. The wheel turns through 3 revolutions in 60 seconds.

$\therefore$ In 45 seconds, the wheel turns $\frac{45}{60} \times 3$ revolutions. [using **proportion**]

$= \frac{45 \times 3 \times 2 \times \pi}{60}$ [since 1 revolution is 2π radians]

= 14.14 radians (2 dp)

2. In 1 minute a point on the circumference travels a distance equal to three times the circumference, which is $3 \times 2\pi r = 6\pi r$ metres.

$\therefore$ its speed is $\frac{6 \times \pi \times 1.5}{60}$ metres per second [speed = $\frac{\text{distance}}{\text{time}}$]

$= 0.5\ \text{ms}^{-1}$ (1 dp)

When are radians used?

In most practical situations, such as building, navigation and engineering, angles are normally measured in degrees.

However, angles are measured in radians in most advanced areas of pure mathematics and science for a number of reasons.

One good reason is that if x is measured in radians, the ratio $\frac{\sin x}{x}$ gets close to 1, as x gets close to 0.

This result is very important and is used to establish many other important results in more advanced mathematics.

Unit 12.2 Activity 1B: Radians

1. Change the following angles given in degrees into radians (take $\pi = 3.14$):

a. 18° **b.** 38° **c.** 156° **d.** 247° **e.** 418°

2. Change the following angles, given in radians, into degrees:

a. 2.6 **b.** 3.4 **c.** 1.18 **d.** 0.57 **e.** 0.03

3. The following angles are given in radians as multiples or fractions of π. Change them into degrees:

a. $\frac{2\pi}{3}$ **b.** $\frac{2\pi}{4}$ **c.** $\frac{5\pi}{9}$ **d.** $\frac{3\pi}{8}$ **e.** $\frac{4\pi}{5}$

f. $\frac{7\pi}{12}$ **g.** $\frac{5\pi}{12}$ **h.** $\frac{\pi}{8}$ **i.** 3π **j.** $\frac{5\pi}{6}$

4. Change these angles, given in degrees, into radians expressed as fractions or multiples of π:

a. 120° **b.** 135° **c.** 315° **d.** 405° **e.** 22.5°

f. 72° **g.** 108° **h.** 84° **i.** 132° **j.** 27°

5. A fly walking round a circle does 5.5 circuits. How many radians will the fly have turned through?

Exact trigonometric values for angles in radians

Angles in radians can be converted to degrees (using the conversion π radians = 180°) in order to find exact trigonometric values.

Example H

Q. Find the value of $\sin \frac{4\pi}{3}$, leaving the answer in surd form.

A. $\frac{4\pi}{3}$ radians $= \frac{4}{3} \times 180°$ [changing $\frac{4\pi}{3}$ radians to degrees]

$= 240°$

$\therefore \sin \frac{4\pi}{3} = \sin 240°$

$= -\sin 60°$ [since 240° is in quadrant III]

$= -\frac{\sqrt{3}}{2}$

Unit 12.2 Activity 1C: Exact trigonometric values for angles in radians

Evaluate each of the following, giving the answer exactly:

1. $\sin \frac{\pi}{2}$ **2.** $\cos \frac{3\pi}{4}$ **3.** $\sin \frac{3\pi}{4}$ **4.** $\tan \frac{3\pi}{4}$

5. $\tan (\frac{-\pi}{3})$ **6.** $\sin (\frac{-\pi}{6})$ **7.** $\cos (\frac{-2\pi}{3})$ **8.** $\cos 2\pi$

9. $\sin \frac{4\pi}{3}$ **10.** $\tan (\frac{7\pi}{6})$

Graphs of the trigonometrical functions

The graph of $y = \sin x$ can be plotted from the points in the table:

$x°$	–360	–270	–180	–90	0	90	180	270	360	450	540
$\sin x°$	0	1	0	–1	0	1	0	–1	0	1	0

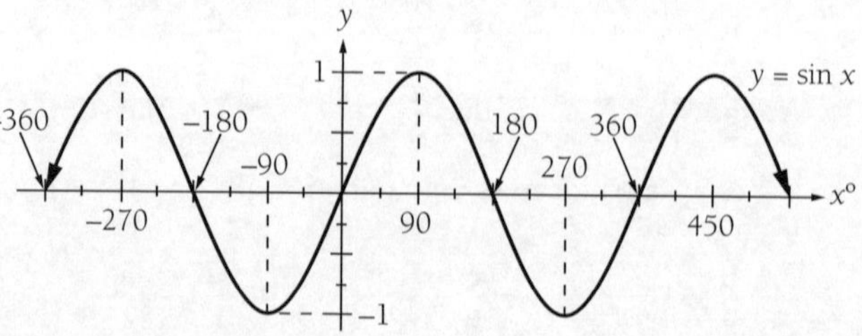

Similarly the graph of $y = \cos x$ is:

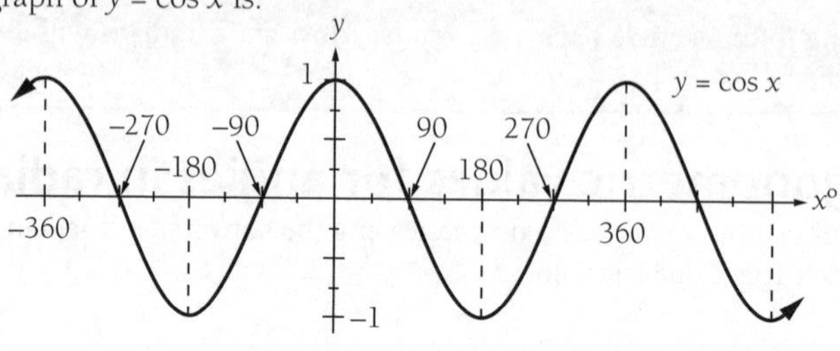

Notes:

1. The graphs of $y = \sin x$ and $y = \cos x$ have a characteristic curved shape (found by calculating intermediate points such as $\sin 30° = 0.5$ or $\cos 45° = 0.71$).
2. The graphs of sine and cosine repeat themselves every 360°. The sine and cosine functions are said to be **periodic** with a **period** of 360°.
3. The **domain** of both sine and cosine is R, the set of all real numbers.
4. The **range** of both sine and cosine is $-1 \le y \le 1$.
5. The **amplitude** of the sine and cosine graphs is the height of the graph above its **mean** line (here the x-axis). It is half the distance between the maximum and **minimum** y-values of the graph. The amplitude of $y = \sin x$ and $y = \cos x$ is 1.

The graph of $y = \tan x$ is quite different from the sine and cosine graphs, as seen by **plotting** the points from the following table:

$x°$	–90	–45	0	45	90	135	180	225	270	315
$\tan x°$	E	–1	0	1	E	–1	0	1	E	–1

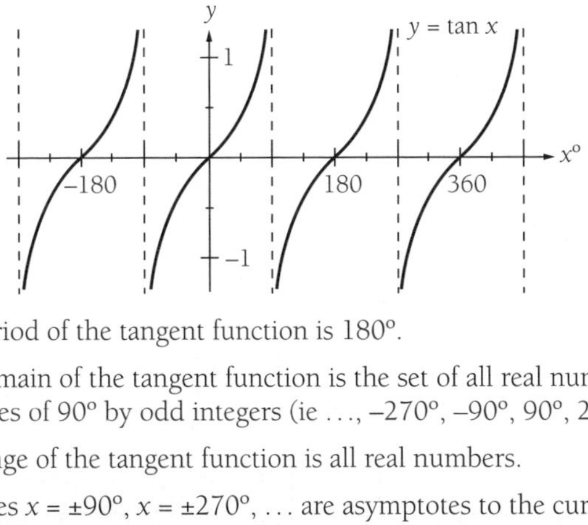

Notes:

1. The period of the tangent function is 180°.
2. The domain of the tangent function is the set of all real numbers except multiples of 90° by odd integers (ie …, –270°, –90°, 90°, 270°, 450°, …).
3. The range of the tangent function is all real numbers.
4. The lines $x = \pm 90°$, $x = \pm 270°$, … are asymptotes to the curve $y = \tan x$.

Unit 12.2 Activity 1D: Basic trigonometric graphs

1. Copy and complete the graph of $y = \cos x$ from 0° to 360°. Draw and label the y-axis clearly.

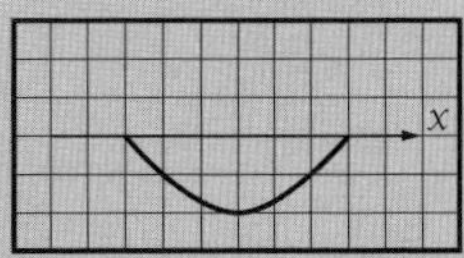

2. Copy and complete the graph of $y = \sin x$ from 0° to 360°. Draw and label the y-axis clearly.

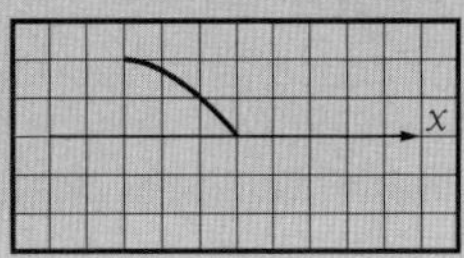

3. A diagram of part of the graph of one of the trigonometric functions is shown.

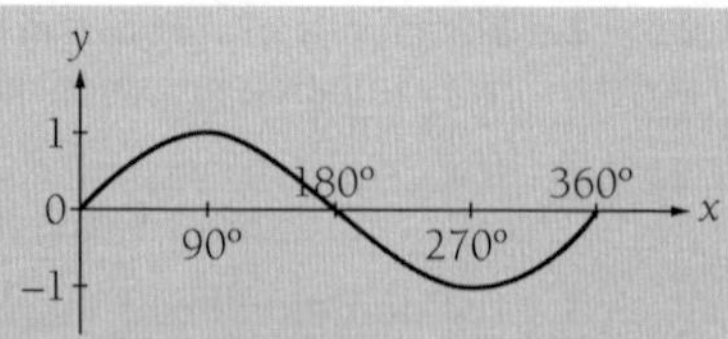

a. Identify the trigonometric function.
b. What is the domain of the function in the diagram?
c. What is the range of the function in the diagram?
d. Does the graph have any symmetries?
e. Are there any asymptotes for this function?

4. The diagram shows the graph of a function $y = f(x)$.

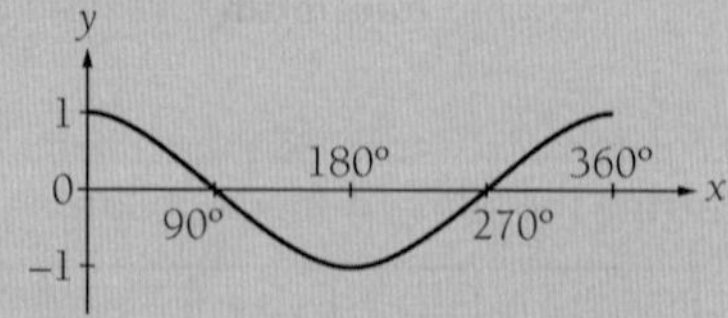

a. What is the value of $f(90°)$?
b. Which value of x satisfies $f(x) = -1$?
c. What is the domain of f?
d. What is the range of f?
e. Which trigonometric function gives the same values as f over this domain?

5. The graph shows a trigonometric function $y = g(x)$ for $0 \le x \le 360°$.

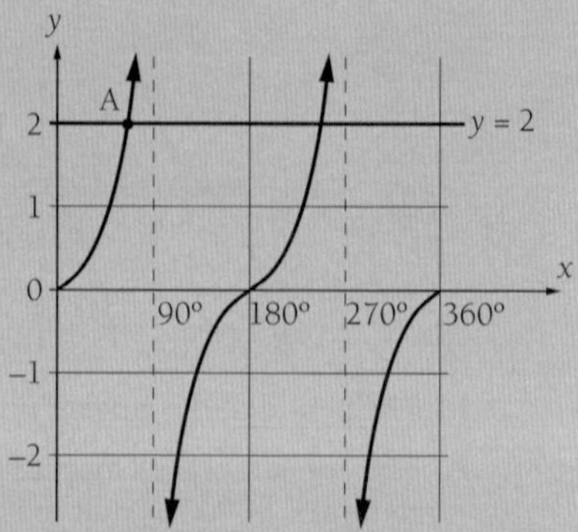

a. Identify the trigonometric function.
b. Which angles x, where $0 \le x \le 360°$, are undefined for $g(x)$?
c. What are the equations of the asymptotes in this diagram?
d. Describe the symmetries shown by the graph of $y = g(x)$ where $0 \le x \le 360°$.
e. The line $y = 2$ cuts the graph of $y = g(x)$ at the point marked A. What is the x-coordinate of A?
f. If the graph of $y = \cos x$ was drawn on the same diagram as $y = g(x)$, how many times would the two graphs intersect?
g. If the graph of $y = \sin x$ was drawn on the same diagram as $y = g(x)$, how many times would the two graphs intersect?

Trigonometric functions with one transformation

The basic trigonometric graphs can be transformed by horizontal or vertical translations, reflections or change of scale (stretch or **compression**). Note carefully the effect each transformation has on the equation of the graph.

Example I

Q. Sketch the graph of $y = \sin(x + 90°)$ and identify its features.

A. A table of values is given below:

$x°$	–180	–90	0	90	180	270	360	450
y	–1	0	1	0	–1	0	1	0

The graph is sketched from the table:

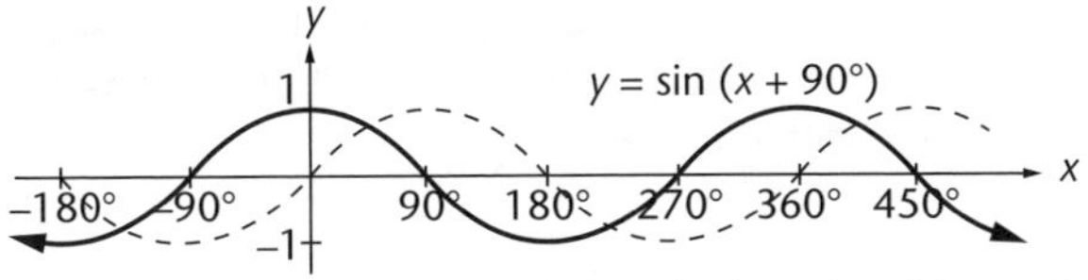

Comparing the graph of $y = \sin(x + 90°)$ with the graph of $y = \sin x$ (shown dotted), it can be seen that $y = \sin x$ has been translated 90° to the left to become $y = \sin(x + 90°)$.

Interesting features of $y = \sin(x + 90°)$ include:

- The graph is periodic with period 360°.
- The amplitude of $y = \sin(x + 90°)$ is 1.
- The lines $x = (180n)°$ are mirror lines for the graph, where n is any integer.
- The graph has point symmetry about every point $(90n°, 0)$ where n is any odd integer.
- The graph of $y = \sin(x + 90°)$ is the same graph as $y = \cos x$.

Example J

Q. Sketch the graph of $y = 3\sin x$, $-180° \le x \le 360°$ and identify its features.

A. A graph is drawn from the table of values below:

$x°$	–180	–90	0	90	180	270	360
y	0	–3	0	3	0	–3	0

In this case, the graph of $y = \sin x$ (shown dotted) has been stretched vertically by a factor of 3, without any horizontal stretch taking place.

- The amplitude of $y = 3\sin x$ is 3.
- By restricting the domain to $-180° \le x \le 360°$ (ie [–180°, 360°]), the graph has only the line $x = 90°$ as an axis of symmetry (no point symmetry).

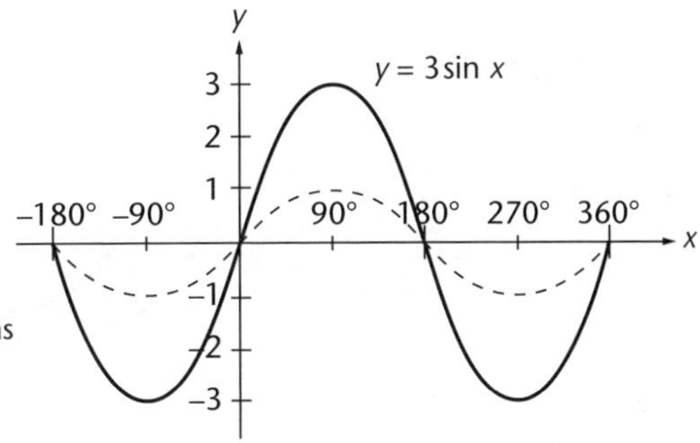

Note: The amplitude of $y = k\sin x$ (or $y = k\cos x$) is $|k|$, ie the size of the number multiplying the sine or cosine is the amplitude of the curve.

Example K

Q. Sketch the graph of $y = \cos x - 2$.

A. By plotting points and joining them, the graph shown below is obtained. It is the **translation** of $y = \cos x$ by the vector $\begin{pmatrix} 0 \\ -2 \end{pmatrix}$

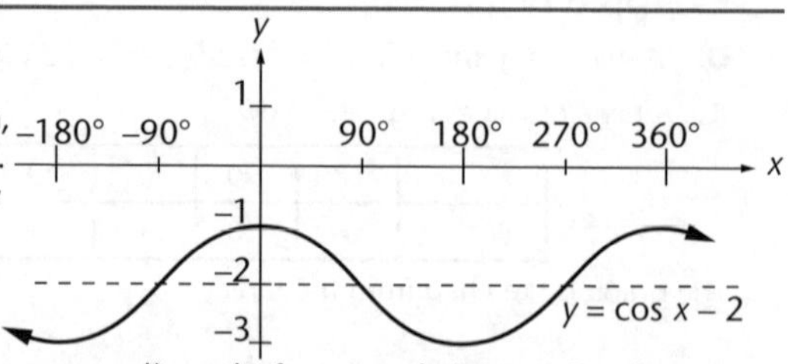

Note: The graph of $y = \sin x + k$ (or $y = \cos x + k$) results from translating $y = \sin x$ (or $y = \cos x$) vertically by k units, ie by the vector $\begin{pmatrix} 0 \\ k \end{pmatrix}$

Example L

Q. Sketch the graph of $y = \sin 3x$, and identify its features.

A. By plotting points and joining them, the graph is as shown below:

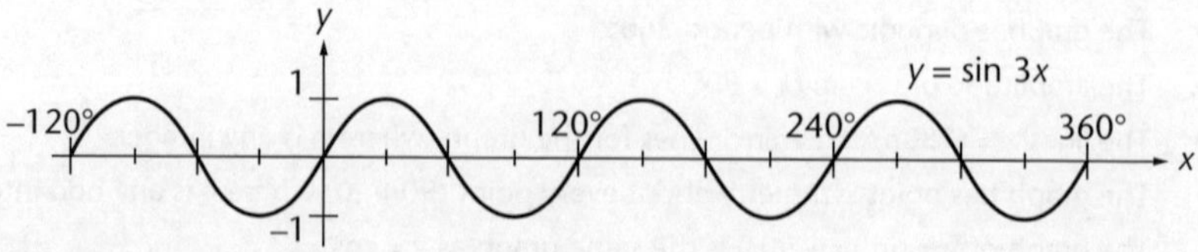

The graph of $y = \sin x$ has been **compressed** horizontally by a factor of 3. The following features can be seen:

- The period of the graph is 120° [which is $\frac{360°}{3}$]
- The amplitude of the graph is 1
- The graph has point symmetry about (0°, 0), (±60°, 0), (±120°, 0), …

Note: The period of $y = \sin kx$ (or $y = \cos kx$) is $\frac{360°}{k}$

By considering the features of a graph, its equation can be written down.

Example M

Q. Write down the equation of the graph shown.

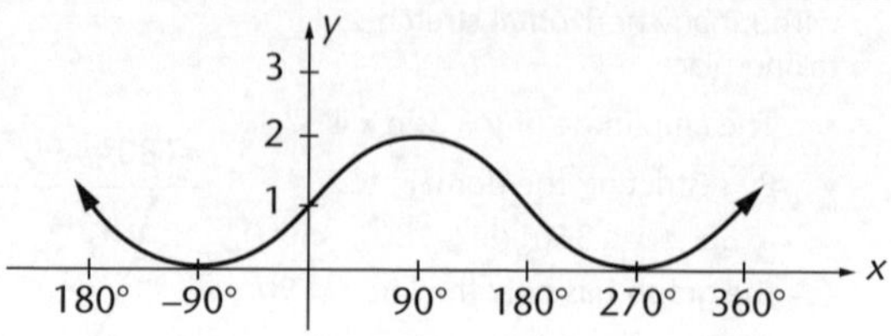

A. The graph of $y = \sin x$ has been translated vertically by 1 unit. The equation of the graph is therefore $y = \sin x + 1$.

Unit 12.2 Activity 1E: Graphs of transformed trigonometric functions

Sketch the following graphs:

1. $y = \cos(x + 90°)$ **2.** $y = \sin(x - 90°)$ **3.** $y = 2\sin x,\ 0 \le x \le 360°$

4. $y = -\cos x,\ -90° \le x \le 270°$ **5.** $y = \sin x + 1$ **6.** $y = \sin(x + 30°)$

7. $y = \sin(x - 30°),\ -150° \le x \le 210°$ **8.** $y = \sin 2x,\ -90° \le x \le 270°$

Write down the equations of the following graphs:

9.

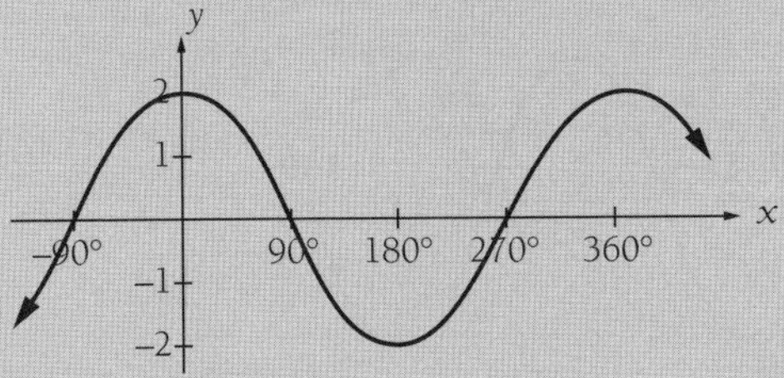

10.

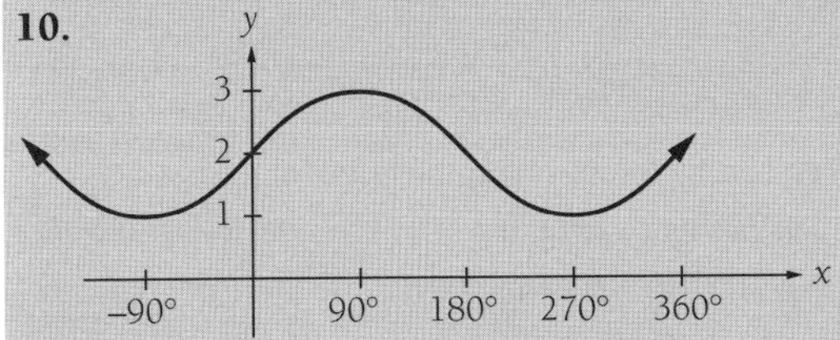

11.

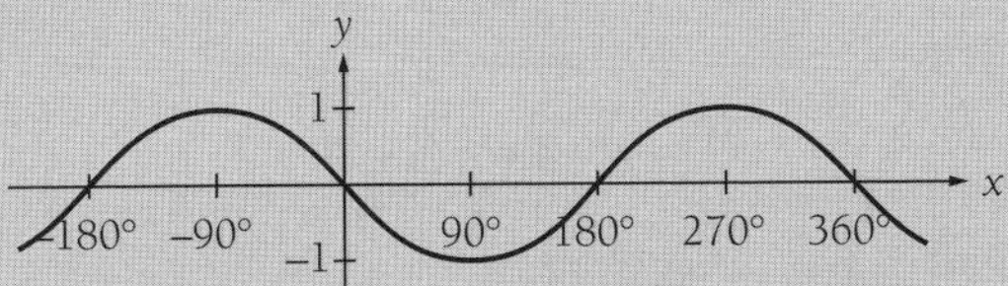

12.

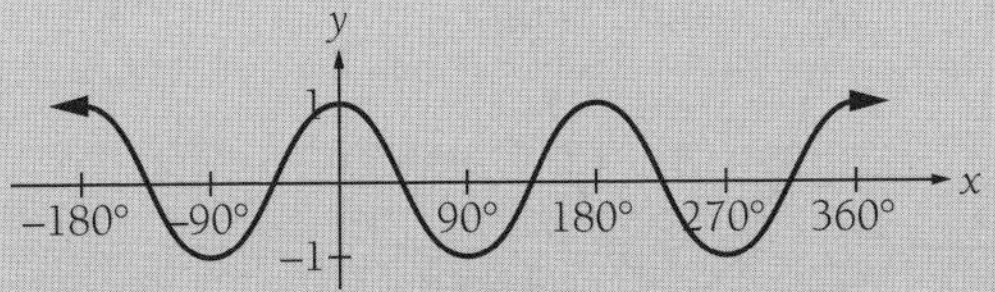

Graphs of trigonometric functions with two transformations

As has been seen, each transformation of $y = \sin x$ (or $y = \cos x$) has an effect on the equation of the graph. By considering each of these changes in turn, the graph of a trigonometric function with more than one transformation can be sketched.

Example N

Q. Sketch the graph of $y = 3 \sin 2x$.

A. $y = 3 \sin 2x$ is sketched by transforming the graph of $y = \sin x$:

- The 3 **enlarges** all vertical values by a factor of 3 (the amplitude is 3).
- The 2 **compresses** the graph horizontally by *halving the period* to 180°.

The graph is shown below:

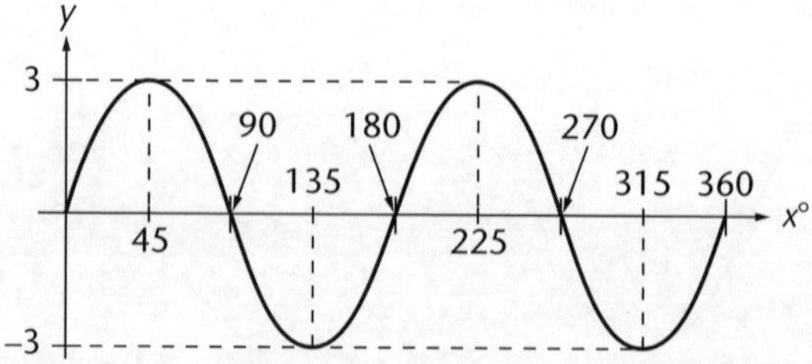

Example O

Q. Sketch the graph of $y = 4 \cos (x + 30°)$.

A. The graph of $y = 4 \cos (x + 30°)$ is sketched by transforming the graph of $y = \cos x$.

- The 4 (the amplitude) stretches the y-values by a factor of 4.
- The 30° horizontally translates the graph by –30° (30° to the left).

The graph is shown below:

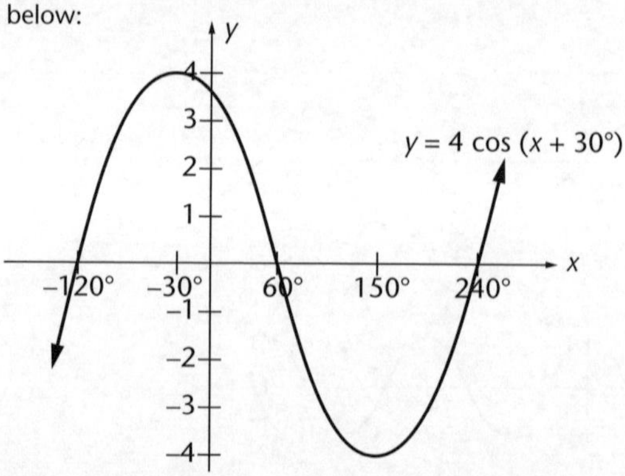

Unit 12.2 Activity 1F: Trigonometric graphs with two transformations

Sketch the following graphs:

1. $y = 3 \sin 2x$	**2.** $y = 2 \sin (x - 60°)$	**3.** $y = -\tan (x - 30°)$
4. $y = \frac{1}{2} \cos x + 1$	**5.** $y = 2 \cos \frac{1}{2}x$	**6.** $y = \sin (x + 90°) - 2$

Write down the equations of the functions represented by these graphs:

7.

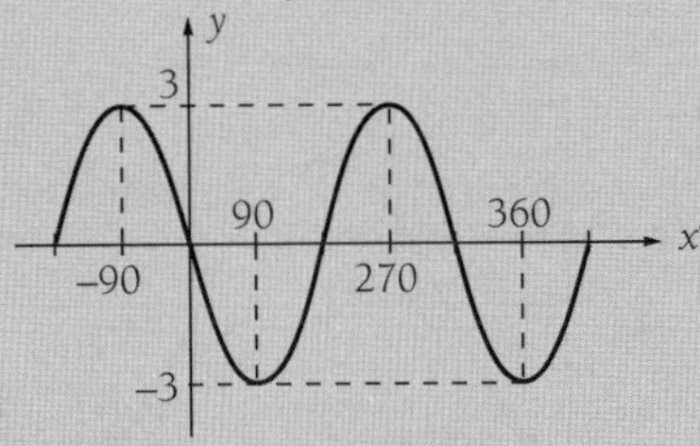

8.

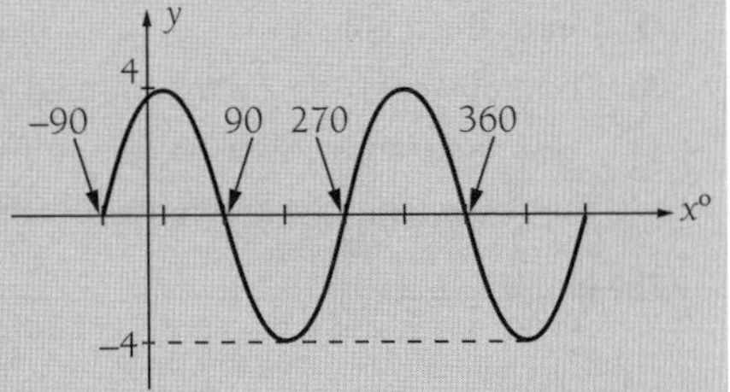

9. The graph below shows which function? Choose from **a** – **e**.

a. $y = -2 \sin x$

b. $y = -\sin 2x$

c. $y = 2 \sin (x + 180°)$

d. $y = 2 \sin (x - 90°)$

e. $y = 2 \sin (x + 90°)$

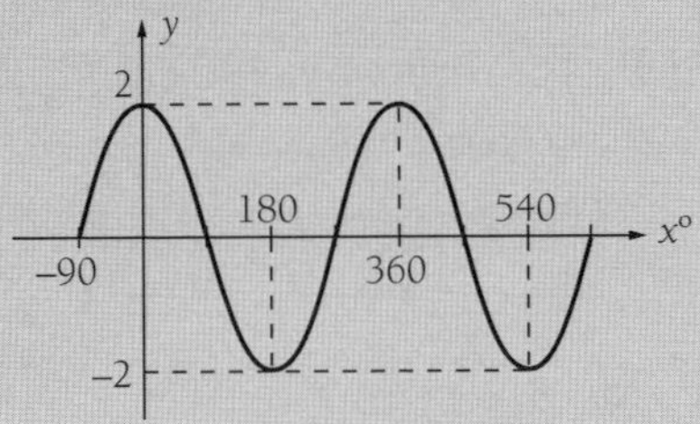

10. Sketch the graphs of the four other functions in question 9.

Trigonometric graphs using radians

Trigonometric graphs can also be drawn with angles measured in radians. Remember to set your calculator to radian mode when calculating trigonometric functions of angles measured in radians.

A small Ⓡ appears on the display (instead of Ⓓ for degrees)

Example P

Q. Sketch the graph of $y = \cos x$, $0 \le x \le 2\pi$.

A. The table lists some important points on the curve.

x	0	$\frac{\pi}{2}$	π	$\frac{3\pi}{2}$	2π
y	1	0	–1	0	1

The graph is as shown.

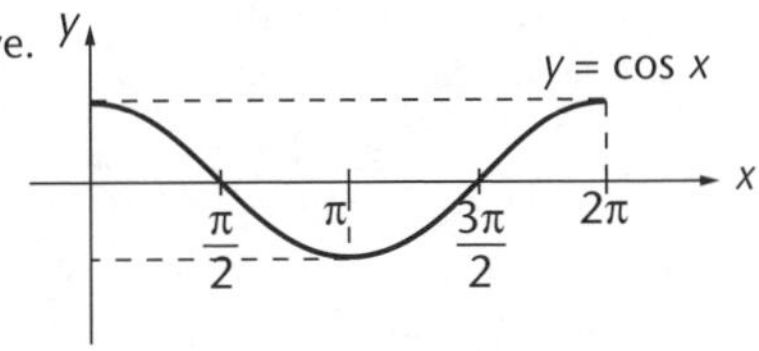

Trigonometric graphs using radians can be transformed in the same way as trigonometric graphs using degrees.

Example Q

Q. Sketch the graph of $y = 2 \sin (x - \frac{\pi}{2})$, $0 \leq x \leq 2\pi$.

A. The graph of $y = \sin x$ is transformed as follows:

- The 2 stretches the y-values (the amplitude is 2).
- The $\frac{\pi}{2}$ translates the graph horizontally by $\frac{\pi}{2}$ units.

The graph is as shown.

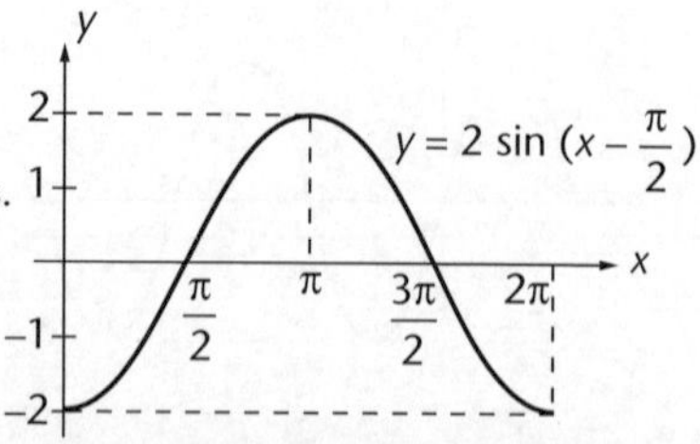

Unit 12.2 Activity 1G: Trigonometric graphs using radians

Sketch graphs of the following functions:

1. $y = 2 \sin x$, $0 \leq x \leq 2\pi$

2. $y = \tan x$, $-\pi \leq x \leq 2\pi$

3. $y = \sin (x - \frac{\pi}{4})$, $-\pi \leq x \leq 2\pi$

4. $y = \sin (x + \frac{\pi}{3})$, $0 \leq x \leq 2\pi$

5. $y = 3 \cos (x - \frac{\pi}{2})$, $0 \leq x \leq 3\pi$

6. $y = -2 \sin (x + \frac{\pi}{3})$, $-\pi \leq x \leq \pi$

7. $y = 2 \sin 3x$, $-\frac{\pi}{2} \leq x \leq \frac{\pi}{2}$

8. $y = 3 \cos 2x$, $-\frac{\pi}{2} \leq x \leq \frac{\pi}{2}$

9. a. Which function is represented by the graph shown?

i. $y = 2 \sin x$

ii. $y = \sin 2x$

iii. $y = \sin x$

iv. $y = 2 \cos x$

v. $y = \cos x$

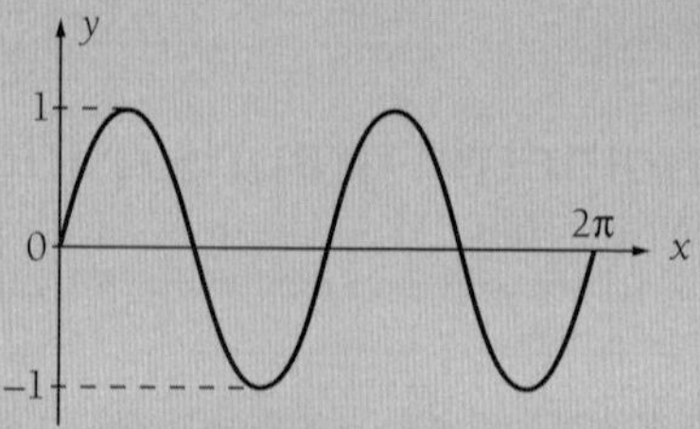

b. Sketch, on two separate pairs of axes, the other four functions.

10. The diagram shows the graphs of two trigonometrical functions g and h.

a. Write down the equation of g and the equation of h.

b. What is the period of h?

c. Using the graphs, or otherwise, find the value(s) of x for which $g(x) = h(x)$ in $0 \leq x \leq 2\pi$.

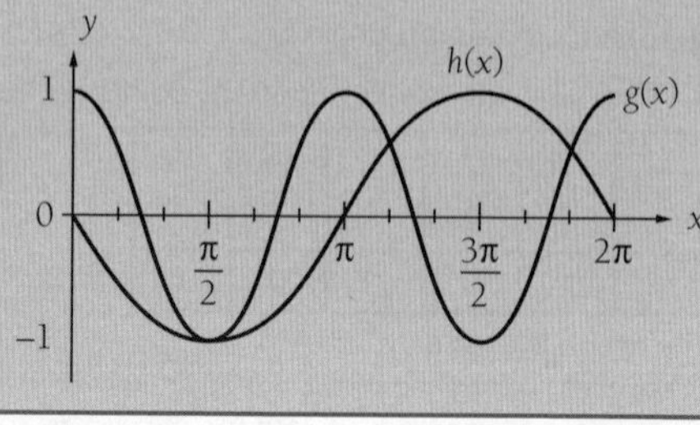

Unit 12.2 Trigonometry and Vectors
Topic 2: Further trigonometric functions

Topic 2 builds on the foundation laid in the previous Topic, solving problems and equations involving trigonometric functions. Topic 2 contains:

- Problems with models involving functions of the form:
  $y = A \sin B(x + C) + D$ $\quad$ $y = A \cos B(x + C) + D$
  $y = A \tan B(x + C) + D$.
- Problems which require knowledge of amplitude, period and frequency.
- Problems which involve finding the values of A, B, C or D.
- Form an equation for a model and use the model to solve problems. Information may be given or collected or provided in the form of data.

Periodic functions

A function $f(x)$ is periodic if there is a constant, h, such that $f(x) = f(x + h)$, for each x in the domain of f. This means that the graph of $f(x)$ repeats itself every h units along the x-axis. The least value of h is said to be the period of $f(x)$. One **cycle** of the graph occurs every h units.

The trigonometric functions are the best-known examples of periodic functions.

Example A

$f(x) = \sin x$ is a periodic function since $\sin x = \sin(x + 2\pi)$, showing that the sine curve repeats itself every 2π radians, ie has a period of 2π.

From the graph of $f(x) = \sin x$ below, $f(x)$ maps onto itself under a translation $\begin{pmatrix} 2\pi \\ 0 \end{pmatrix}$.

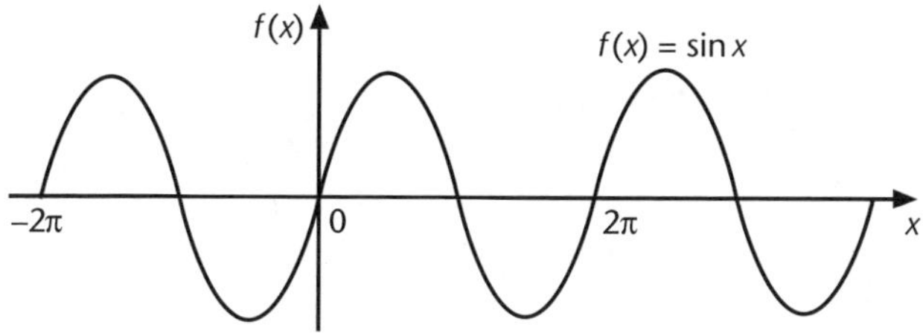

Period of a function

If the equation of a curve is $y = \sin Bx$ (or $y = \cos Bx$) then the period of the curve (the horizontal length of one cycle) is $\frac{2\pi}{B}$ radians, or $\frac{360}{B}$ degrees.

Example B

The period of $y = \sin 3x$ is $\frac{2\pi}{3}$ radians (or 120°) $\quad$ [substituting $B = 3$]

The period of $y = \cos\left(\frac{1}{2}x\right)$ is $\frac{2\pi}{\frac{1}{2}} = 4\pi$ radians (or 720°) $\quad$ [substituting $B = \frac{1}{2}$]

If the equation of a curve is $y = \tan Bx$ then the period of the curve is $\frac{\pi}{B}$ radians or $\frac{180}{B}$ degrees.

Example C

The period of $y = \tan 2x$ is $\frac{\pi}{2}$ radians (or 90°) [substituting $B = 2$]

Graphs of the trigonometric functions

The graphs of the six trigonometric functions illustrate important properties of the functions including domain (x-values for which the function is defined) and range (y-values for which the function is defined), symmetry and whether the function is even or odd.

The graphs are drawn by plotting the points from a suitable table of values ($\in$ means 'belongs to').

1. $f(x) = \sin x$

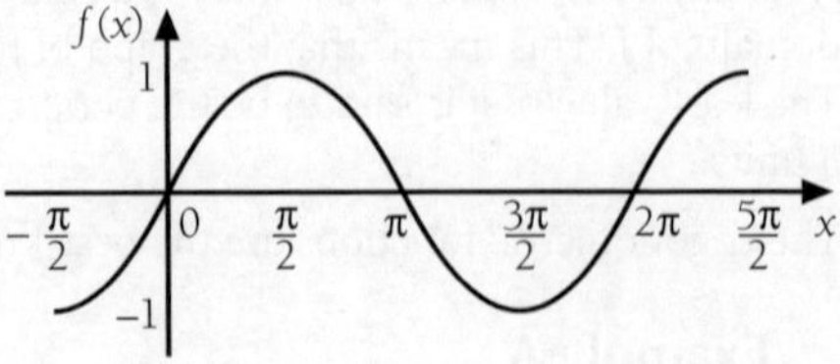

Domain: $x \in R$
Range: $[-1, 1]$

The period of the curve is 2π because $\sin x = \sin (x + 2n\pi)$, $n \in I$ (integers).

In other words, the curve repeats itself every 2π radians.

An **odd function** has the property that $f(-x) = -f(x)$.

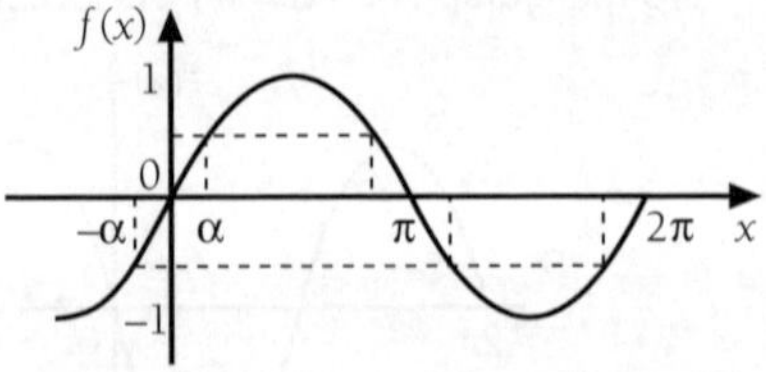

Since $\sin (-x) = -\sin x$, $f(x) = \sin x$ is an odd function. Therefore, the curve has half-turn rotational symmetry about the origin.

By symmetry, the graph alongside shows:

$$\sin (-\alpha) = -\sin \alpha$$
$$\sin (\pi - \alpha) = \sin \alpha$$
$$\sin (\pi + \alpha) = -\sin \alpha$$
$$\sin (2\pi - \alpha) = -\sin \alpha$$

2. $f(x) = \cos x$

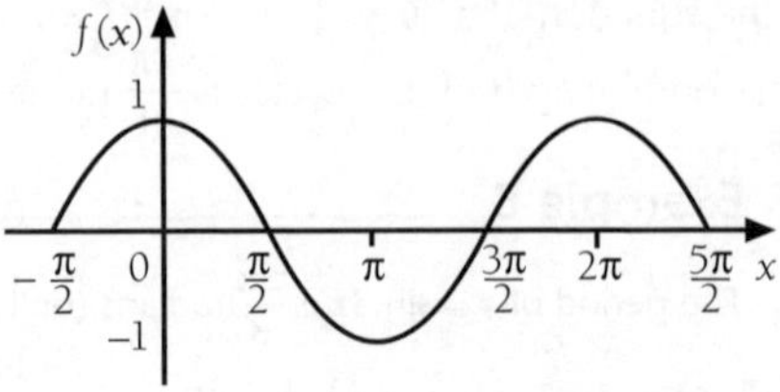

Domain: $x \in R$
Range: $[-1, 1]$

The period of the curve is 2π because $\cos x = \cos (x + 2n\pi)$, $n \in I$.

An **even function**, f, has the property that $f(-x) = f(x)$.

$f(x) = \cos x$ is an even function because $\cos(-x) = \cos x$ and therefore the curve is symmetrical about the y-axis.

From the graph alongside:

$\cos(-\alpha) = \cos\alpha$

$\cos(\pi - \alpha) = -\cos\alpha$

$\cos(\pi + \alpha) = -\cos\alpha$

$\cos(2\pi - \alpha) = \cos\alpha$

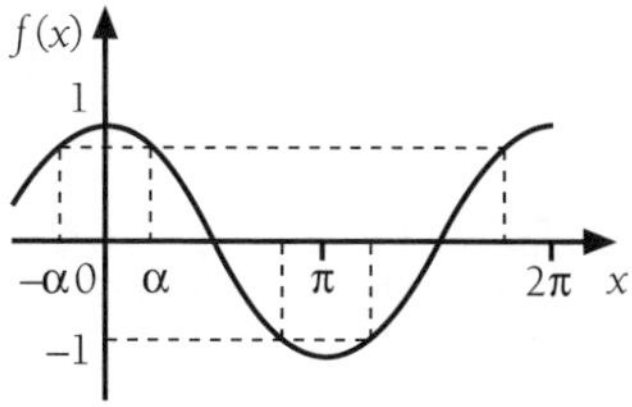

3. $f(x) = \tan x$

Domain: $x \in R, x \neq \ldots, \frac{-\pi}{2}, \frac{\pi}{2}, \frac{3\pi}{2}, \ldots$

Range: $y \in R$

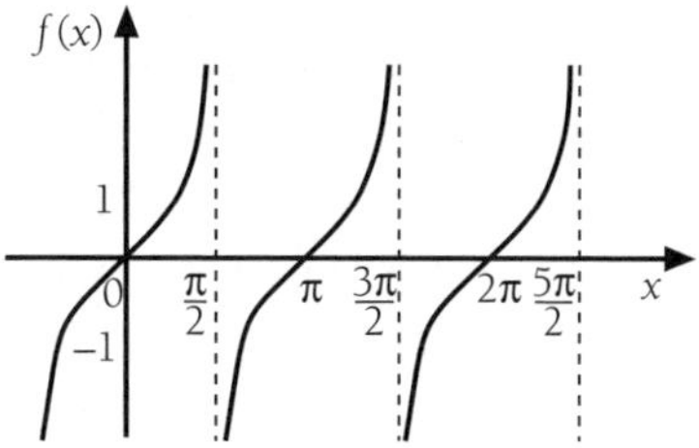

The dotted lines are asymptotes, which are lines the curve gets closer and closer to but never actually touches.

The domain could be written:

$x \in R, x \neq (2n + 1)\frac{\pi}{2}, n \in I$, which would exclude all values of x for which $\tan x$ is undefined (odd multiples of $\frac{\pi}{2}$).

The period of the curve is π because $\tan x = \tan(x + n\pi), n \in I$.

Since $\tan(-x) = -\tan x$, $f(x) = \tan x$ is an *odd* function so that $\tan x$ has half-turn rotational symmetry about the origin.

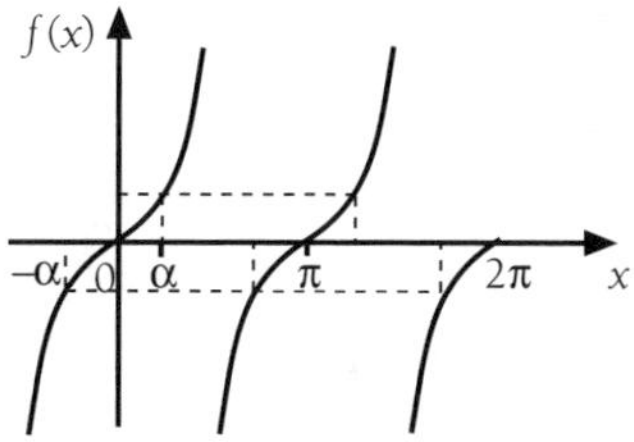

From the graph alongside:

$\tan(-\alpha) = -\tan\alpha$

$\tan(\pi - \alpha) = -\tan\alpha$

$\tan(\pi + \alpha) = \tan\alpha$

$\tan(2\pi - \alpha) = -\tan\alpha$

The reciprocal trigonometric ratios

The three **reciprocal** trigonometric ratios are **cosecant** (**cosec**), **secant** (**sec**) and **cotangent** (**cot**).

$$\text{cosec}\,\theta = \frac{1}{\sin\theta} = \frac{r}{y} \qquad \sec\theta = \frac{1}{\cos\theta} = \frac{r}{x} \qquad \cot\theta = \frac{1}{\tan\theta} = \frac{\cos\theta}{\sin\theta} = \frac{x}{y}$$

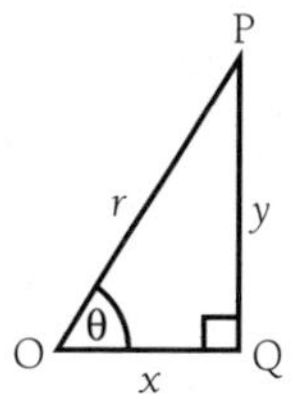

Plotting points for these reciprocal functions:

1. $f(x) = \sec x$

A suitable table of values for drawing the graph is:

x (in radians)	–1.5	–1.0	–0.5	0.0	0.5	1.0	1.5	2.0	2.5	3.0	3.5	4.0	4.5	5.0
sec x	14.1	1.9	1.1	1.0	1.1	1.9	14.1	–2.4	–1.2	–1.0	–1.1	–1.5	–4.7	3.5

Domain: $x \in \mathrm{R}, x \neq \ldots, \frac{-\pi}{2}, \frac{\pi}{2}, \frac{3\pi}{2}, \ldots$

Range: $y \in \mathrm{R}, y \leq -1$ or $y \geq 1$.

The period of the curve is 2π and sec x is an *even* function so that it is symmetrical about the y-axis.

The dotted lines $y = \pm\frac{\pi}{2}, y = \pm\frac{3\pi}{2}, \ldots$ are asymptotes for the curve.

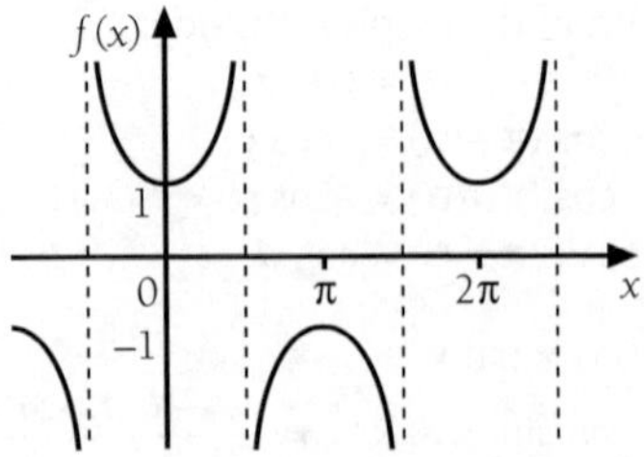

2. $f(x) = \text{cosec } x$

From a suitable table of values, a graph can be drawn.

This is the curve $y = \sec x$ given a translation $\begin{pmatrix} \frac{\pi}{2} \\ 0 \end{pmatrix}$.

Domain: $x \in \mathrm{R}, x \neq \ldots, -\pi, 0, \pi, 2\pi, \ldots$

Range: $y \in \mathrm{R}, y \leq -1$ or $y \geq 1$

The period of the curve is 2π and it is an *odd* function so that it has half-turn rotational symmetry about the origin.

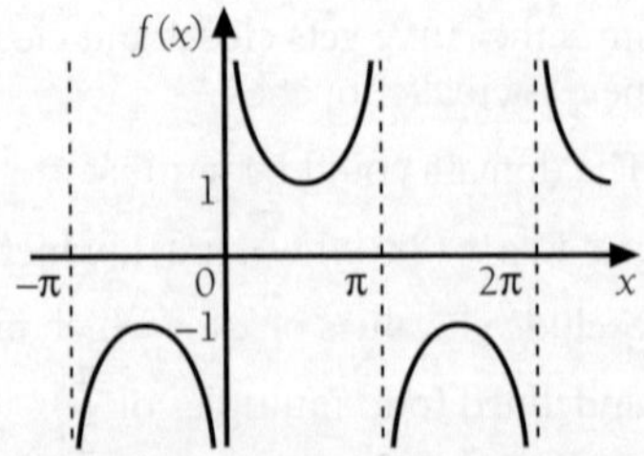

3. $f(x) = \cot x$

Domain: $x \in \mathrm{R}, x \neq \ldots -\pi, 0, \pi, 2\pi, \ldots$

Range: $y \in \mathrm{R}$

The period of the curve is π and it is an *odd* function so that it has half-turn rotational symmetry about the origin.

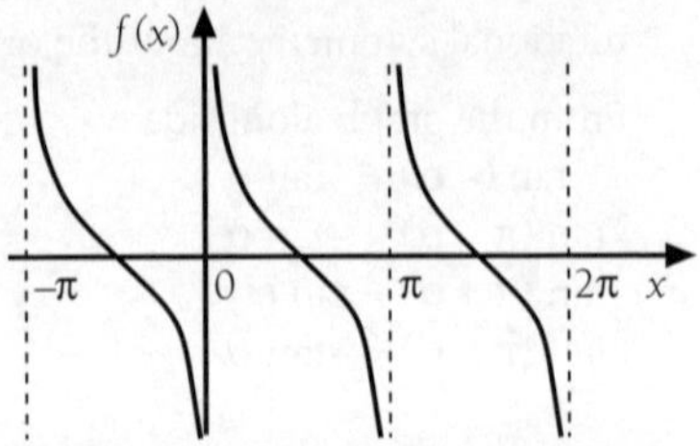

Questions can be set using the periodic properties of trigonometric curves.

Example D

Q. Find a formula connecting α and θ in the graph of $y = \cos x$ shown below.

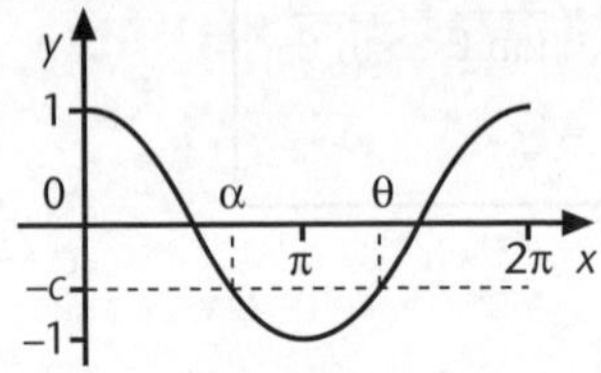

A. In the figure the shaded areas are congruent, so

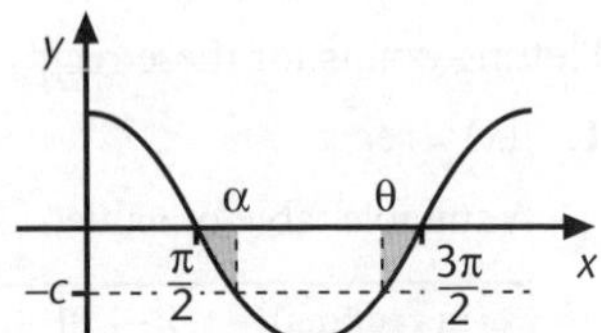

$\alpha - \frac{\pi}{2} = \frac{3\pi}{2} - \theta$ [equating equal lengths on x-axis]

$\alpha + \theta = \frac{3\pi}{2} + \frac{\pi}{2}$ [rearranging equation]

$\alpha + \theta = 2\pi$

Unit 12.2 Activity 2A: Basic trigonometry

1. Find the exact value of each of the following (the angles are in radians):

a. $\cos\frac{3\pi}{4}$ **b.** $\sin\frac{7\pi}{6}$ **c.** $\tan\frac{5\pi}{3}$ **d.** $\sec\frac{-\pi}{4}$ **e.** $\operatorname{cosec}\frac{8\pi}{3}$ **f.** $\cot\frac{4\pi}{3}$

g. $\cos\frac{5\pi}{6}$ **h.** $\sin\frac{5\pi}{4}$ **i.** $\tan\frac{4\pi}{3}$ **j.** $\sec\frac{7\pi}{6}$ **k.** $\operatorname{cosec}\frac{3\pi}{4}$ **l.** $\cot\frac{7\pi}{6}$

m. $\operatorname{cosec}\frac{5\pi}{3}$ **n.** $\sec\frac{15\pi}{4}$ **o.** $\cot\frac{-3\pi}{4}$ **p.** $\sec^2\pi$ **q.** $\operatorname{cosec}^2\left(\frac{-9\pi}{4}\right)$

2. If $f(x) = 2\sin x\cos x + \sin x$, evaluate $f\left(-\frac{\pi}{4}\right)$.

3. If $f(x) = \frac{\sin x}{\cos x}$, evaluate:

a. $f\left(\frac{-7\pi}{4}\right)$ **b.** $f\left(\frac{-7\pi}{4}\right)$

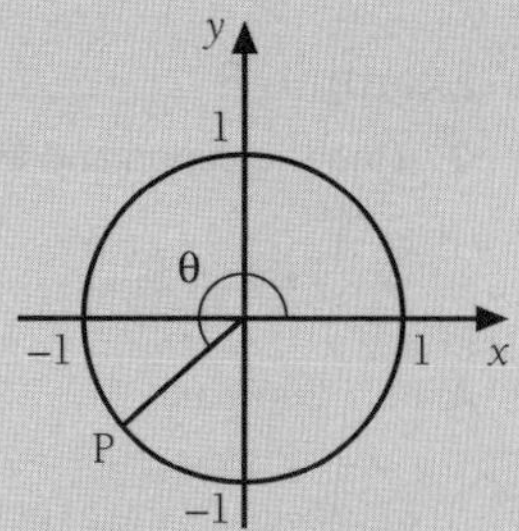

4. Evaluate $f\left(\frac{\pi}{2}\right) - f\left(\frac{-3\pi}{4}\right)$ where $f(x) = \sin x$.

5. Evaluate $f\left(\frac{\pi}{2}\right) - f(0)$ where $f(x) = -x\sin x + \cos x$.

6. The radius of a circle is one unit and the coordinates of P are $\left(\frac{-\sqrt{3}}{2}, \frac{-1}{2}\right)$. Find the *exact* values of:

a. θ in radians **b.** $\sec\theta$ **c.** $\tan\theta$

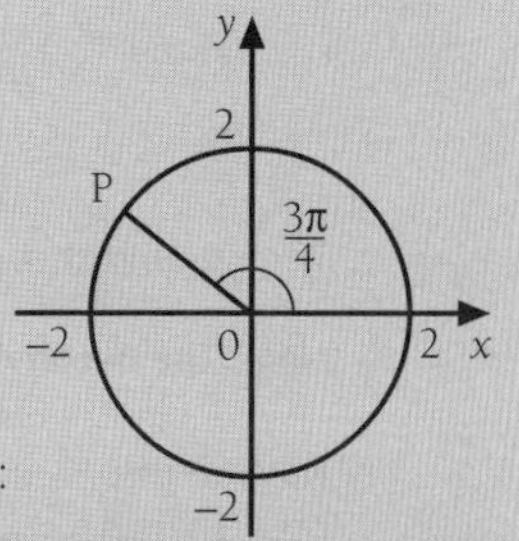

7. The radius of the circle shown alongside is two units. Find the *exact* coordinates of P.

8. Find a formula connecting α and θ in each of the following:

a.

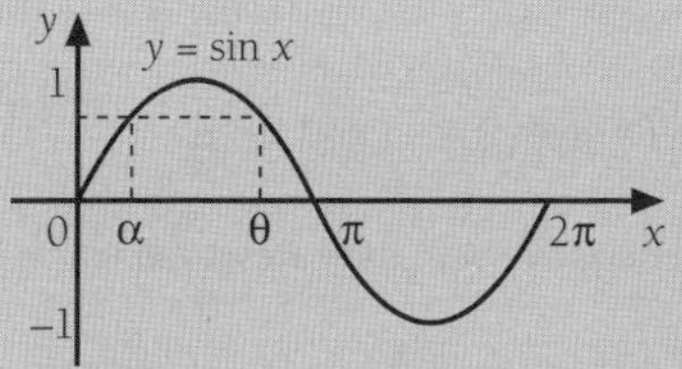

b.

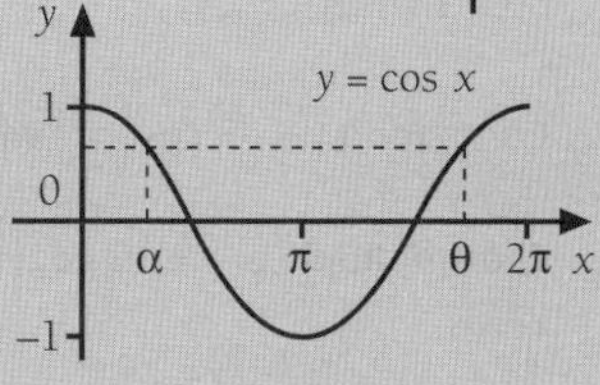

c.

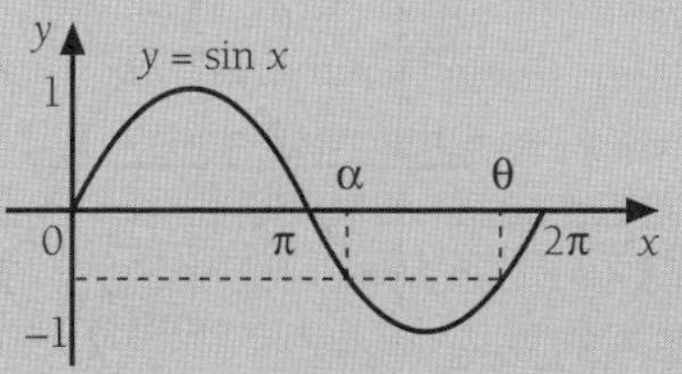

d.

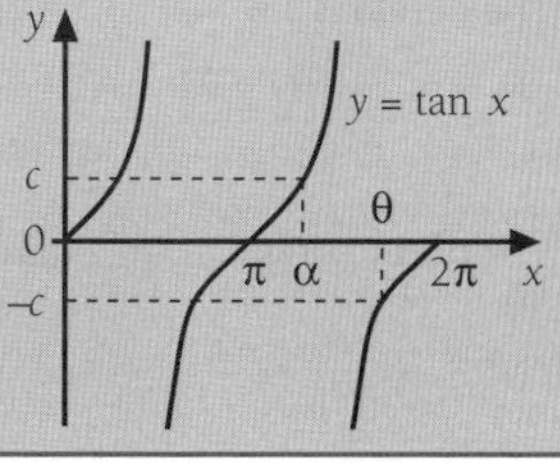

Sketching sine and cosine curves

Sine and cosine curves can be drawn if the following information is known:

- The amplitude (half the distance between the maximum and minimum y-values).
- The period (which is denoted by T).
- If **reflection in the x-axis** has taken place.
- If translation has taken place.

Consider the sine curve $y = A \sin Bx$, and the cosine curve, $y = A \cos Bx$ where A and B are positive. Both graphs have amplitude A and period $\frac{2\pi}{B}$.

If the graph of $y = A \sin Bx$ is reflected in the x-axis, the graph of $y = -A \sin Bx$ is obtained. A similar rule applies for reflecting the cosine graph.

Example E

Q. Find the period and amplitude of each curve and sketch its graph showing a complete cycle.

1. $y = 2 \sin\left(\frac{1}{2}x\right)$ **2.** $y = -3 \cos\left(\frac{\pi x}{2}\right)$

A. 1. Comparing $y = 2 \sin\left(\frac{1}{2}x\right)$ with $y = A \sin (Bx)$ gives $A = 2$ and $B = \frac{1}{2}$.

The amplitude of $y = 2 \sin\left(\frac{1}{2}x\right)$ is $A = 2$, and the period is

$T = \frac{2\pi}{\frac{1}{2}}$ [using $T = \frac{2\pi}{B}$]

$= 4\pi$

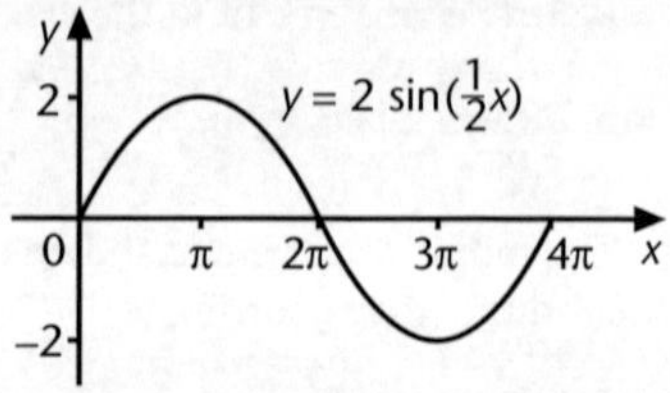

The graph is shown alongside.

Note: The maximum value of $2 \sin\left(\frac{1}{2}x\right)$ is 2 and this occurs when $x = \pi$; the minimum value of $2 \sin\left(\frac{1}{2}x\right)$ is -2, and this occurs when $x = 3\pi$.

2. Comparing $y = -3 \cos\left(\frac{\pi x}{2}\right)$ with $y = A \cos Bx$ gives $A = -3$ and $B = \frac{\pi}{2}$.

The amplitude of $y = -3 \cos\left(\frac{\pi x}{2}\right)$ is 3; the negative sign indicates a reflection in the x-axis.

The period is $T = \frac{2\pi}{\frac{\pi}{2}}$ [using $T = \frac{2\pi}{B}$]

$= 4$

The dotted graph in the figure is $y = 3 \cos\left(\frac{\pi x}{2}\right)$.

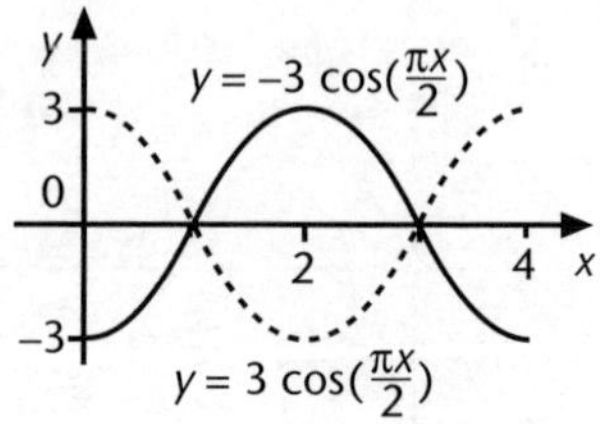

The graph of $y = -3\cos\left(\frac{\pi x}{2}\right)$ is drawn in full and is obtained after a reflection of $y = 3\cos\left(\frac{\pi x}{2}\right)$ in the x-axis.

General equations of sine and cosine curves

The general equations of sine and cosine curves are:

$$\boxed{y = A\sin B(x + C) + D} \quad \text{and} \quad \boxed{y = A\cos B(x + C) + D}$$

These curves are obtained by translating the graphs of $y = A\sin Bx$ and $y = A\cos Bx$ by the vector $\begin{pmatrix}-C\\D\end{pmatrix}$. (As before, the amplitudes are A and the periods are $\frac{2\pi}{B}$)

The following examples explain the processes involved in sketching the graphs of the general equations $y = A\sin B(x + C) + D$ and $y = A\cos B(x + C) + D$.

Note: When sketching the graphs, the last step is the translation. Note also the placings of the brackets.

Example F

Q. Sketch the graph of $y = 3\cos(x - \frac{\pi}{2})$ showing one complete cycle.

A. $A = 3$, $B = 1$, $C = -\frac{\pi}{2}$, $D = 0$ [comparing $y = 3\cos(x - \frac{\pi}{2})$ with $y = A\cos B(x + C) + D$]

The amplitude is $A = 3$.

The period is $T = \frac{2\pi}{1}$ [using $T = \frac{2\pi}{B}$]

$= 2\pi$

The translation is $\begin{pmatrix}\frac{\pi}{2}\\0\end{pmatrix}$ [using the translation $\begin{pmatrix}-C\\D\end{pmatrix}$]

The graph of $y = 3\cos x$ is drawn dotted, and then translated by $\begin{pmatrix}\frac{\pi}{2}\\0\end{pmatrix}$ to obtain the graph of $y = 3\cos(x - \frac{\pi}{2})$, which is drawn in full.

After the translation, the periodic nature of the curve is used to extend it into the region $0 \le x \le \frac{\pi}{2}$.

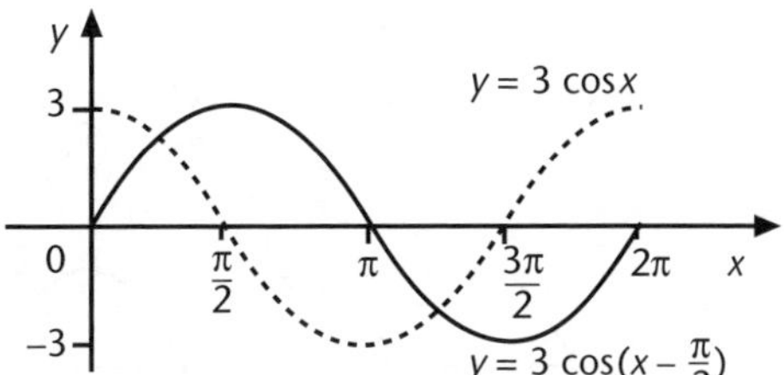

Note: The graph of $y = 3\cos(x - \frac{\pi}{2})$ is the same as the graph of $y = 3\sin x$. This can be explained when the trigonometric identities have been studied in Topic 5 (p.123).

Example G

Q. Sketch the graph of $y = \frac{1}{2}\sin 2x + 1$, showing one complete cycle.

A. $A = \frac{1}{2}$, $B = 2$, $C = 0$, $D = 1$ [comparing $y = \frac{1}{2}\sin 2x + 1$ with $y = A\sin B(x + C) + D$]

Amplitude is $A = \frac{1}{2}$

The period is $T = \frac{2\pi}{2}$ [using $T = \frac{2\pi}{B}$]

$= \pi$

The translation is $\begin{pmatrix} 0 \\ 1 \end{pmatrix}$ [using a translation of $\begin{pmatrix} -C \\ D \end{pmatrix}$]

The graph of $y = \frac{1}{2}\sin 2x$ is drawn dotted.

This graph is then translated by $\begin{pmatrix} 0 \\ 1 \end{pmatrix}$ to obtain the graph of $y = \frac{1}{2}\sin 2x + 1$, which is drawn in full.

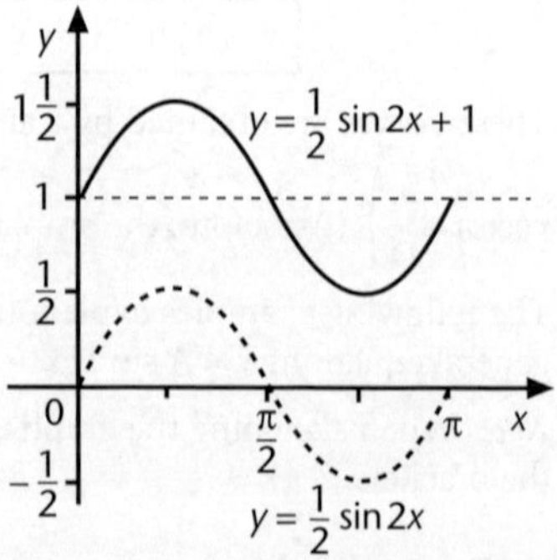

Example H

Q. Sketch the graph of $y = 2\sin 3(x + 180°)$.

A. $A = 2$, $B = 3$, $C = 180°$, $D = 0$ [comparing $y = 2\sin 3(x + 180°)$ with $y = A\sin B(x + C) + D$]

The amplitude is $A = 2$

The period is $T = \frac{360°}{3}$ [using $T = \frac{360°}{B}$, in degree form]

$= 120°$

The translation is $\begin{pmatrix} -180° \\ 0 \end{pmatrix}$ [using a translation of $\begin{pmatrix} -C \\ D \end{pmatrix}$]

One cycle of $y = 2\sin 3x$ is drawn dotted. This graph is then translated $\begin{pmatrix} -180° \\ 0 \end{pmatrix}$ to obtain the graph of $y = 2\sin 3(x + 180°)$, which is drawn in full.

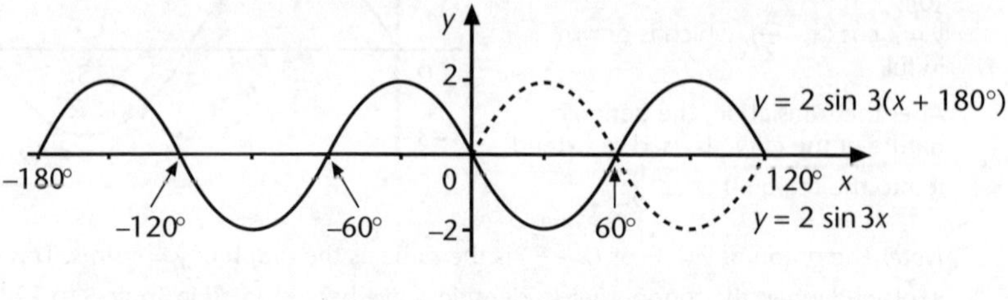

Note: **1.** Extra space is needed to the left of the *y*-axis so that the curve $y = 2\sin 3x$ can be translated $\begin{pmatrix} -180° \\ 0 \end{pmatrix}$. One cycle of the curve is then continued to the right of the *y*-axis.

2. If a graphic calculator is used to draw this graph, extra brackets may have to be inserted. For example, the graph above would be entered as $y = 2 \sin(3(x + 180°))$.

Care must be taken with the equations of some graphs to ensure that their form matches that of the general equations. In the example that follows, **factorising** was necessary in order to express the equation in the form of the general equation.

Example I

Q. Sketch the graph of $y = -\cos(2x - \frac{\pi}{2})$.

A. $y = -\cos(2x - \frac{\pi}{2})$

$y = -\cos 2(x - \frac{\pi}{2})$ [factorising $2x - \frac{\pi}{2}$ (using the coefficient of x as the 'common' factor)]

This equation can now be directly compared with the general equation for a cosine curve.

$A = -1, B = 2, C = -\frac{\pi}{4}, D = 0$ [comparing $y = -\cos 2(x - \frac{\pi}{4})$ with $y = A \cos B(x + C) + D$]

The amplitude is $A = 1$, and the negative sign indicates a reflection in the x-axis.

The period is $T = \frac{2\pi}{2}$ [using $T = \frac{2\pi}{B}$]

$= \pi$

The translation is $\begin{pmatrix} \frac{\pi}{4} \\ 0 \end{pmatrix}$ [using a translation of $\begin{pmatrix} -C \\ D \end{pmatrix}$]

There are 3 stages in sketching this graph.

Stage 1: Sketch $y = \cos 2x$, which has an amplitude of $A = 1$ and a period of $T = \pi$.

Stage 2: Reflect this curve in the y-axis. This is the graph of $y = -\cos 2x$.

Stage 3: Translate the curve from Stage 2 by $\begin{pmatrix} \frac{\pi}{4} \\ 0 \end{pmatrix}$. This is the required graph.

The graph drawn with a full line is the answer. Note that the translation is done *last*.

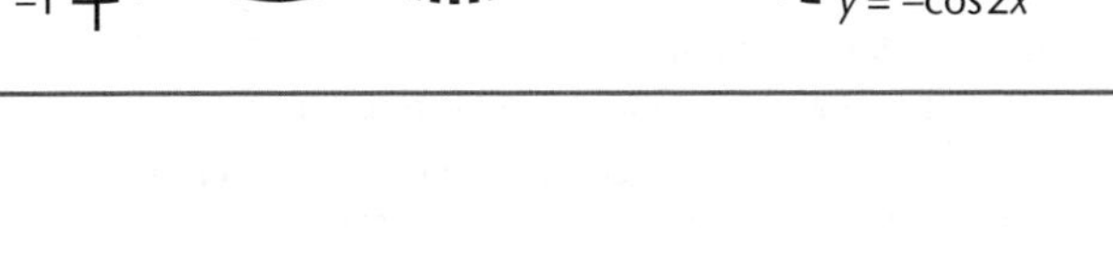

The following example illustrates a practical use of sketching sine and cosine graphs:

Example J

Q. A fishing boat goes aground in a thick fog at a place where the water level, relative to its mean level, is given by $h = -3\cos\frac{\pi t}{6}$, where t is in hours and h is in metres.

1. Sketch the graph of h, if low tide first occurs at 10 am.

2. What is the depth of water at high tide, and when does it occur?

3. How many hours pass between low tides?

4. If the boat can only be refloated when the water level is at least 2.8 metres above mean level, find the time when this first occurs.

A. 1. $A = -3$, $B = \frac{\pi}{6}$, $C = 0$, $D = 0$ [comparing $h = -3\cos\frac{\pi t}{6}$ with $y = A\cos B(x + C) + D$]

Since $A = -3$, the amplitude is 3 metres, and the negative sign indicates a reflection in the t-axis.

The period is $T = \dfrac{2\pi}{\frac{\pi}{6}}$ [using $T = \frac{2\pi}{B}$]

$= 12$ hours

There is no translation involved. The graph is shown in the figure, with the time axis labelled in hours as time of day, as well as units of time.

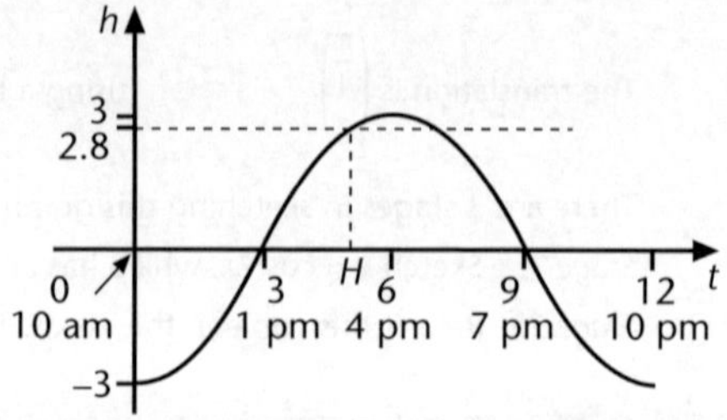

2. High tide is 3 metres and occurs at 4 pm (6 hours after 10 am).

3. The period of the graph is 12 hours, and is the time between low tides.

4. A horizontal line is drawn on the graph at 2.8 metres, so the first time that refloating occurs is when $t = H$ hours. To find t, solve the equation

$$2.8 = -3\cos\frac{\pi H}{6}$$

$$-0.9333 = \cos\frac{\pi H}{6} \quad \text{[dividing both sides by } -3]$$

$$\frac{\pi H}{6} = 2.7744 \text{ hours (4 dp)} \quad [\text{taking } \cos^{-1}(0.9333)]$$

$$H = \frac{6 \times 2.7744}{\pi} \quad \text{[rearranging equation]}$$

$H = 5.299$ hours (3 dp)

$H = 5$ hours and 18 minutes (to nearest minute)

The time that refloating can first occur is 18 minutes past 3 pm.

Investigations: Practical trigonometry

1. Many houses have corrugated roofing iron which has a cross-section which resembles a sine curve. Find the amplitude and the period of roofing iron and write the equation in the form $x = a \sin \omega t$.

2. The equation $h = H \cos\left(\frac{\pi t}{6}\right)$ models tides which are 12 hours apart, where the maximum height of the tide above its mean level is H.

   For a particular day in your local area, modify the equation $h = H \cos (K\pi t)$ by finding the value of K so that the equation gives the local period of time between high tides, and find the value of H for the height of the tide.

   In what way(s) is this modified equation not true for the following week (month) in your area?

   If another town has a high tide 3 hours later than your local area, what will be its 'tide equation' as a transformation of your equation?

   Discuss the limitations of using a cosine curve to model the motions of tides.

3. Find a long flexible spring, or a piece of elastic, and suspend it from a beam with a mass hanging from it. When the mass is at rest in the equilibrium position, pull it down a further distance and release it. The mass will oscillate up and down.

   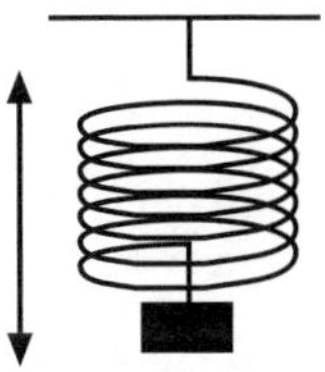

   Measure the amplitude, and the time for a complete oscillation. Use these figures to find 'a' and 'ω' in the formula $x = a \cos (\omega t)$, which represents the motion of the mass, where x is the distance measured in cm from the point of release of the mass and t the time in seconds so that when $x = 0$, $t = 0$.

   a. Draw a graph of the motion of the mass and find its frequency. Write a formula for the motion of the mass using sine instead of cosine to obtain the same graph.

   b. Discuss the long-term accuracy of modelling the motion of the mass with the equation $x = a \cos \omega t$.

4. The general equation of a sine curve is $y = \text{A} \cos \text{B}(x + \text{C}) + \text{D}$. Find the values of A, B, C and D so that the general equation above has the same graph as each of the following:

   a. $y = \cos x$ **b.** $y = -\cos x$ **c.** $y = \cos 2x$

   d. $y = 2 \cos x$ **e.** $y = -2 \cos 2x$ **f.** $y = \frac{1}{2} \cos (3x - \frac{\pi}{2})$

   g. $y = -3 \cos 2(x - \frac{\pi}{2}) - 3$ **h.** $y = 0.5 \cos (x + \pi) - 2$

Frequency

The **frequency** of a curve is the number of cycles that occur in a given unit of time (often one second).

The frequency of a curve is the reciprocal of its period.

The units of frequency are cycles per second or cycles per minute, etc.

Example K

The period of the curve is 8 seconds, so the frequency will be the reciprocal of 8 ie frequency is $\frac{1}{8}$ cycles per second.

Note: Since it is known that the amplitude of the curve is 2 and the period is 8, the equation of the curve can be written down:

$y = 2 \sin Bx$ where $\frac{2\pi}{B} = 8$ and $B = \frac{\pi}{4}$,

ie $y = 2 \sin \frac{\pi x}{4}$.

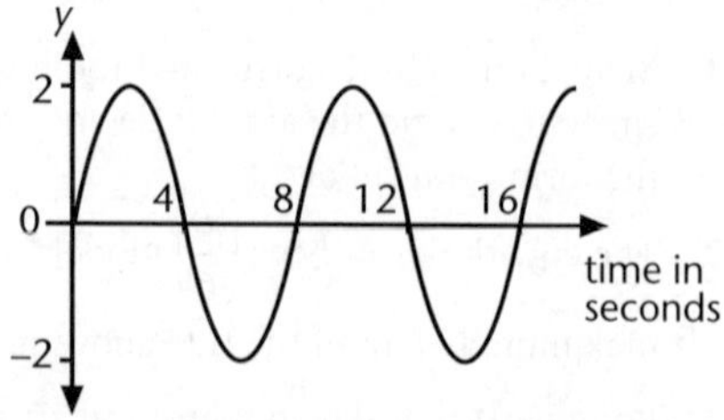

Unit 12.2 Activity 2B: Transformations of trigonometric graphs

1. Sketch the graph of each of the following functions, and write down the period and amplitude. Show at least one full cycle of each graph.
   - **a.** $y = 2 \cos x$
   - **b.** $y = -\sin x$
   - **c.** $y = \sin 2x + 1$
   - **d.** $y = 3 \sin \frac{1}{2}x$
   - **e.** $y = 2 \cos 3x$
   - **f.** $y = \sin (x + \pi)$
   - **g.** $y = 2 \cos (x - \frac{\pi}{2})$
   - **h.** $y = \frac{-1}{2} \sin (x - \pi)$

2. Find B in radians so that the graph of $y = \cos x$ is the same as the graph of $y = \sin (x + B)$, explaining your working with a graph.

3. The height of the sea water (due to tides) above the mean sea level is given by the formula:
   $h = H \cos \left(\frac{t\pi}{6}\right)$, where t is the time in hours after high tide and H is the height of the high tide in metres. Suppose a high tide of 3.0 metres occurs at 10 am.
   - **a.** Sketch the graph of $h = H \cos \left(\frac{t\pi}{6}\right)$ for $0 \le t \le 12$.
   - **b.** Using your graph, show that low tide occurs 6 hours later at 4 pm.
   - **c.** Find the times of day when the height of the tide is 2.5 m.

4. $f(x) = 3 \cos \left(\frac{2\pi x}{3}\right)$ and $g(x) = 2 \sin 2(x - \frac{\pi}{6})$. Find:
   - **a.** The maximum and minimum values of f and g.
   - **b.** The period of f and g.
   - **c.** If the units on the x-axis are in hours, find the frequency of f and g.

5. Suppose the temperature T degrees Celsius during a certain day is modelled by the sine curve $T(t) = -10 \sin \left(\frac{\pi t}{12}\right) + 15$, where t is the time in hours since midnight (when $t = 0$ it is midnight).

a. Sketch the graph of the curve.
b. Find the maximum and minimum temperatures during the 24 hours, and the times of day when they occur.
b. Find the amplitude and frequency of the curve.

6. Sketch the following trigonometric graphs and write down the equation of each. There may be more than one possible answer; find as many different answers as you can.
a. The period is 360°, amplitude is 2, and $f(0) = 0$.
b. The period is 720°, amplitude is 0.5, and it is even.
c. The period is 360°, amplitude is 2, and the minimum value of $f(x)$ is zero.
d. $f(x + 90°) = f(x)$ for all values of x, $f(x)$ is odd, and $f(x) \leq 0$, for $0 \leq x \leq 45°$.
e. A cosine curve with an amplitude of 2 and a frequency of 3 cycles per second.

7. **a.** Find the period and amplitude of the graphs of these functions:
i. $y = 2\cos(\frac{2}{3}x)$ **ii.** $y = \frac{1}{2}\sin\left(\frac{\pi x}{2}\right)$
b. If π radians is represented by 10 seconds, find the frequency of each graph.
c. Sketch the graphs showing one complete cycle.

8. Sketch the following graphs and state the period of each. Indicate the asymptotes clearly with dotted lines.
a. $y = \tan(x - \frac{\pi}{2})$ **b.** $y = \tan(2x)$ **c.** $y = -\tan\left(\frac{\pi x}{4}\right)$ **d.** $y = \tan(2x + \pi)$

9. A Ferris wheel rotates clockwise about an axis through the point A. Passengers board the wheel at B, and the horizontal ground is 0t. Suppose a chair starts at B and t seconds later it is at P which has a vertical height h above the ground, where $h = -10\cos\left(\frac{\pi t}{15}\right) + 11$.

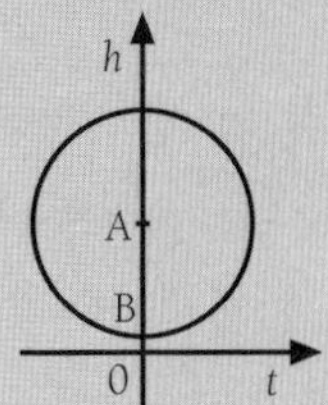

a. What is the radius of the Ferris wheel?
b. What is the height of B above the ground?
c. How long does a complete revolution of the wheel take, assuming it does not stop?
d. How high above the ground is the highest position of a chair?
e. Find the height above the ground of a chair 10 seconds after it started at B.
f. A chair starts at B and t seconds later it is 13 metres above the ground. Find two possible values for t.

10. A ski slope AB is in the shape of a sine curve, with axes shown.
a. Find the equation of the curve.
b. Find the point on the curve where the ski slope is steepest.

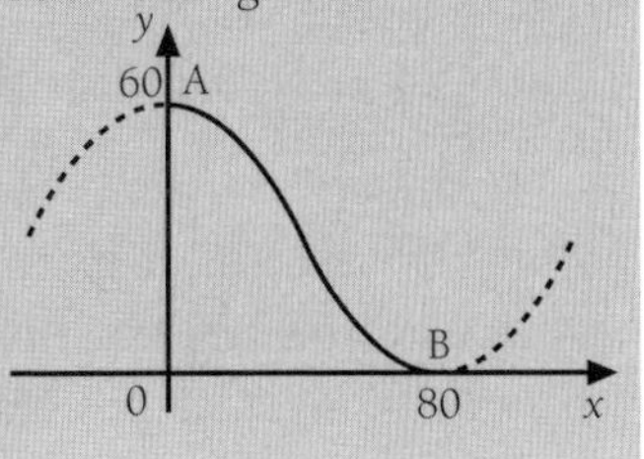

11. Each of the following graphs has an equation of the form $y = A \sin Bx$ or $y = A \cos Bx$. Choose which equation is more suitable and find A and B. Write the equation of the graph in full.

a.

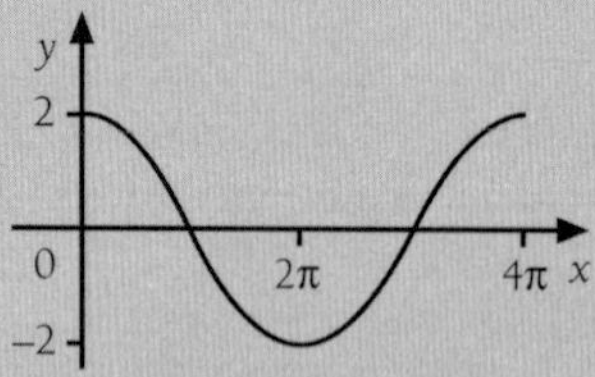

b.

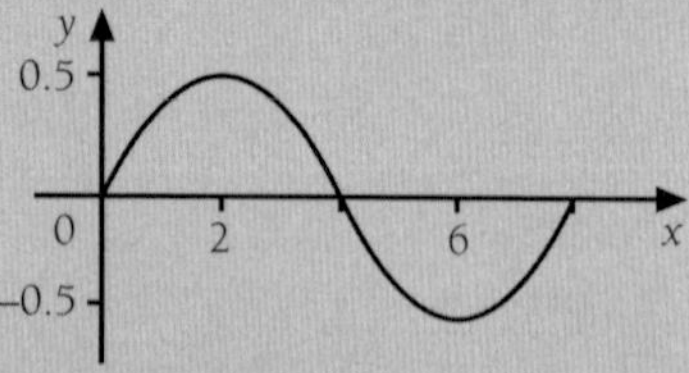

c.

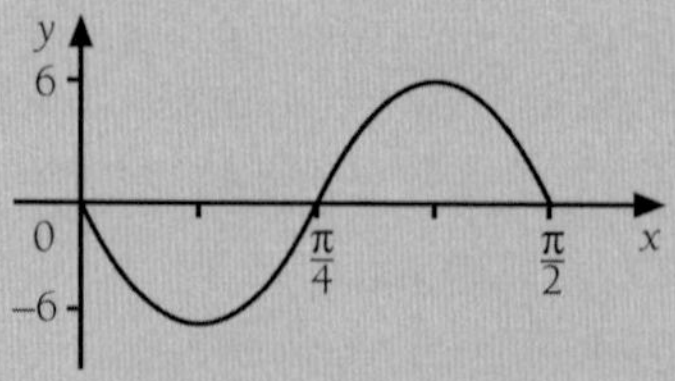

d.

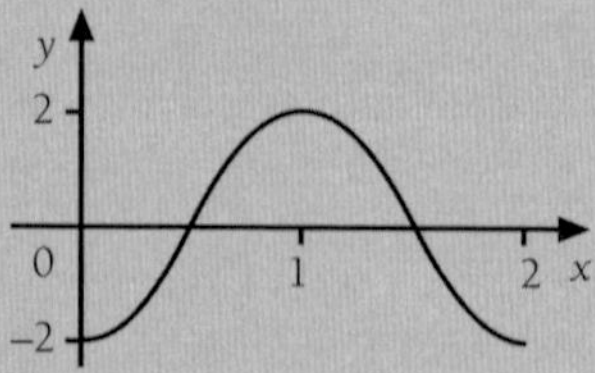

e.

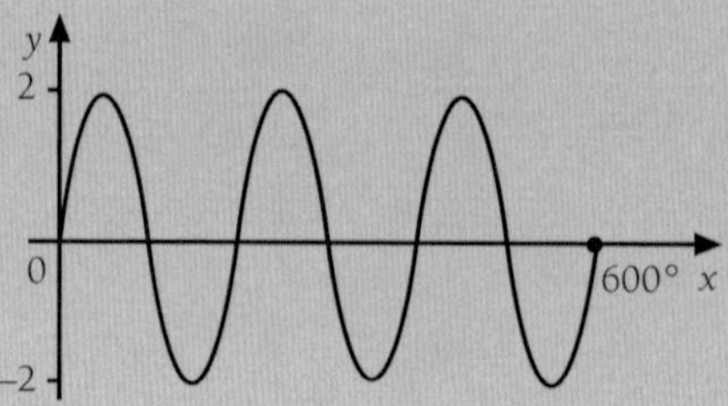

f.

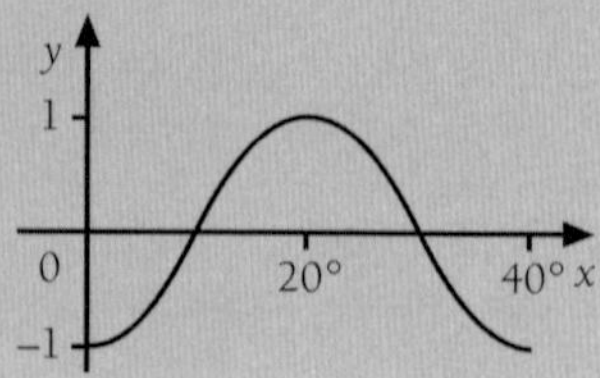

12. The figure shows the graph of $y = a \cos bx$.

a. What are the values of the amplitude and period of this function?

b. What are the values of a and b in the equation of the curve $y = a \cos bx$?

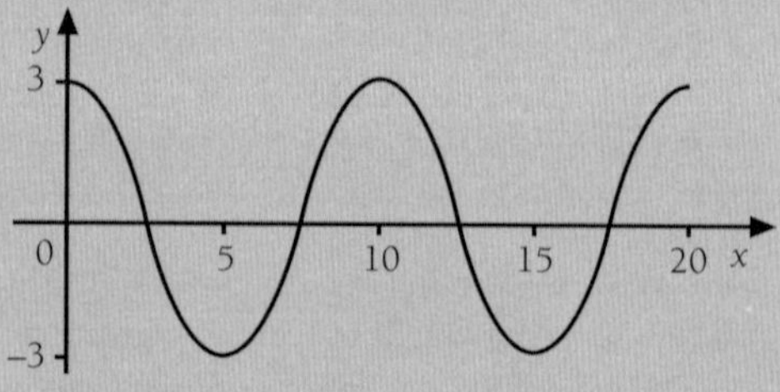

13. The graph shown is a model for the tide variation in a harbour. The equation of the curve can be represented by either $y = A \sin B(x + C) + D$ or $y = A \cos B(x + C) + D$.

a. Choose which of these two equations best represents the curve, and find A, B, C and D.

b. If a boat needs a depth of at least 8 metres of water, for how long during the 12-hour period shown is it able to enter the harbour?

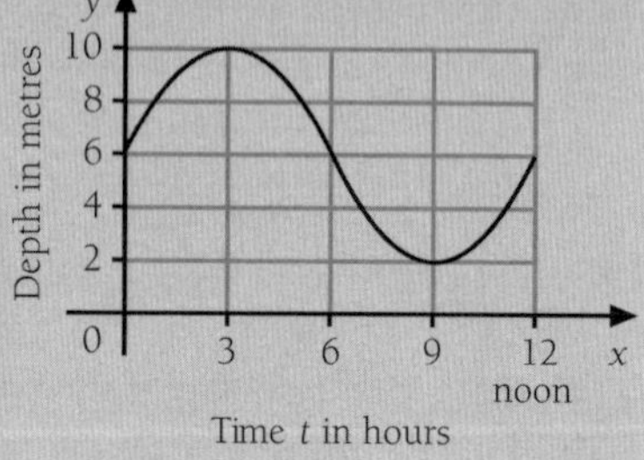

14. An alternator generates electricity on a certain day. The size of the current is modelled by a sine curve, where A is the current in amperes and t is the time in seconds since midnight. The greatest current during a cycle of 4 seconds is 370 amps. The amplitude of the curve is 160 amps.

a. Sketch the graph and find its equation.

b. What is the lowest number of amps generated by this alternator?

c. A turbo alternator generates twice the current of this alternator but in half the period. Write down the equation of the graph of the turbo alternator, assuming it is a sine curve.

15. Research has revealed that human beings undergo rhythmic changes in their behaviour. These biological rhythms can be measured in body temperature, blood pressure, mental alertness, physical strength, and other body functions. These rhythms take place over roughly a 24-hour cycle and are called circadian rhythms. Our internal body clock is reset daily at sunrise. Certain times during the 24-hour cycle, often night time, are for repair and recovery of our body functions.

Suppose John's circadian rhythms are studied with regard to his 'percentage likelihood of success in problem solving' S, and his 'physical work rate measured in watts' W. The sine curve representing S has the equation $S = -25 \sin\left(\frac{\pi t}{6}\right) + 65$, and the sine curve representing W has the equation $W = -30 \sin\left(\frac{\pi t}{12}\right) + 200$, where t is the time in hours starting at midnight.

a. Sketch the two graphs on the same axes so they can be compared.

b. Find the best time of the day for John to solve a complex problem that requires both physical effort and problem-solving skills. Explain how you obtained your answer.

c. Discuss a suitable transformation of the curve representing W so that it has a better match with the curve representing S.

d. In general terms, discuss the limitations of using a sine curve to model John's physical work rate.

Unit 12.2 Trigonometry and Vectors

Topic 3: Straightforward trigonometric equations

Topic 3 builds on the content of the previous two Topics, with a focus on solving straightforward trigonometric equations.

- Solve trigonometric equations in degrees or radians, including the use of trigonometric graphs.
- Solve problems in context.

Introduction

Previously, the solution to an equation such as $\sin x = 0.5$ was found by using the **inverse** sine function. Thus, the solution was the single value $x = \sin^{-1}(0.5) = 30°$.

In most of the trigonometric equations that follow, the student will be given a domain from which to find all values that satisfy the equation.

Example A

Q. Solve $\cos x = 0.5$, where $0° \le x \le 360°$.

A. The domain is $0° \le x \le 360°$ (this is often written using interval notation as $[0°, 360°]$). It will be found by substitution that both $x = 120°$ and $x = 240°$ satisfy the equation $\cos x = -0.5$ and lie in the domain $0° \le x \le 360°$.

Note: From the work which follows it will be seen that 120° and 240° are the only possible solutions over this domain.

Solution of trigonometric equations by graph

The properties of the graphs of trigonometric functions can be used to find solutions to trigonometric equations.

Example B

Q. Solve $\sin x = 0.8$, $-360° \le x \le 360°$.

A. The graph of $y = \sin x$ is drawn for $-360° \le x \le 360°$, along with the line $y = 0.8$.

By calculator, $\sin^{-1}(0.8) = 53.1°$, which gives the solution in the first quadrant.

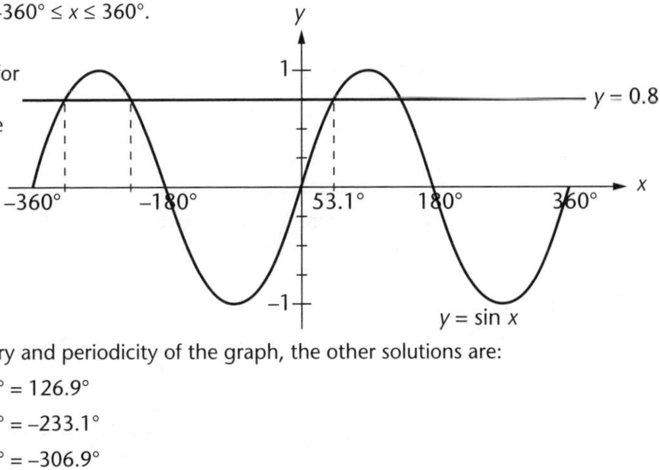

Using the symmetry and periodicity of the graph, the other solutions are:

$180° - 53.1° = 126.9°$

$-180° - 53.1° = -233.1°$

and $-360° + 53.1° = -306.9°$

Solutions to simple trigonometric equations in the domain $0° \le x \le 360°$

Since the period of $y = \sin x$ and $y = \cos x$ is 360°, a common domain for solutions is $0° \le x \le 360°$.

Example C

Q. Find all solutions to the equation $\sin x = 0.5$, $0° \le x \le 360°$.

A. Using a calculator,
$\sin^{-1}(0.5) = 30°$ is one solution.

By symmetry of the graph of $y = \sin x$, the other solution is $180° - 30° = 150°$.

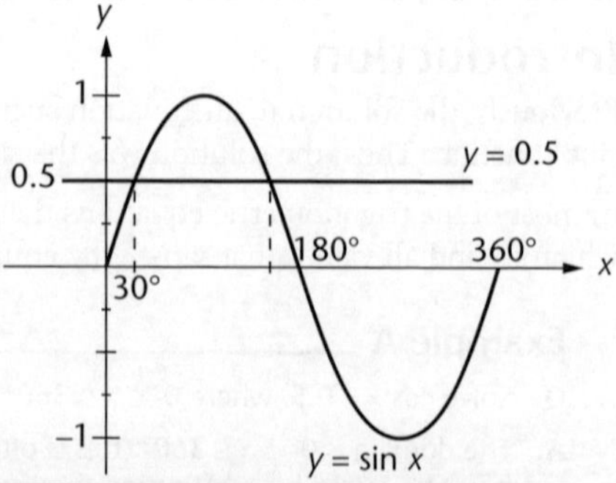

Note: In general, if $\sin x = c$ for $0° \le x \le 360°$, where $0 \le c \le 1$, then $\alpha° = \sin^{-1}(c)$ is one solution. The other solution is $180° - \alpha°$.

When solving $\sin x = c$ for negative c, it is best to find $\alpha° = \sin^{-1}(|c|)$, where $|c|$ is the **absolute value** of c (the (positive) size of c). Since sine is negative in the third and fourth quadrants, the two solutions will be $180° + \alpha°$, $360° - \alpha°$.

Example D

Q. Solve $\sin x = -0.3$, $0° \le x \le 360°$.

A. By calculator,

$\sin^{-1}(0.3) = 17.5°$ [since $|-0.3| = 0.3$]

$\therefore$ Solutions are:

$180° + 17.5° = 197.5°$

and

$360° - 17.5° = 342.5°$ [see graph]

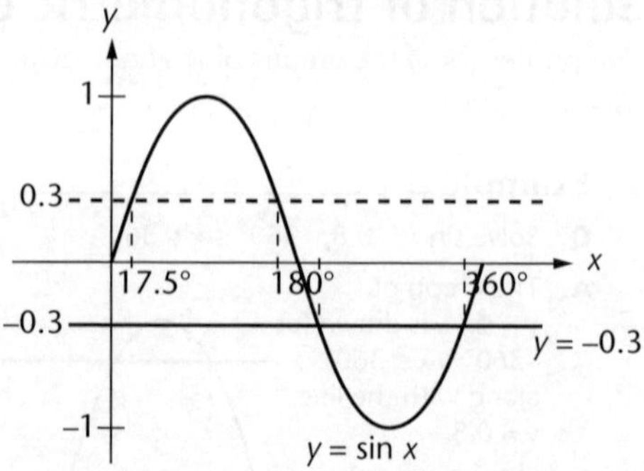

Similar techniques are used for solving $\cos x = c$ where $-1 \le c \le 1$ (the cosine graph is positive only in the first and fourth quadrants).

Example E

Q. Find all solutions to the equation $\cos x = 0.5$, $0° \le x \le 360°$.

A. By calculator,

$\cos^{-1}(0.5) = 60°$

By symmetry, the other solution is $360° - 60° = 300°$

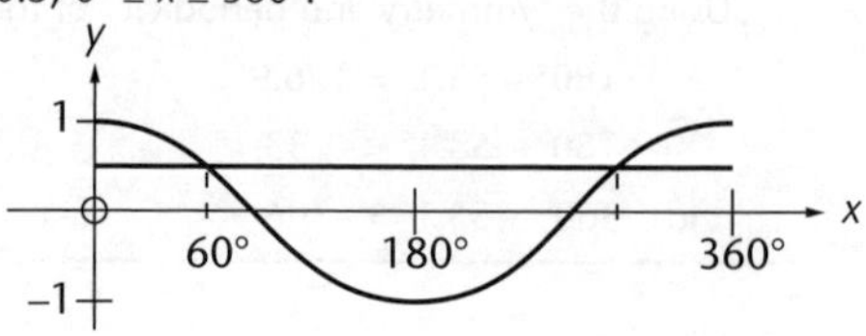

Example F

Q. Find all solutions to the equation $\tan x = \sqrt{3}$, $0° \leq x \leq 360°$.

A.

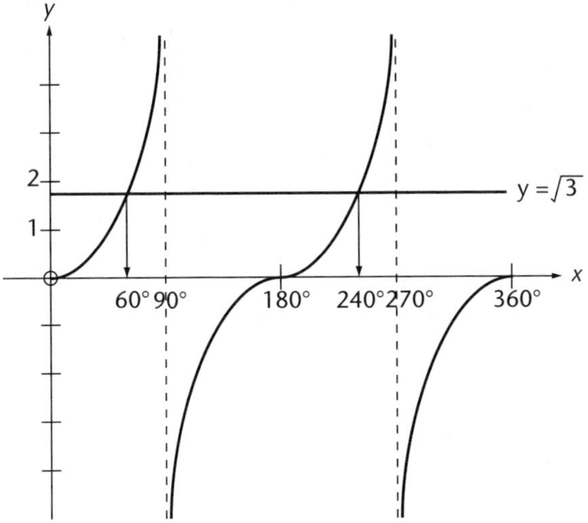

$\tan^{-1}(\sqrt{3}) = 60°$ so one solution is $x = 60°$. From the graph, the other solution is 240°.

Note: In general, if $\cos x = c$ for $0° \leq x \leq 360°$ and $\alpha° = \cos^{-1}(c)$ is one solution, then the other solution is $360° - \alpha°$. (This method works for both positive and negative c.)

The following points can be confirmed by examining the graph of $y = \sin x$:

- $\sin x = c$, $0° \leq x \leq 360°$ has exactly two solutions when $-1 < c < 0$ or $0 < c < 1$
- $\sin x = 1$, $0° \leq x \leq 360°$ has one solution only ($x = 90°$)
- $\sin x = -1$, $0° \leq x \leq 360°$ has one solution only ($x = 270°$)
- $\sin x = 0$, $0° \leq x \leq 360°$ has three solutions ($x = 0°$, $x = 180°$, $x = 360°$)

Similar results can be derived for $\cos x = c$ and for $\tan x = c$. Since the graph of $y = \tan x$ is considerably different from those of $y = \sin x$ and $y = \cos x$, it should be noted that:

- $\tan x = c$, $0° \leq x \leq 360°$ has exactly two solutions for $c \neq 0$.
- $\tan x$ can take any real value.

Unit 12.2 Activity 3A: Solving straightforward trigonometric equations

1. Using the graph of $y = \sin x$ (see Example B), solve the following trigonometric equations for the given domains:

a. $\sin x = 0.8$, $0° \leq x \leq 360°$

b. $\sin x = 0.8$, $0° \leq x \leq 180°$

c. $\sin x = 0.8$, $-360° \leq x \leq 0°$

d. $\sin x = 0.8$, $-360° \leq x \leq -180°$

e. $\sin x = 0.8$, $180° \leq x \leq 360°$

2. The graph shows $y = \sin x$ and $y = 0.6$, $-360° \leq x \leq 360°$.
   a. Use your calculator to solve $\sin x = 0.6$ in degrees ($0° \leq x \leq 90°$).
   b. Using the result from **2.a.** and the graph, solve the following equations over the given domains:
      i. $\sin x = 0.6$, $-360° \leq x \leq 360°$
      ii. $\sin x = 0.6$, $0° \leq x \leq 360°$
      iii. $\sin x = 0.6$, $0° \leq x \leq 180°$
      iv. $\sin x = 0.6$, $180° \leq x \leq 360°$

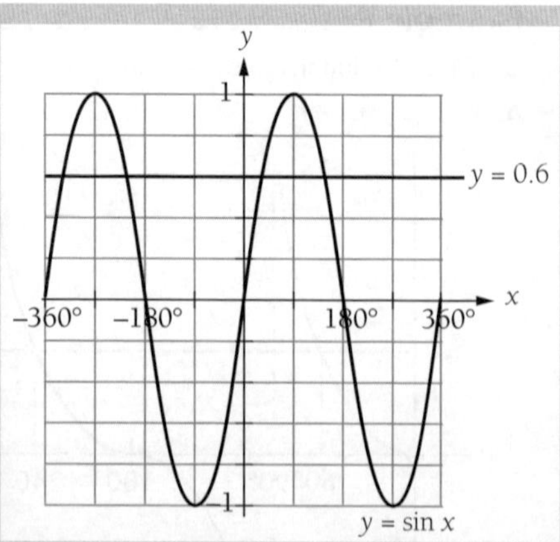

3. The graph shows $y = \cos x$ and $y = -0.4$, over the domain $[-360°, 360°]$.
   a. Use your calculator to solve $\cos x = -0.4$ in degrees.
   b. Using the result from **3.a.**, solve the following equations over the given domains:
      i. $\cos x = -0.4$, $-360° \leq x \leq 360°$
      ii. $\cos x = -0.4$, $0° \leq x \leq 360°$
      iii. $\cos x = -0.4$, for x in $[-360°, 0°]$
      iv. $\cos x = -0.4$, for ° in $[-180°, 180°]$

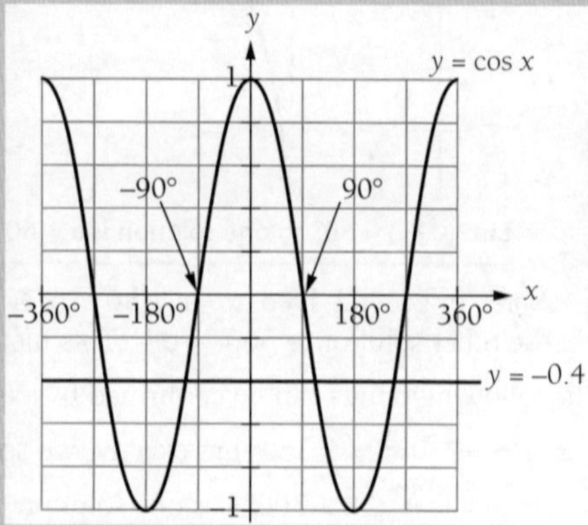

4. Find the solutions to $\sin x = 0.3$, $0° \leq x \leq 360°$ as follows:
   a. Sketch a graph of $y = \sin x$, $0° \leq x \leq 360°$.
   b. On the same diagram, sketch the straight line $y = 0.3$.
   c. Solve the equation $\sin x = 0.3$ using your calculator (ie by working out $\sin^{-1}(0.3)$).
   d. Use the periodicity and symmetry of the sine graph to find all solutions of the equation $\sin x = 0.3$, $0° \leq x \leq 360°$.

5. In a similar manner to that of question **4.**, find all solutions to $\cos x = 0.7$, $0° \leq x \leq 360°$.

6. Find all solutions to the equation $\cos x = -0.3$, $0° \leq x \leq 360°$.

7. a. Find all solutions to $\sin x = -0.2$, for x in $[-180°, 360°]$.
   b. If the domain of the equation was extended to $[-180°, 720°]$, what other solutions to the equation would there be?

8. a. Make a neat sketch of $y = \tan x$, $0° \leq x \leq 360°$.
   b. Sketch the line $y = 1$ on the same diagram.
   c. Find all solutions to the equation $\tan x = 1$, $0° \leq x \leq 360°$.

9. How many solutions does $\cos x = 0$, $0° \leq x \leq 360°$ have?

10. How many solutions does $\cos x = 1$, $0° \leq x \leq 360°$ have?

Solving trigonometric equations using radians

Trigonometric equations can also be solved using angles measured in radians. Similar techniques are used for solving these equations in radians.

Example G

Q. Find all solutions of $\sin x = 0.6$, $0 \leq x \leq 2\pi$.

A. The graphs of $y = \sin x$ and $y = 0.6$ are drawn for $0 \leq x \leq 2\pi$ (domain is in radians).

Using a calculator in radian mode, $\sin^{-1}(0.6) = 0.6435$

By symmetry, the other solution is $\pi - 0.6435 = 2.4981$

Thus, the solutions to $\sin x = 0.6$, $0 \leq x \leq 2\pi$ are 0.6435 and 2.4981.

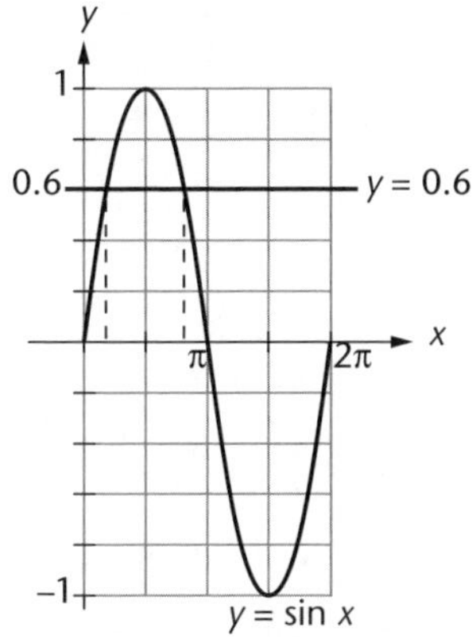

When exact solutions are required (see table on page 76), it is usually best to find the solutions in degrees first. These values are then converted to radians (using the conversion rule π radians = 180°). In this way answers are given as fractions of π.

Example H

Q. Solve $\sin x = -0.5$, $0 \leq x \leq 2\pi$, giving solutions as exact values (fractions of π).

A. Working in degrees, using the methods of the previous section, the solutions are 210° and 330°. Converting these solutions to radians gives:

$$210° = \frac{210 \times \pi}{180} \text{ radians} = \frac{7\pi}{6} \text{ radians}$$

$$330° = \frac{330 \times \pi}{180} \text{ radians} = \frac{11\pi}{6} \text{ radians}$$

So the exact solutions to $\sin x = -0.5$ are $\frac{7\pi}{6}$ and $\frac{11\pi}{6}$.

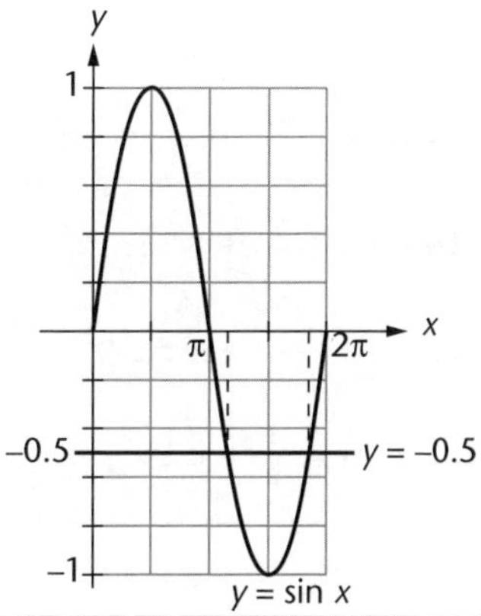

Unit 12.2 Activity 3B: Solving trigonometric equations using radians

1. $\sin x = 0.3, 0 \leq x \leq 2\pi$

2. $\sin x = 0.7, 0 \leq x \leq 2\pi$

3. $\cos x = 0.4, 0 \leq x \leq 2\pi$

4. $\sin x = -0.1, 0 \leq x \leq 2\pi$

5. $\sin x = 0.4, 0 \leq x \leq 2\pi$

6. $\cos x = -0.3, 0 \leq x \leq 2\pi$

7. $\tan x = 0.8, 0 \leq x \leq 2\pi$

8. $\tan x = \frac{-2}{3}, 0 \leq x \leq 2\pi$

9. $\tan x = 0.6, 0 \leq x \leq 2\pi$

10. $\tan x = \frac{-1}{2}, 0 \leq x \leq 2\pi$

11. Which of the following is a solution to $\sin x = 0.5$?

a. $\frac{\pi}{2}$ **b.** $\frac{\pi}{3}$ **c.** $\frac{\pi}{4}$ **d.** $\frac{\pi}{6}$ **e.** 0

12. Which of the following is a solution to $\cos x = \frac{-\sqrt{3}}{2}$?

a. $\frac{\pi}{3}$ **b.** $\frac{2\pi}{3}$ **c.** $\frac{5\pi}{6}$ **d.** $\frac{\pi}{6}$ **e.** $\frac{4\pi}{3}$

13. Which of the following sets contains the solutions to $\sin x =$, $0 \leq x \leq 2\pi$?

a. $\left\{\frac{\pi}{4}, \frac{7\pi}{4}\right\}$ **b.** $\left\{\frac{\pi}{4}, \frac{3\pi}{4}\right\}$ **c.** $\left\{\frac{3\pi}{4}, \frac{5\pi}{4}\right\}$ **d.** $\left\{\frac{5\pi}{4}, \frac{7\pi}{4}\right\}$ **e.** $\left\{\frac{3\pi}{4}, \frac{7\pi}{4}\right\}$

14. Which of the following sets contains the solutions to $\sin x = -\frac{1}{\sqrt{2}}, 0 \leq x \leq 2\pi$?

a. $\left\{\frac{\pi}{4}, \frac{7\pi}{4}\right\}$ **b.** $\left\{\frac{\pi}{4}, \frac{3\pi}{4}\right\}$ **c.** $\left\{\frac{3\pi}{4}, \frac{5\pi}{4}\right\}$ **d.** $\left\{\frac{5\pi}{4}, \frac{7\pi}{4}\right\}$ **e.** $\left\{\frac{3\pi}{4}, \frac{7\pi}{4}\right\}$

15. Which of the following sets contains the solutions to $\tan x = 1, 0 \leq x \leq 2\pi$?

a. $\left\{\frac{\pi}{4}, \frac{7\pi}{4}\right\}$ **b.** $\left\{\frac{\pi}{4}, \frac{5\pi}{4}\right\}$ **c.** $\left\{\frac{\pi}{4}, \frac{3\pi}{4}\right\}$ **d.** $\left\{\frac{5\pi}{4}, \frac{7\pi}{4}\right\}$ **e.** $\left\{\frac{3\pi}{4}, \frac{5\pi}{4}\right\}$

Find the solutions for the following trigonometrical equations. (Give answers in radians as multiples of π):

16. $\sin x = \frac{-1}{2}, 0 \leq x \leq 2\pi$

17. $\cos x = \frac{1}{2}, -\pi \leq x \leq 2\pi$

18. $\sin x = \frac{1}{\sqrt{2}}, 0 \leq x \leq 2\pi$

19. $\tan x = \sqrt{3}, -\pi \leq x \leq \pi$

20. $\tan x = -1, 0 \leq x \leq 2\pi$

21. $\sin x = \frac{-\sqrt{3}}{2}, -\pi \leq x \leq \pi$

22. $\cos x = \frac{1}{\sqrt{2}}, -\pi \leq x \leq \pi$

Trigonometric modelling (problems in context)

Since radians are unit-free (they are treated as numbers) they are often used when practical situations are modelled using trigonometry.

Example I

Q. A wave pattern on water is modelled by the sine function, $y = \sin t$ where y is the height of the water in centimetres above or below the 2 m depth the water has when still, and t is measured in minutes (regarded as radians).

a. Find the depth of the water after $\frac{\pi}{2}$ seconds.

b. What is the lowest depth?

c. When is the depth 2.005 m in the first 6 minutes?

A. **a.** After $\frac{\pi}{2}$ seconds, the depth is

$2 + (\sin \frac{\pi}{2}) \times 0.01$ [working in metres]

$= 2.01$ m

b. The lowest depth occurs when y has its least value, which is -1×0.01 metres $= -0.01$ m.

$\therefore$ lowest depth is $2 - 0.01 = 1.99$ m.

c. 6 is close to 2π so the equation is solved in the domain $[0, 2\pi]$.

The depth at time t is $2 + 0.01\sin t$.

Setting this equal to 2.005, gives:

$2 + 0.01\sin t = 2.005$

$0.01\sin t = 0.005$ [subtracting 2]

$\sin t = 0.5$ [dividing by 0.01]

Solving this equation over the domain $[0, 2\pi]$ gives:

$t = 0.5236$ minutes or $t = 2.6180$ minutes.

Unit 12.2 Activity 3C: Practical problems using trigonometry

1. A wave pattern on water is modelled by the cosine function $y = \cos t$, where y is the height of water (in centimetres) above its usual 1 m depth when still, and t is the time in minutes (regarded as radians).

a. What is the depth of the water after $\frac{\pi}{2}$ seconds?

b. What is the lowest depth of the water?

c. In the first 6 minutes, when is the depth 1.005 m?

2. A logo is being drawn using the cosine function $y = \cos x$, $-180° \leq x \leq 180°$ coloured red, the x-axis corresponding to this domain coloured blue, and the y-axis corresponding to this range coloured black.

a. Sketch the logo.

b. Which angles have a red value of 0.3?

3. On the right is a diagram of a wheel. It is turning in an anti-clockwise direction about A. The height of a point P on the circumference, above A, can be modelled by the equation $h = \sin t$, where t is the time in seconds and h is the height in centimetres. How long does it take the point P to reach a height of 0.5 cm above A, if the wheel is turning through 1° per second?

4. A machine generates force according to the formula $F(t) = \tan(t°)$, $25 \leq t \leq 75$, where $F(t)$ is the size of the force (in newtons) t seconds after midday on Tuesday (1° corresonds to 1 second). Fince the length of time the force exceeds 1 newton.

5. A baby watches a toy. The distance, in centimetres, of the toy above a point one centimetre from the floor can be modelled by the equation $d = \cos t$, where values of t are to be taken such that 1 radian corresponds to one second. When is the second time the toy is 0.6 cm above the point?

Unit 12.2 Trigonometry and Vectors
Topic 4: Trigonometric equations including multi-step problems

Topic 4 continues the focus on solving straightforward trigonometric equations:
- Solve trigonometric equations.
- Solve multi-step trigonometric problems, including complex manipulations of trigonometric equations.
- Solve problems in context.

Basic trigonometric equations

A trigonometric equation is given in terms of a circular function. The solution is an angle, expressed in degrees or radians. Examples A, B and C illustrate the basic techniques for solving trigonometric equations.

Example A

Q. Solve $\sin x = 0.363$ for $0° \leq x \leq 360°$.

A. There will be two different angles in the set $0° \leq x \leq 360°$ for which $\sin x = 0.363$. These will be in the first and second quadrants because sine corresponds to the y-coordinate and is positive only in these two quadrants.

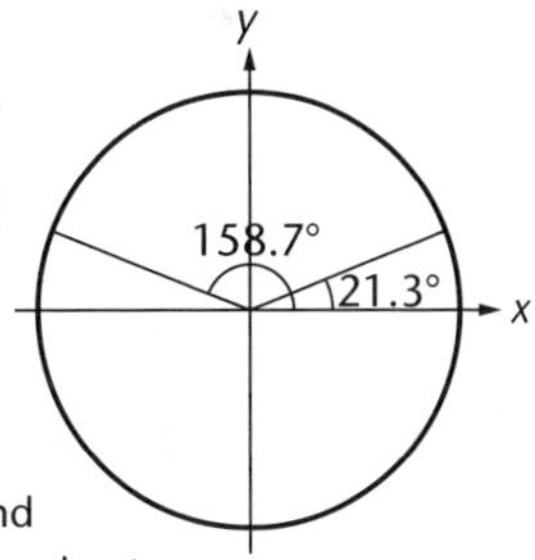

The $\sin^{-1}$ key on a calculator gives 21.3 (1 dp) as follows:

shift $\sin^{-1}$ · 3 6 3 =

The two angles required are: 21.3° in the 1st quadrant and
$180° - 21.3° = 158.7°$ in the 2nd quadrant.

Example B

Q. Solve the equation $\cos x = -0.4721$ for $0° \leq x \leq 360°$.

A. $\cos^{-1}(0.4721) = 61.8°$ [using the $\cos^{-1}$ key, ignoring the negative.]

Cosine, measuring the *x*-coordinate, is negative in the 2nd and 3rd quadrants so the angles required are:

$180° - 61.8° = 118.2°$ in the 2nd quadrant and

$180° + 61.8° = 241.8°$ in the 3rd quadrant.

Example C

Q. Solve the equation $\tan \theta = -1.42$ for $0° \le \theta \le 360°$.

A. $\tan^{-1}(1.42) = 54.8°$ [using the $\tan^{-1}$ key, ignoring the negative]

Tangent is negative in the 2nd and 4th quadrants and hence the solutions to the above equation are:

$180° - 54.8° = 125.2°$ in the 2nd quadrant and

$360° - 54.8° = 305.2°$ in the 4th quadrant.

Transformations of basic trigonometric equations

Various transformations of the basic trigonometric equations can take place. One type is illustrated below.

Example D

Q. Solve the equation $\sin(x + 37°) = -0.832$ for $0° \le x \le 360°$.

A. $\sin^{-1}(0.832) = 56.3°$ [using the $\sin^{-1}$ key, ignoring the negative]

Sine is negative in the 3rd and 4th quadrants:

$\therefore\ x + 37° = 180° + 56.3°$ or $360° - 56.3°$

$= 236.3°$ or $303.7°$ [simplifying]

$\therefore\ x = 199.3°$ or $266.7°$. [subtracting 37°]

Note: Both answers are part of the domain $0° \le x \le 360°$.

Negative angles

Sometimes a *negative angle* results when solving trigonometric equations.

Negative angles are the angles obtained by rotating in a *clockwise* direction about the unit circle, rather than in an anti-clockwise direction. Thus $-x$ corresponds to the same position on the unit circle as $360° - x$.

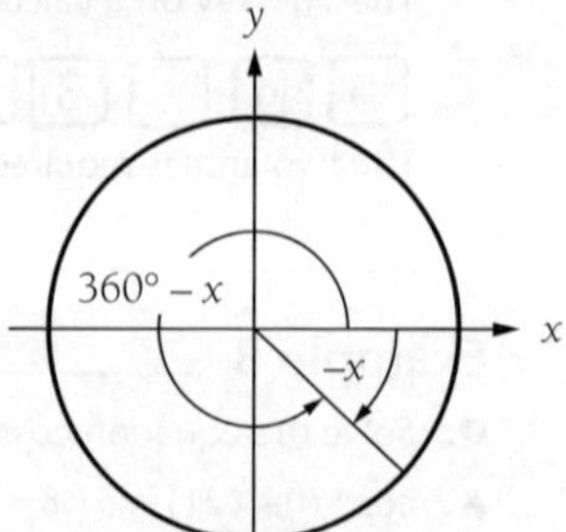

Example E

Q. Solve the equation, $\cos(x + 49°) = 0.913$ for $0° \le x \le 360°$.

A. $\cos^{-1}(0.913) = 24.1°$ [using $\cos^{-1}$ key on calculator]

$\therefore\ x + 49° = 24.1°$ [the angle in quadrant I]

or $x + 49° = 360° - 24.1°$ [the angle in quadrant IV]

$\therefore\ x = -24.9°, 286.9°$

The solution $x = -24.9°$ is not part of the domain $0° \le x \le 360°$, so this solution is written as $x = 360° - 24.9° = 335.1°$ instead.

$\therefore$ The solutions are 286.9° or 335.1°.

Unit 12.2 Activity 4A: Simple trigonometric equations

Solve these basic trigonometric equations where $0° \leq x \leq 360°$:

1. $\sin x = 0.3172$
2. $\cos x = 0.4314$
3. $\tan x = -1.21$
4. $\cos x = -0.4621$
5. $\sin x = -0.8623$
6. $\tan x = 0.6821$
7. $\cos x = 0.3164$
8. $\sin x = 0.4839$
9. $\sin y = -0.8134$
10. $\tan x = -2.137$
11. $\cos x = 0.135$
12. $\tan x = 2.3$
13. $\sin x = \frac{-1}{\sqrt{2}}$
14. $\sin x = \frac{\sqrt{3}}{2}$
15. $\tan x = \frac{-1}{\sqrt{3}}$
16. $\sin x = -1$
17. $\cos x = \frac{1}{\sqrt{2}}$
18. $\tan x = \sqrt{3}$
19. $\sin x = \frac{-1}{2}$
20. $2 \sin^2 x = 1$

Solve these transformed trigonometric equations, where $0° \leq x \leq 360°$:

21. $\sin (x + 34°) = 0.8634$
22. $\cos (x - 53°) = -0.7144$
23. $\sin (x - 90°) = 0.4$
24. $\tan (x + 100°) = -2$
25. $\cos (x + 45°) = \frac{-1}{\sqrt{2}}$
26. $\sin (x + 60°) = \frac{\sqrt{3}}{2}$
27. $\cos (x - 30°) = \frac{-1}{2}$
28. $2 \sin (x - 37°) = 0.8613$
29. $3 \tan (x + 35°) = 5$
30. $4 \cos (x + 30°) = 0.83$

Equations where the unknown has a coefficient ≠ 1

When the unknown has a coefficient other than 1, care must be taken to ensure that *all solutions* are found. There are usually two solutions for each rotation of 360° from the positive horizontal axis. Thus, if $0° \leq x \leq 360°$, $3x$ takes values in the set $0° \leq 3x \leq 1\,080°$. This involves three rotations from the positive horizontal axis, so there will be six solutions to an equation involving $3x$, such as $\tan 3x = 0.5913$.

Example F

Q. Solve $\tan 3x = 0.5913$ for $0° \leq 3x \leq 360°$.

A. $\tan^{-1} (0.5913) = 30.6°$ [using a calculator]

Since $0° \leq x \leq 360°$, $3x$ lies in the set $0° \leq 3x \leq 1\,080°$. [$3 \times 360° = 1\,080°$]

There are two solutions for $3x$ in the set 0° to 360°:

$3x = 30.6°$ and $3x = 180° + 30.6°$ [tan positive in quadrants I and III]

$= 210.6°$

$\therefore\ x = 10.2°$ and $70.2°$ [dividing by 3]

Two other solutions occur when $3x$ lies in the interval [360°, 720°]. These are obtained by solving:

$3x = 360° + 30.6° = 390.6°$ $\quad \therefore\ x = 130.2°$ [dividing by 3]

$3x = 360° + 210.6° = 570.6°$ $\quad \therefore\ x = 190.2°$ [dividing by 3]

The last two solutions occur when $3x$ lies in the interval [720°, 1 080°]. These are:

$3x = 720° + 30.6° = 750.6°$ $\quad \therefore x = 250.2°$ $\quad$ [dividing by 3]

$3x = 720° + 210.6° = 930.6°$ $\quad \therefore x = 310.2°$ $\quad$ [dividing by 3]

$\therefore$ Solutions are $x = 10.2°, 70.2°, 130.2°, 190.2°, 250.2°, 310.2°$

Example G

Q. Solve $\sin 2x = \frac{-1}{\sqrt{2}}$, $0° \le x \le 360°$.

A. $\sin^{-1}\left(\frac{1}{\sqrt{2}}\right) = 45°$ $\quad$ [exact trig. value or by using a calculator]

Since $2x$ lies in the set 0° to 720° and sin is negative in quadrants III and IV:

$2x = 180° + 45°, 360° - 45°, 360° + (180° + 45°), 360° + (360° - 45°)$

$\therefore x = 112.5°, 157.5°, 292.5°, 337.5°$ $\quad$ [dividing by 2]

Unit 12.2 Activity 4B: Trigonometric equations where coefficient of $x \ne 1$

1. Solve the following equations for $0° \le x \le 360°$:

a. $\sin 3x = 0.5613$ **b.** $\cos 2x = 0.2132$ **c.** $\tan 3x = 1.3342$

d. $\tan 2x = -0.6345$ **e.** $\cos 4x = -0.5632$ **f.** $\sin 2x = -0.4763$

g. $\tan 2x = 0.7634$ **h.** $\sin 4x = 0.8632$ **i.** $\sin \frac{x}{2} = 0.1342$

j. $\cos \frac{x}{3} = 0.4152$ **k.** $\cos 2x = \frac{\sqrt{3}}{2}$ **l.** $\sin 3x = 1$

2. Solve the following equations for the domains given:

a. $\sin 2x = 0.8329, 0° \le x \le 180°$ **b.** $\cos 3x = 0.3625, 0° \le x \le 180°$

c. $\tan \frac{x}{2} = 5.6713, 0° \le x \le 720°$ **d.** $\sin 3x = -0.1618, -180° \le x \le 180°$

e. $\cos 2x = -0.8127, 0° \le x \le 540°$ **f.** $\tan 4x = -9.5144, 0° \le x \le 180°$

g. $\sin 2x = \frac{1}{2}, 0° \le x \le 90°$ **h.** $2 \sin \frac{1}{2}x = -0.4618, 0° \le x \le 720°$

i. $3 \tan 2x = 1, 0° \le x \le 270°$ **j.** $5 \sin 3x = 2 \sin 3x + 0.48, 0° \le x \le 180°$

k. $3 \cos 3x = 2.3416, -180 \le x \le 180°$ **l.** $\sqrt{3} \tan 2x = 3, 0° \le x \le 180°$

Trigonometrical equations using radians

Similar techniques to those described above are used when solving trigonometric equations with angles measured in radians.

Example H

Q. Solve $\cos 2x = \frac{1}{\sqrt{2}}$, $0 \le x \le 2\pi$

A. Since x lies in the interval $[0, 2\pi]$, $2x$ lies in the interval $[0, 4\pi]$.

There are two solutions for $2x$ in the set $[0, 2\pi]$:

$2x = \frac{\pi}{4}$ or $2x = 2\pi - \frac{\pi}{4}$ [cosine positive in quadrants I, IV]

$\therefore\ x = \frac{\pi}{8}$ or $x = \frac{7\pi}{8}$ [simplifying and dividing by 2]

There are also two solutions for $2x$ in the set $[2\pi, 4\pi]$:

$2x = 2\pi + \frac{\pi}{4}$ or $2x = 2\pi + \frac{7\pi}{4}$

$\therefore\ x = \frac{9\pi}{8}$ or $x = \frac{15\pi}{8}$ [simplifying and dividing by 2]

$\therefore$ Solutions are $\frac{\pi}{8}, \frac{7\pi}{8}, \frac{9\pi}{8}, \frac{15\pi}{8}$.

Note: This equation can also be solved by using the period of cos $2x$, which is π. The first two solutions are $\frac{\pi}{8}$ and $\frac{7\pi}{8}$, then further solutions are found by adding multiples of the period until values are found outside the domain $[0, 2\pi]$.

Example I

Q. Solve the equation $\sin (x + \frac{\pi}{2}) = \frac{\sqrt{3}}{2}$, $0 \le x \le 2\pi$. Give exact solutions (as fractions of π).

A. $\sin^{-1} (\frac{\sqrt{3}}{2}) = 60°$ [using a calculator]

$60° = \frac{60 \times \pi}{180}$ radians [changing to radians]

$= \frac{\pi}{3}$ radians

Sine is positive in quadrants I and II, so the two solutions are:

$\therefore\ x + \frac{\pi}{2} = \frac{\pi}{3}$ or $x + \frac{\pi}{2} = \pi - \frac{\pi}{3}$

$\therefore\ x = \frac{-\pi}{6}$ or $\frac{\pi}{6}$ [rearranging and solving]

Since $x = \frac{-\pi}{6}$ is not in the domain $0 \le x \le 2\pi$, this solution is expressed as $2\pi - \frac{\pi}{6} = \frac{11\pi}{6}$, so the solutions to $\sin (x + \frac{\pi}{2}) = \frac{\sqrt{3}}{2}$ are $x = \frac{\pi}{6}$ or $x = \frac{11\pi}{6}$.

Unit 12.2 Activity 4C: Trigonometric equations using radians

1. Solve the following trigonometric equations (give answers in radians as multiples of π):

a. $\sin (x + \frac{\pi}{4}) = \frac{\sqrt{3}}{2}$, $0 \le x \le 2\pi$

b. $\cos (x - \frac{\pi}{3}) = \frac{1}{2}$, $-\pi \le x \le \pi$

c. $\sin 2x = \frac{\sqrt{3}}{2}$, $0 \le x \le 2\pi$

d. $\cos 3x = \frac{-1}{\sqrt{2}}$, $0 \le x \le 2\pi$

e. $\sin 3x = 1$, $\pi \le x \le 2\pi$

f. $\tan 2x = 1$, $0 \le x \le 2\pi$

g. $\sin \frac{x}{2} = \frac{1}{2}$, $0 \le x \le 2\pi$

h. $\cos \frac{x}{3} = \frac{-1}{\sqrt{2}}$, $0 \le x \le \pi$

2. Find solution sets for the following trigonometric equations.
Express answers in radians.

a. $\cos(x + \frac{\pi}{4}) = \frac{-1}{2}, 0 \leq x \leq 2\pi$

b. $\sin(x - \frac{\pi}{3}) = 0.7614, 0 \leq x \leq 2\pi$

c. $3 \sin x = 1, 0 \leq x \leq 2\pi$

d. $4 \cos x = -3, 0 \leq x \leq 2\pi$

e. $8 \sin x = 7, 0 \leq x \leq 2\pi$

f. $-2 \tan x = 3, 0 \leq x \leq 2\pi$

g. $7 \tan x = -8, 0 \leq x \leq 2\pi$

h. $-\cos x = 0.1513, 0 \leq x \leq 2\pi$

Trigonometric modelling (problems in context)

Example J

Q. Some stars have a brightness that periodically increases and decreases. The brightness (B) of one such star can be modelled by the equation:

$B = 3.9 + 0.33\sin\left(\frac{2\pi t}{5.6}\right)$, where t is measured in days.

1. What is the maximum brightness of the star?

2. What is the minimum brightness of the star?

3. How many days elapse between successive times of maximum brightness?

4. When was the first occasion when the brightness was 4.065?

A. **1.** The maximum brightness occurs when $\sin\left(\frac{2\pi t}{5.6}\right)$ is 1. Thus, the maximum brightness is $3.9 + 0.33 \times 1 = 4.23$.

2. The minimum brightness occurs when $\sin\left(\frac{2\pi t}{5.6}\right)$ is –1. Thus, the minimum brightness is $3.9 - 0.33 \times 1 = 3.57$.

3. A period will elapse between times of maximum brightness.

Thus, the period $= 2\pi \div \frac{2\pi}{5.6} = 5.6$ days.

4. Let the first occasion when brightness is 4.065 be after x days. To find x, solve:

$3.9 + 0.33 \sin\left(\frac{2\pi x}{5.6}\right) = 4.065$

$\therefore\ 0.33 \sin\left(\frac{2\pi x}{5.6}\right) = 0.165$ [subtracting 3.9]

$\therefore\ \sin\left(\frac{2\pi x}{5.6}\right) = 0.5$ [dividing by 0.33]

The solution is found using the inverse sine function on a calculator.

$\therefore\ \frac{2\pi x}{5.6} = \sin^{-1} 0.5 = \frac{\pi}{6}$ [or 0.5236 by calculator in radian mode]

$\therefore\ x = \frac{5.6\pi}{6 \times 2\pi}$ [rearranging]

$= \frac{5.6}{12}$

$= 0.4667$ days (4 dp)

Unit 12.2 Activity 4D: Applications of trigonometry

1. The height above the ground of a point on a turning wheel is given by the formula, $H = 5\cos(60t)^\circ + 10$, where t is in seconds and H is in metres.

a. What is the height above the ground of the point after 2 seconds?

b. What is the greatest height of the point above the ground?

c. What is the least height of the point above the ground?

d. How long is the period of the motion?

e. Give two times when the height of the point is 12.5 m above the ground.

2. The height above the sea floor of a cork on the surface of water in a bay is given by the formula, $D = 2\cos(4t)^\circ + 4$, where t is in seconds and D is in metres.

a. What is the least distance of the cork from the sea floor?

b. Find the height of the cork above the sea floor when the time is 15 seconds.

c. At what times in the first minute is the height of the cork above the sea floor 6 m?

3. A mass oscillates on the end of a spring suspended from a ceiling. The depth of the mass below the ceiling is given by $d = 4\sin(24t)^\circ + 12$ where d is measured in centimetres and t is measured in seconds.

a. What is the maximum depth of the mass below the ceiling?

b. What is the minimum depth of the mass below the ceiling?

c. How long does it take for the mass to move between its lowest and highest points?

d. Give the first two times when the depth of the mass below the ceiling is 14 cm.

4. A mark on a steadily revolving Ferris wheel is at a height h metres above the ground after t seconds. The height h is given by:

$$h = 8 + 7\sin\frac{\pi t}{12}$$

a. How high above the ground is the mark after 2 seconds?

b. After how many seconds is the mark at its greatest height? What is this height?

Multi-step problems

A trigonometric model may involve a **piecewise** function, in which one function applies for part of the domain, and another function (or functions) applies for the other values in the domain.

Example K

Q. For the first 10 minutes of recording, a wave has its height described by the formula, $h(t) = 5\cos\left(\frac{2\pi t}{5}\right)$ where t is in seconds and h is in centimetres. After that time, a more suitable formula for the height is, $h(t) = A\cos\left(\frac{4\pi t}{5}\right)$ as the period is halved.

a. How many waves had passed before the change of formula?

b. What is the value of A? Give a reason for your answer.

c. How much time elapses from the time of the change until the wave has a height of 0 cm?

A. a. 10 minutes had passed before the change, which is 600 seconds. The period of the motion is $2\pi \div \frac{2\pi}{5} = 5$ seconds.

$\therefore$ The number of cycles (or waves) was $600 \div 5 = 120$ waves.

b. $A = 5$ since the wave is **continuous**, the height (amplitude) of the wave does not change.

c. Let the required time be x.

$\therefore\ 5\cos\left(\frac{4\pi x}{5}\right) = 0$

$\therefore\ \cos\left(\frac{4\pi x}{5}\right) = 0$ [dividing by 5]

$\therefore\ \frac{4\pi x}{5} = \frac{\pi}{2}$ [since $\cos^{-1} 0 = \frac{\pi}{2}$]

$\therefore\ x = \frac{5}{8}$ seconds [rearranging and solving for x]

Unit 12.2 Activity 4E: Piecewise trigonometric functions

1. A wave has its height described by:

$$h(t) = \begin{cases} 4\sin\dfrac{2\pi t}{81}, & 0 \le t \le 170 \\ 2 + A\sin\dfrac{8\pi t}{81}, & t > 170 \end{cases}$$

where t is in seconds and h is in centimetres.

a. Find the value of A.

b. Find the first time after 170 seconds that the height of the wave is 2 cm.

2. A pattern with line symmetry is drawn following the rule,

$$y = \begin{cases} 1 - \dfrac{1}{2}\sin 2x & 0 \le x \le \dfrac{\pi}{2} \\ 1 - 2\cos x & \dfrac{\pi}{2} \le x \le \dfrac{3\pi}{2} \\ 1 + \dfrac{1}{2}\sin 2x & \dfrac{3\pi}{2} \le x \le 2\pi \end{cases}$$

For what values of x is the height of the pattern, y, equal to:

a. 0.75?

b. 2?

Unit 12.2 Trigonometry and Vectors

Topic 5: Trigonometric identities

The material in this Topic is about solving problems and equations involving trigonometric functions, and contains:

- Problems with trigonometric manipulation, such as proving trigonometric identities which may involve the following:
  - reciprocal relationships
  - Pythagorean identities
  - compound angle formulae
  - double angle formulae
  - sum and product formulae

  and combinations of these.
- More complex problems, where students will be expected to choose and apply appropriate trigonometric relationships in a chain of reasoning to solve the problem. This may include:
  - proof
  - developing a formula
  - rewriting trigonometric expressions in other forms
  - knowledge of sine and cosine rules.

Trigonometric identities

An **identity** is an equation that is true for all values of the unknown, whereas an equation is usually true for one or some values only.

Example A

1. $3x - 15 = 0$ is an equation that is true only for $x = 5$.
2. The equation $\sin\theta = \cos\theta$ is true for certain values of θ, namely $\theta = \ldots \frac{\pi}{4}, \frac{5\pi}{4}, \frac{9\pi}{4}, \ldots$.
3. The equation $x^2 - x = x(x - 1)$ is true whatever the value of x and so is an identity.
4. The equation $\frac{\sin\theta}{\cos\theta} = \tan\theta$ is an identity which is true for all values of θ except when $\tan\theta$ is undefined, which occurs when $\theta = (2n+1)\frac{\pi}{2}$, $n \in \mathrm{I}$.

Equations can be solved, but identities cannot be solved since they are true for all values of the variable (except those values that make the expression undefined).

In this Topic the main trigonometric identities are verified and then used to prove other identities. Trigonometric identities will also be used to solve equations in Topic 7.

The following identities are already known:

$$\tan\theta = \frac{\sin\theta}{\cos\theta} \qquad \sec\theta = \frac{1}{\cos\theta} \qquad \operatorname{cosec}\theta = \frac{1}{\sin\theta} \qquad \cot\theta = \frac{1}{\tan\theta} = \frac{\cos\theta}{\sin\theta}$$

Pythagorean identities

Three more identities are now developed, often called **Pythagorean Identities**.

The first Pythagorean identity is derived from Pythagoras' theorem.

Beginning with a circle of radius r and any angle θ as in the diagram:

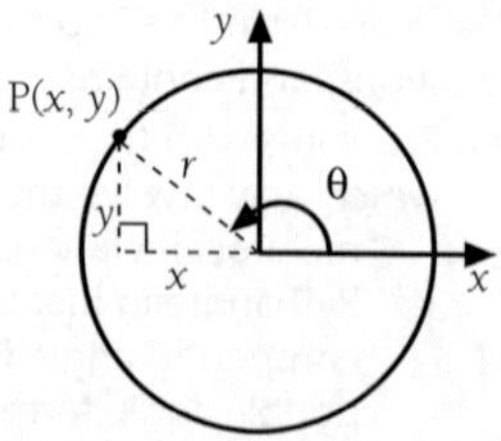

$$x^2 + y^2 = r^2 \quad \text{[using Pythagoras' theorem]}$$

$$\frac{x^2}{r^2} + \frac{y^2}{r^2} = 1 \quad \text{[dividing both sides by } r^2\text{]}$$

$$\text{ie } \left(\frac{x}{r}\right)^2 + \left(\frac{y}{r}\right)^2 = 1 \quad \left[\text{since } \frac{a^2}{b^2} = \left(\frac{a}{b}\right)^2\right]$$

But $\frac{x}{r} = \cos\theta$, $\frac{y}{r} = \sin\theta$ (see original definition), so $(\cos\theta)^2 + (\sin\theta)^2 = 1$.

Writing $(\cos\theta)^2$ as $\cos^2\theta$ and $(\sin\theta)^2$ as $\sin^2\theta$ gives:

$$\boxed{\sin^2\theta + \cos^2\theta = 1}$$

From $\sin^2\theta + \cos^2\theta = 1$, two more identities can be established:

$$\frac{\sin^2\theta}{\cos^2\theta} + \frac{\cos^2\theta}{\cos^2\theta} = \frac{1}{\cos^2\theta} \quad \text{[dividing by } \cos^2\theta\text{]}$$

$$\left(\frac{\sin\theta}{\cos\theta}\right)^2 + 1 = \left(\frac{1}{\cos\theta}\right)^2 \quad \left[\text{since } \frac{a^2}{b^2} = \left(\frac{a}{b}\right)^2\right]$$

$$(\tan\theta)^2 + 1 = (\sec\theta)^2 \quad \left[\text{using } \frac{\sin\theta}{\cos\theta} = \tan\theta \text{ and } \frac{1}{\cos\theta} = \sec\theta\right]$$

Hence, the second Pythagorean identity can be stated as:

$$\boxed{\tan^2\theta + 1 = \sec^2\theta}$$

Similarly:

$$\frac{\sin^2\theta}{\cos^2\theta} + \frac{\cos^2\theta}{\sin^2\theta} = \frac{1}{\sin^2\theta} \quad \text{[dividing } \sin^2\theta + \cos^2\theta = 1 \text{ by } \sin^2\theta\text{]}$$

$$1 + \left(\frac{\cos\theta}{\sin\theta}\right)^2 = \left(\frac{1}{\sin\theta}\right)^2 \quad \left[\text{since } \frac{a^2}{b^2} = \left(\frac{a}{b}\right)^2\right]$$

$$1 + (\cot\theta)^2 = (\operatorname{cosec}\theta)^2 \quad \left[\text{using } \frac{\cos\theta}{\sin\theta} = \cot\theta \text{ and } \frac{1}{\sin\theta} = \operatorname{cosec}\theta\right]$$

Hence, the third Pythagorean identity can be stated as:

$$\boxed{1 + \cot^2\theta = \operatorname{cosec}^2\theta}$$

Proving trigonometric identities

The three Pythagorean identities, along with $\frac{\sin\theta}{\cos\theta} = \tan\theta$ and the definitions of $\sec\theta$, $\operatorname{cosec}\theta$ and $\cot\theta$, can be applied to a variety of other identities to establish their validity.

To prove an identity, start with either the left-hand side (LHS) or the right-hand side (RHS), choosing the side that looks more readily able to be simplified. Re-express the terms using identities so that it takes the form of the other side.

Example B

Q. Prove that $\dfrac{\cot^2\theta - 1}{\cot^2\theta + 1} = 1 - 2\sin^2\theta$.

A. Starting with LHS

$\dfrac{\cot^2\theta - 1}{\cot^2\theta + 1} = \dfrac{\cot^2\theta - 1}{\csc^2\theta}$ [substituting $1 + \cot^2\theta = \csc^2\theta$]

$= \dfrac{\cot^2\theta + 1 - 2}{\csc^2\theta}$ [re-expressing to form identity]

$= \dfrac{\csc^2\theta - 2}{\csc^2\theta}$ [substituting $\csc^2\theta$ for $1 + \cot^2\theta$]

$= \dfrac{\csc^2\theta}{\csc^2\theta} - \dfrac{2}{\csc^2\theta}$ [separating into two fractions]

$= 1 - 2\sin^2\theta$ [since $\dfrac{1}{\csc^2\theta} = \sin^2\theta$]

$=$ RHS

Example C

Q. Prove that $\dfrac{1 - \sin\theta}{1 + \sin\theta} = (\sec\theta - \tan\theta)^2$

A. Starting with RHS

$(\sec\theta - \tan\theta)^2 = \left(\dfrac{1}{\cos\theta} - \dfrac{\sin\theta}{\cos\theta}\right)^2$ [using definitions of $\sec\theta$, $\tan\theta$]

$= \dfrac{(1 - \sin\theta)^2}{\cos^2\theta}$ [squaring **numerator** and **denominator**]

$= \dfrac{(1 - \sin\theta)^2}{1 - \sin^2\theta}$ [since $\sin^2\theta + \cos^2\theta = 1$]

$= \dfrac{\cancel{(1 - \sin\theta)}(1 - \sin\theta)}{\cancel{(1 - \sin\theta)}(1 + \sin\theta)}$ [factorising and cancelling]

$=$ LHS

Unit 12.2 Activity 5A: Trigonometric identities

1. Prove the following identities:
   a. $4\sin^2 x + 7\cos^2 x = 4 + 3\cos^2 x$
   b. $3\sin^2 x - 2\cos^2 x = 5\sin^2 x - 2$
   c. $2\sec^2 x + \tan^2 x = 3\sec^2 x - 1$
   d. $3\csc^2 x + 5\cot^2 x = 8\cot^2 x + 3$
   e. $\cos^2 x + \tan^2 x = \sec^2 x - \sin^2 x$
   f. $\tan x + \cot x = \sec x \csc x$

2. Prove each of the following identities:
   a. $\tan\theta\sec\theta = \sin\theta(1 + \tan^2\theta)$
   b. $(\csc^2\theta - 1)(1 + \tan^2\theta) = \csc^2\theta$
   c. $(\sec\theta - 1)(\sec\theta + 1)\cot\theta = \tan\theta$
   d. $\sec x - \cos x = \sin x \tan x$
   e. $\dfrac{1}{1 - \cos\theta} + \dfrac{1}{1 + \cos\theta} = 2\csc^2\theta$
   f. $\dfrac{2\tan\theta}{1 + \tan^2\theta} = 2\sin\theta\cos\theta$
   g. $a^2\cos^2\theta + b^2\sin^2\theta = a^2 + (b^2 - a^2)\sin^2\theta$
   h. $\cos^2 x \csc x \tan^2 x = \sin x$

i. $\dfrac{\cos x}{1-\tan x}+\dfrac{\sin x}{1-\cot x}=\sin x+\cos x$

j. $\dfrac{\tan^2\theta-1}{\tan^2\theta+1}=1-2\cos^2\theta$

3. Eliminate θ from each of the following pairs of equations:

a. $x = 2\cos\theta,\ y = 2\sin\theta$

b. $x = b\tan\theta,\ y = a\sec\theta$

c. $x = \tan\theta,\ y = a\sin\theta$

d. $x = \sin\theta + \cos\theta,\ y = \sin\theta - \cos\theta$

4. Suppose $u = \dfrac{1+\sin\theta}{\cos\theta}$.

a. Prove that $\dfrac{1}{u}=\dfrac{1-\sin\theta}{\cos\theta}$.

b. Deduce a formula for $\sin\theta$ in terms of u.

Investigation: Compound angle identities

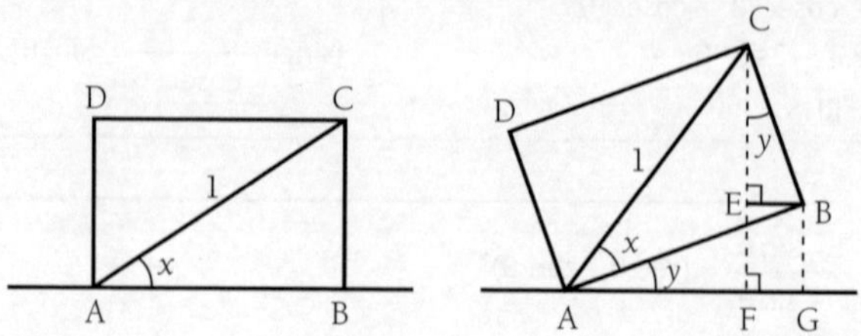

The figure on the left shows a rectangle ABCD with a diagonal of length one unit, and an angle of size x. The rectangle is rotated anticlockwise about the point A, through an angle y, as shown by the figure on the right. Write down the lengths of the following lines:

BC, AB, CE, EF (using triangle ABG) and CF (using triangle CAF).

Using this information explain why $\sin(x+y) = \sin x\cos y + \cos x\sin y$

In a similar way, use a clockwise rotation of the rectangle ABCD about the point A to verify that $\sin(x-y) = \sin x\cos y - \cos x\sin y$

Compound angle identities

The investigation above verifies these two identities:

$$\sin(A+B) = \sin A\cos B + \cos A\sin B$$
$$\sin(A-B) = \sin A\cos B - \cos A\sin B$$

There are similar identities for $\cos(A \pm B)$ which are given below:

$$\cos(A+B) = \cos A\cos B - \sin A\sin B$$
$$\cos(A-B) = \cos A\cos B + \sin A\sin B$$

Note: It is important to understand that the expression $\sin(A+B)$ does not mean $\sin \times (A+B)$, but means the sine of the angle $(A+B)$. Therefore it should be clear brackets are *not* expanded using the distributive law ie $\sin(A+B) \neq \sin A + \sin B$.

These four identities now combine to derive identities involving tangents.

$$\frac{\sin(A+B)}{\cos(A+B)} = \frac{\sin A \cos B + \cos A \sin B}{\cos A \cos B - \sin A \sin B}$$

Now $\dfrac{\sin x}{\cos x} = \tan x$ so $\tan(A+B) = \dfrac{\sin A \cos B + \cos A \sin B}{\cos A \cos B - \sin A \sin B}$

Dividing the numerator and denominator of the RHS by $\cos A \cos B$ gives:

$$\tan(A+B) = \frac{\tan A + \tan B}{1 - \tan A \tan B}$$

Replacing B by $-B$ in the above identity, and using $\tan(-B) = -\tan B$, gives the sixth identity in this group:

$$\tan(A-B) = \frac{\tan A - \tan B}{1 + \tan A \tan B}$$

Using the sign $\mp$ to denote the opposite of $\pm$ (when one sign takes the positive value the other sign takes the negative value), these results can be summarised as follows:

$$\sin(A \pm B) = \sin A \cos B \pm \cos A \sin B$$
$$\cos(A \pm B) = \cos A \cos B \mp \sin A \sin B$$
$$\tan(A \pm B) = \frac{\tan A \pm \tan B}{1 \mp \tan A \tan B}$$

Note: You will be more successful in applying these identities if you learn them first rather than constantly referring to them. When you have learnt the trigonometric identities your confidence will improve, as will your success in solving problems.

Investigation

In the figure, OA and OB are the spokes of a wheel of radius R. Express $\sin(\angle AOB)$ in terms of a, b, and R.

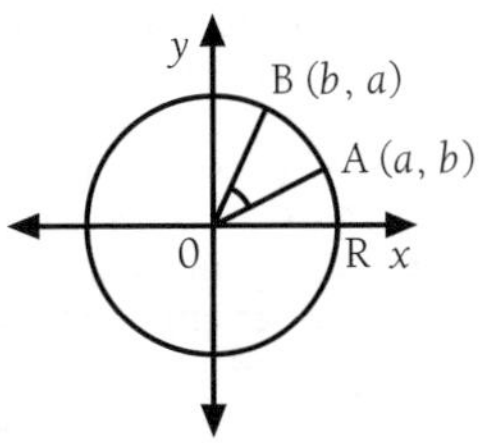

The following worked examples illustrate some applications of these identities:

Example D

Q. Simplify $\sin(\frac{\pi}{2} + x)$

A. Using the identity $\sin(A+B) = \sin A \cos B + \cos A \sin B$

$$\sin(\frac{\pi}{2} + x) = \sin\frac{\pi}{2}\cos x + \cos\frac{\pi}{2}\sin x$$
$$= 1.\cos x + 0.\sin x$$
$$= \cos x$$

Problem solving in trigonometry

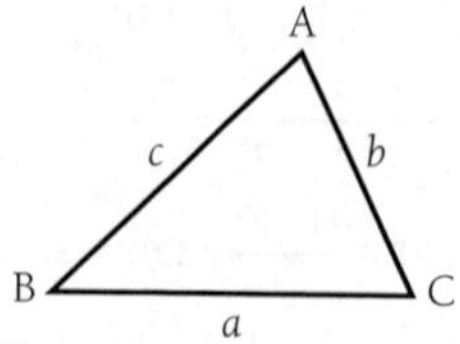

Practical problems can be solved using various identities along with the previously learnt sine rule and the cosine rule.

In the triangle shown:

The **sine rule** states:

$$\frac{a}{\sin A} = \frac{b}{\sin B} = \frac{c}{\sin C}$$

The **cosine rule** states:

$$a^2 = b^2 + c^2 - 2bc \cos A \quad \text{or} \quad \cos A = \frac{b^2 + c^2 - a^2}{2bc}$$

Sometimes the formula for the **area of a triangle** is useful:

$$\text{Area} = \frac{1}{2} bc \sin A$$

In a circle of radius r, where θ is the angle at the centre (in radians) the **arc length**, s, is given by the formula

$$s = r\theta$$

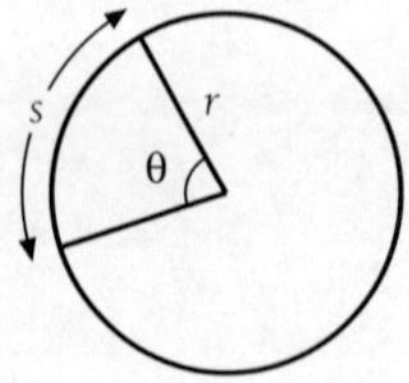

The formula for the **area of a sector** of a circle is:

$$\text{Area} = \frac{1}{2} r^2\theta$$

There is a whole range of formulae and identities available to solve practical questions. As a general guide try the most simple and obvious methods first.

Example E

Q. Show that if L is the length of each side of a regular polygon of n sides, and r is the length of the radius of the *inscribed* circle, then the length of the radius is given by $r = \frac{1}{2} L \cot\left(\frac{\pi}{n}\right)$.

A. A diagram is drawn with a few sides and details shown.
The length of one side of the polygon is

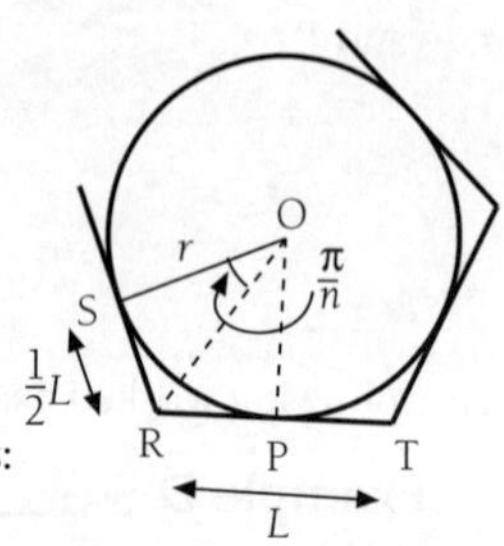

$RT = L$

$SR = \frac{1}{2}L$ [since $SR = RP = \frac{1}{2}RT$]

In a full turn there are 2π radians. Since the polygon has n sides:

$$\angle POS = \frac{2\pi}{n} \text{ radians.}$$

$$\therefore \angle ROS = \frac{\pi}{n} \text{ radians} \qquad \text{[since } \angle ROS = \tfrac{1}{2}\angle POS \text{ by symmetry]}$$

In ΔORS: $\dfrac{r}{\frac{1}{2}L} = \cot\left(\dfrac{\pi}{n}\right)$ [since rad $\perp$ tan, $\angle OSR = 90°$]

$$r = \frac{1}{2} L \cot\left(\frac{\pi}{n}\right) \qquad \text{[multiplying by } \tfrac{1}{2}L\text{]}$$

Investigations

1. OAB is a **sector** of a circle, centre O, radius R. AT is a tangent to the circle at the point A. Angle AOB = x radians.

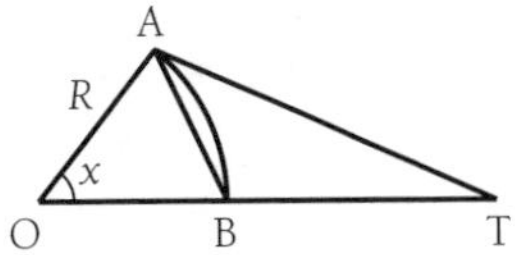

   a. Find the following areas:

      i. Sector OAB. ii. ΔOAB. iii. ΔOAT.

   b. Using the three areas in **a**, prove that $1 \le \dfrac{x}{\sin x} \le \dfrac{1}{\cos x}$.

   c. Hence work out $\lim\limits_{x \to 0} \dfrac{\sin x}{\cos x}$.

2. Angle BAC = x radians, and the radius of the circle is 5 cm. Find the size of angle x so that the area of the ΔABC is equal to the area of the minor **segment** BCD.

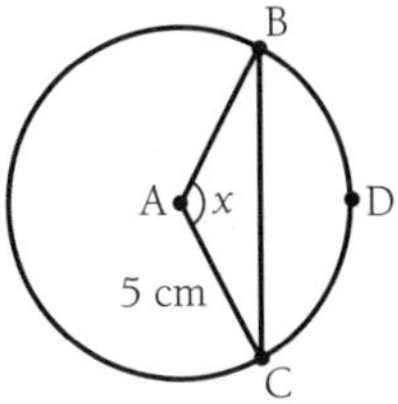

Unit 12.2 Activity 5B: Problem solving in trigonometry

1. From a window, W, which is 75 metres above the ground, the angle of elevation of the top, T, of a nearby building is 30°. From a point, G, on the ground directly below the window, the angle of elevation of T is 55°. Find the height, h, of the building.

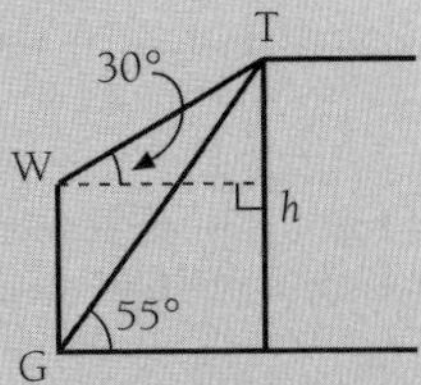

2. A man at sea is paddling his canoe towards a lighthouse, BC. From a certain point, A, he finds that the angle of elevation of the top of the lighthouse is α. After paddling a further distance d metres towards the lighthouse he finds that the angle of elevation of its top is β.

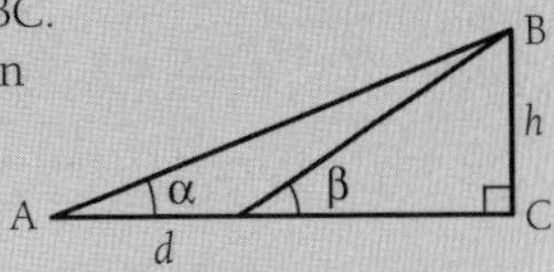

   If h metres is the height of the lighthouse, prove that $h = \dfrac{d \sin \alpha \sin \beta}{\sin (\beta - \alpha)}$.

3. A determined snail climbs 5 metres up a vertical wall, and after a rest climbs a further x metres up the wall. Use the diagram alongside to find an expression for sin θ in terms of x.

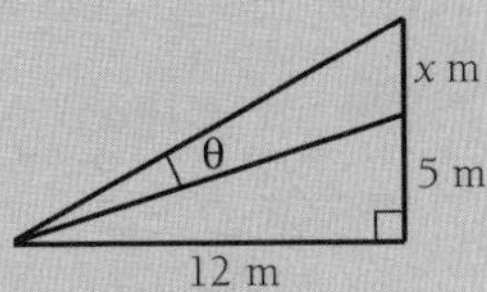

4. Show for the ΔPQR that $p = \dfrac{q \sin \alpha}{\cos (\theta - \alpha)}$.

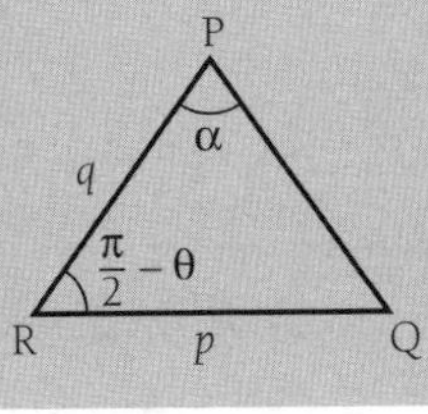

5. ABCD is a parallelogram in which the length of AB is a and the length of BC is b. If angle ABC = θ, prove that the difference in the squared lengths of the diagonals of the parallelogram is $4ab \cos \theta$.

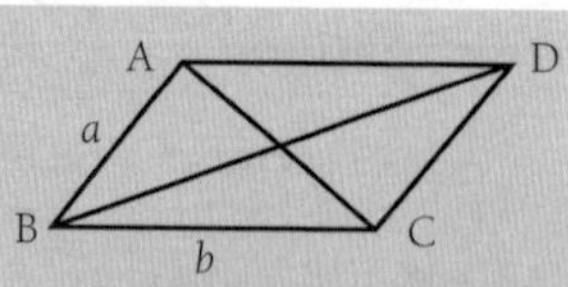

6. In ΔABC, $b = 12$ cm, $c = 10$ cm and C = 45°. Find the length of the third side of the triangle. (There are two possible answers.)

7. A lighthouse is situated at a point L. Two ships A and B are sighted from the lighthouse. Ship A is 10 km away on a bearing of 300°, and ship B is 6 km away on a bearing of 040°. How far apart are the ships, and what is the bearing of ship A from ship B?

8. In triangle ABC, D is a point on BC such that AD = AB. Show that $BD = \dfrac{ac^2}{c^2 + a^2 - b^2}$.

9. Using the cosine rule in any triangle ABC, show that $\dfrac{b^2 - a^2}{abc} = \dfrac{\cos A}{a} - \dfrac{\cos B}{b}$.

10. The diagram shows two parallel tangents AB and XY to a circle, centre O, radius r. AX is the diameter and angle XOP = α. Find an expression for the perpendicular distance of P from XY.

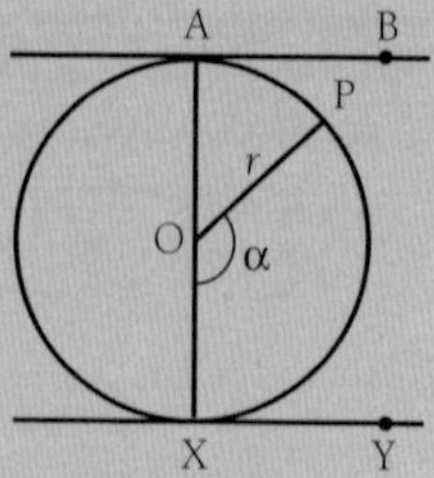

11. Show that the length h of the perpendicular from C to the side AB of any triangle ABC can be found by
$h = \dfrac{c}{\cot \alpha + \cot \beta}$
where $\alpha = \angle$CAB and $\beta = \angle$CBA.

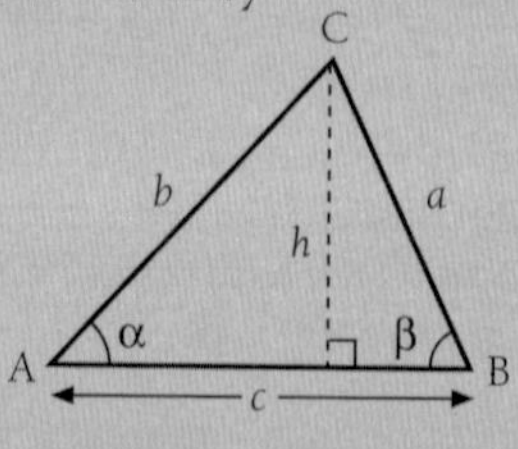

12. A rectangular piece of paper ABCD is w cm wide and L cm long. It is folded so that corner C touches the opposite side at K as shown.
If BE = x cm, prove that
$x = \dfrac{w}{2 \sin \theta \cdot \cos^2 \theta}$.

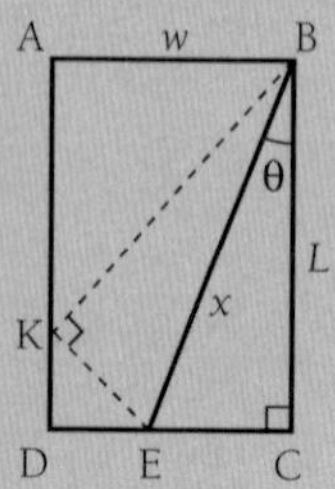

13. A and B are two beads that slide round a vertical wire circle, centre O as shown. OM is horizontal, θ = angle between the vertical and OA, and α = angle between OA and OB, and R is the radius of the circle.
Prove that the perpendicular height of A above B is $R (\cos \theta - \cos (\alpha + \theta))$.

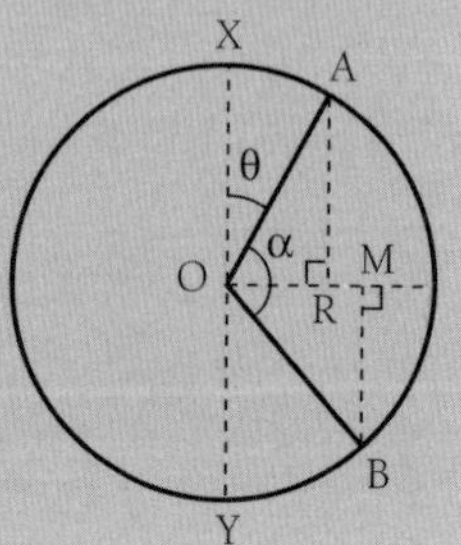

14. Joanne works in the packing department of a chocolate factory. She is packing circular after-dinner mints in flat rectangular boxes, with 24 chocolates to each box.

If arrangement A fits into a box 24 cm by 16 cm, how much area is saved in the box by using arrangement B?

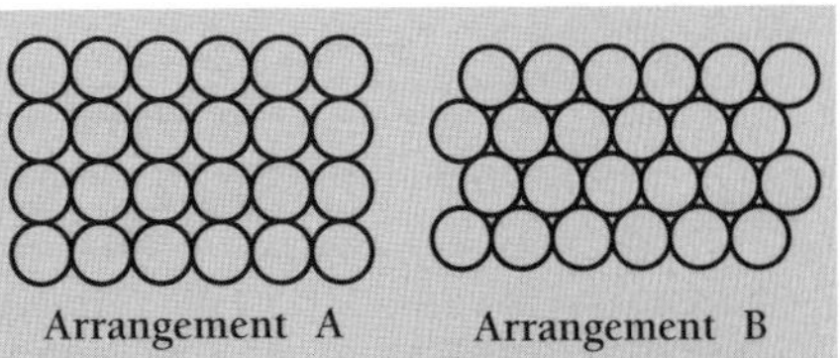

15. A wheel of radius 2 metres rotates on its axis, which passes through the point A. P is a point which is initially at the top of the wheel, and as the wheel rotates on its axis P finishes up at P′. Suppose the angle turned through by the wheel is x radians. If P′ is at height h metres above the lowest point on the wheel,

a. Find a formula for h in terms of x and graph it.

Suppose the wheel is now a Ferris wheel which has a radius of 10 metres, and the height of the lowest point of the wheel above the ground is 1 metre.

b. Find a formula for $h(x)$ and graph it.

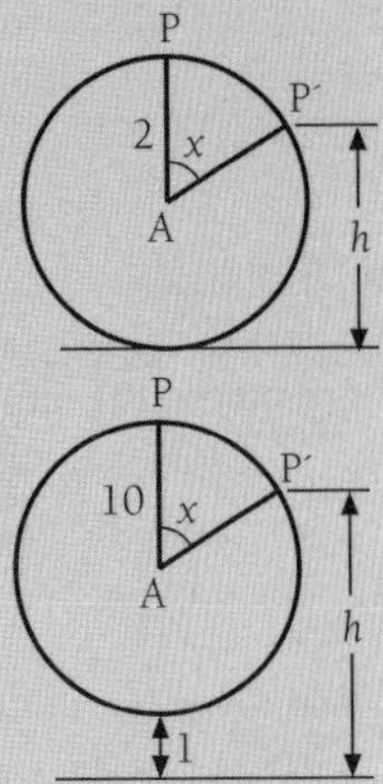

Unit 12.2 Activity 5C: Revision questions

1. Sketch the graphs of the following three functions for $0 \le x \le 2\pi$. Use the same set of axes for all three graphs.

a. $y = 2 \sin x$ **b.** $y = \sin 2x$ **c.** $y = \sin (x - \pi)$

2. Verify the following identities:

a. $\sec^2 \theta + \operatorname{cosec}^2 \theta = \sec^2 \theta \operatorname{cosec}^2 \theta$ **b.** $\tan\left(\theta + \frac{\pi}{4}\right) = \frac{\cos \theta + \sin \theta}{\cos \theta - \sin \theta}$

3. a. A surveyor wishes to calculate the horizontal distance x from a point A to a point X directly below the top of a hill. She measures the angle of elevation α of the top of the hill from A and the angle of elevation β from a point B, a horizontal distance d metres further away from the hill, as in the diagram.

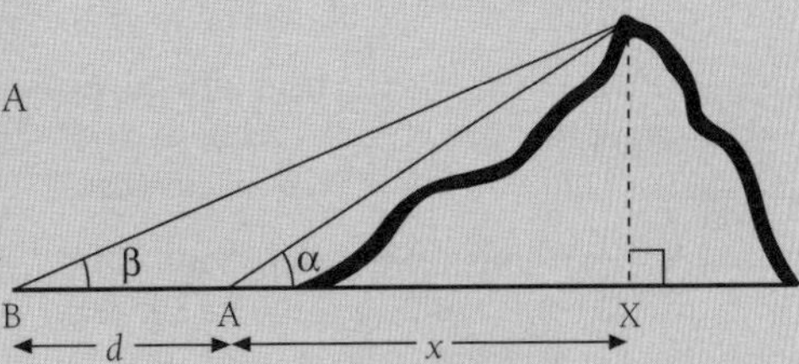

i. The distance x is given by $x = \frac{d \tan \beta}{\tan \alpha - \tan \beta}$. Use a calculator to evaluate x to one decimal place if $d = 240$ m, $\alpha = 24.37°$, $\beta = 18.41°$.

ii. Prove that $x = \dfrac{d \tan \beta}{\tan \alpha - \tan \beta}$. (*Hint:* Express the height of the hill in terms of both α and β using the two right-angled triangles.)

iii. Prove the identity $\dfrac{\tan \beta}{\tan \alpha - \tan \beta} = \dfrac{\cos \alpha \sin \beta}{\sin (\alpha - \beta)}$.

b. If the distance from A to B is no longer horizontal, but slopes up at an angle θ as shown, find an expression for x in terms of α, β, θ, d.

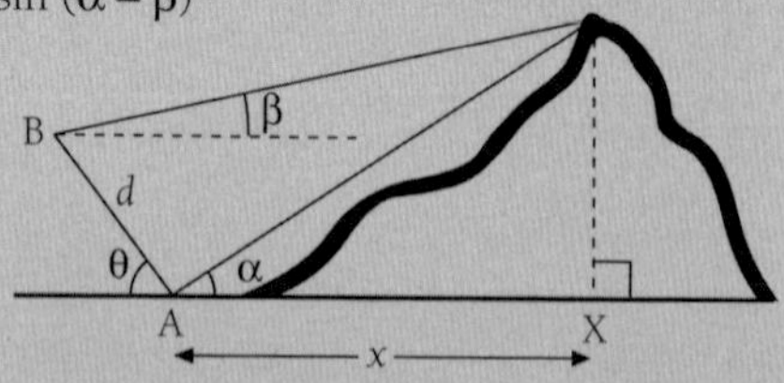

4. The diagram shows two angles a and b that are related by $\sin a = \sin b$. Find another relationship between the angles a and b.

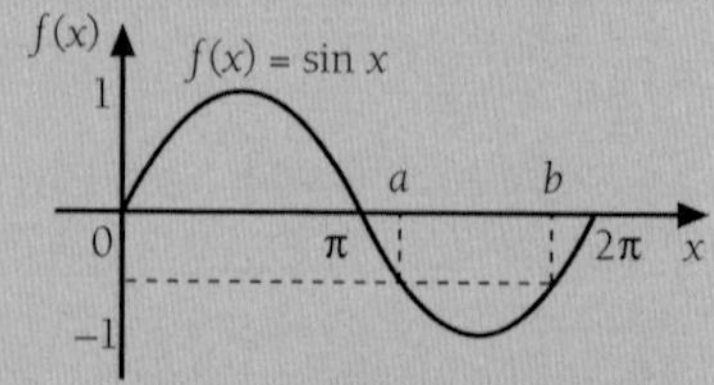

5. The voltage in an electronic circuit is given by $v = \sin^2 t - 2 \sin t + 7$. In order to find the minimum voltage this could be rewritten as $v = x^2 - 2x + 7$ by letting $x = \sin t$.

a. Rearrange $x^2 - 2x + 7$ in completed square form (ie in the form $(x - a)^2 + b$).

b. For what value of x is the voltage a minimum?

c. For what values of t is the voltage a minimum?

d. What is the minimum voltage?

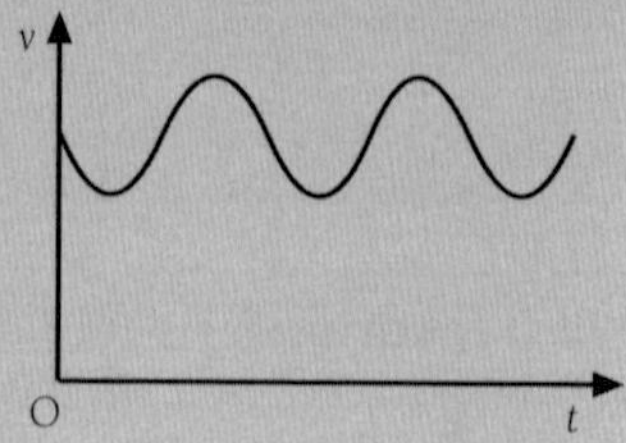

6. Prove the identity $\dfrac{\sin 2x}{\sin x} - \dfrac{\cos 2x}{\cos x} = \sec x$.

7. a. As a bicycle is ridden along a road the pedal height, h cm, above the road changes.

The graph shows the height of the pedal above the road at time t seconds.
The equation is $h = 28 + 18 \cos (\pi t)$.

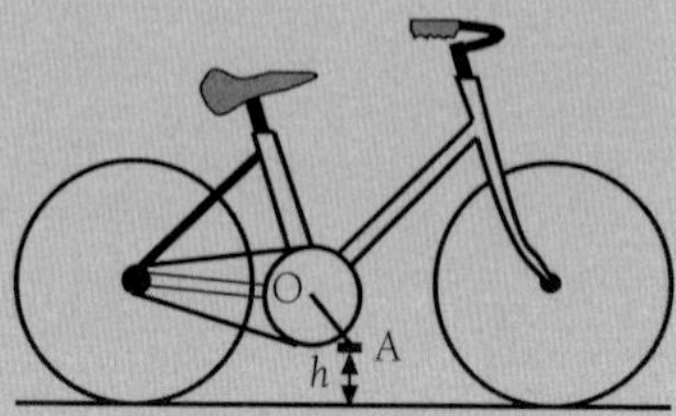

i. How long is the pedal crank OA?

ii. What is the least distance between the pedal and the ground?

iii. What is the value of T (the time marked on the graph) which is the time for the pedal to make one revolution?

iv. What would be the effect on the graph if the pedals were moved round faster? (The gears are not changed.)

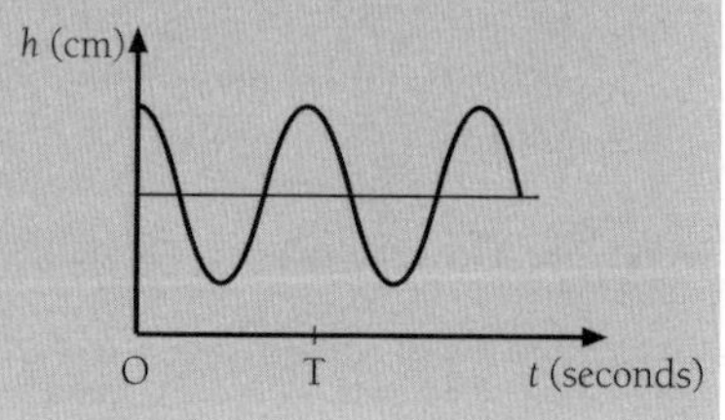

b. A machine has a part that moves somewhat like a pedal and crank of the bicycle, but its motion is represented by the function

$$f(\theta) = \cos\theta \cos\frac{\pi}{3} - \sin\theta \sin\frac{\pi}{3} + 1.$$

Give the value of $f(0.790)$ correct to 3 decimal places.

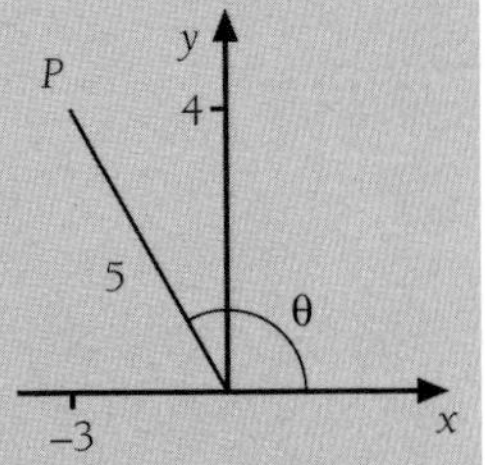

8. The point P in the diagram has Cartesian coordinates $(-3, 4)$ and polar coordinates $(5, \theta)$.

a. Give the values of $\tan\theta$ and $\operatorname{cosec}\theta$.

b. Find the size of the angle θ given in the diagram.

9. Express $\dfrac{\sin 3A}{\sin 2A - \sin 3A}$ in terms of $\cos A$. Simplify your answer as far as possible.

10. Prove that $\dfrac{1 - \sin A}{1 + \sec A} - \dfrac{1 + \sin A}{1 - \sec A} = 2\cot A\,(\cos A + \operatorname{cosec} A)$.

Unit 12.2 Trigonometry and Vectors

Topic 6: Properties of vectors

A vector is a quantity with both magnitude and direction, which is represented by an arrowed line. The length of the line segment represents the magnitude of the vector and the arrow on the line shows the direction of the vector. Examples of vector quantities include velocity, acceleration and force. Quantities such as time, volume and length are not vector quantities because there is no direction associated with them. Such quantities are called scalar quantities.

Topic 6 and the remaining Topics in this Unit cover the content listed in the bullet points in the Syllabus (pp. 20–21):

- Identifying physical quantities as scalar and vector quantities.
- Representing equal, opposite and multiple vectors diagrammatically.
- Discussing unit and position vectors.
- Solving vector addition problems.
- Applying scalar products of two vectors.
- Calculating angles between two vectors.
- Applying vector geometry to real-life problems.
- Solving application problems involving triangle velocities.

Properties of vectors

Vector quantities are quantities that require both **magnitude** (size) and **direction** to be fully specified.

Vector quantities include **velocity**, acceleration and force. To describe vector quantities, a direction as well as a magnitude must be given.

For example, when describing a wind (velocity) it is not sufficient to say 'a wind of 20 km per hour', we must also give a direction such as 'from the west'.

- Quantities that do not require a direction, such as time, length, **speed** or volume, are called **scalar quantities**.
- Vector quantities are represented by vectors, which are directed line segments where the length of the line represents the magnitude of the vector quantity and the orientation of the line represents the direction and the arrowhead represents the direction of action.

Example A

A vector representing a wind velocity of 20 km/hr from the west could be represented by a **line segment** 4 cm long (1 cm = 5 km/hr):

Vector notation

A vector that starts at a point A and ends at a point B can be written as $\overrightarrow{AB}$

Otherwise vectors are usually represented by lower-case letters with a *tilde* underneath: $\underset{\sim}{a}$

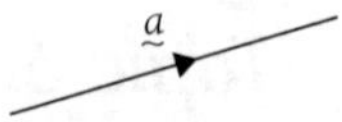

In some textbooks vectors are given in bold print:

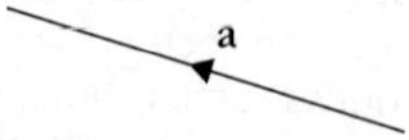

Free vectors have no particular line of action or point of origin but are free to be moved around, as long as they retain their length, direction and sense of direction (indicated by the arrowhead):

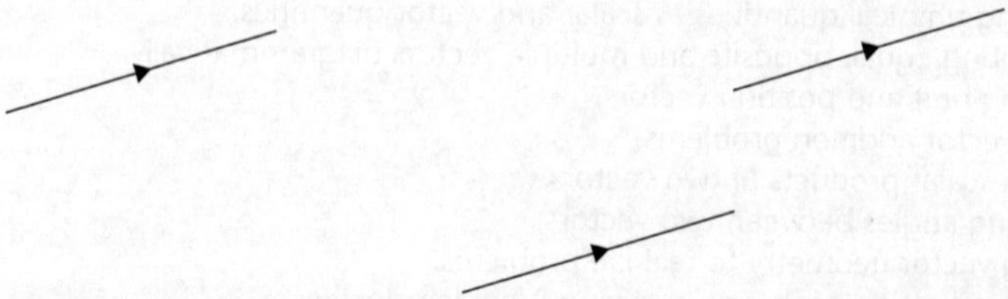

Equality of vectors

Vectors are equal if they have the same length, direction and sense of direction:

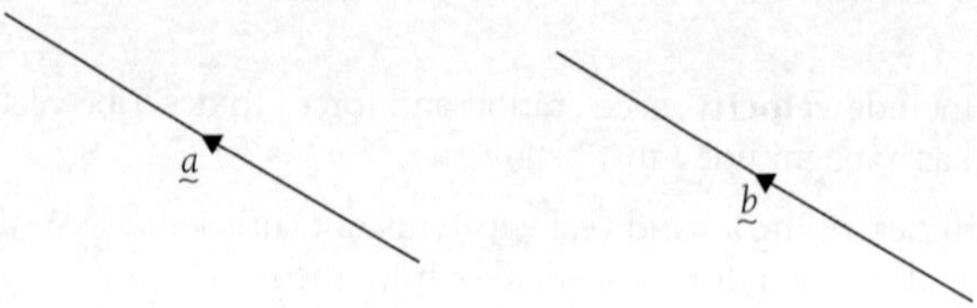

Multiplication of a vector by a scalar

A vector can be multiplied by a scalar (a constant number). Multiplying a vector by a scalar changes the length of the vector; the direction remains the same.

Multiplying a vector by a negative number changes the magnitude and the sense of direction (but not the direction).

Note: The vectors *a* and –*a* differ only in their sense of direction.

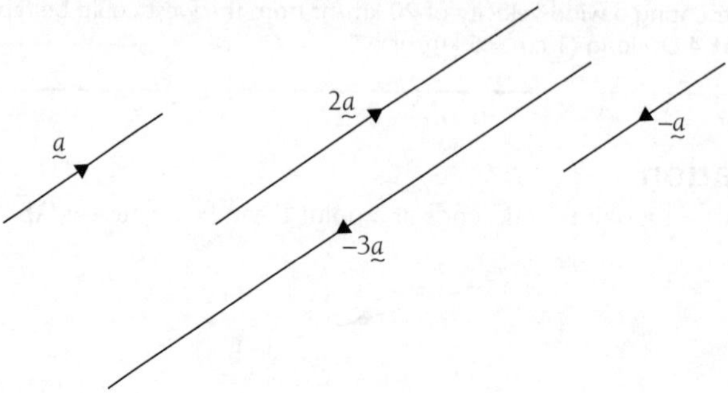

Addition of vectors

The addition of vectors a and b is shown on the diagram below. The vectors are placed 'head to tail' and the addition vector, called the **resultant**, goes from the un-joined tail to the un-joined head.

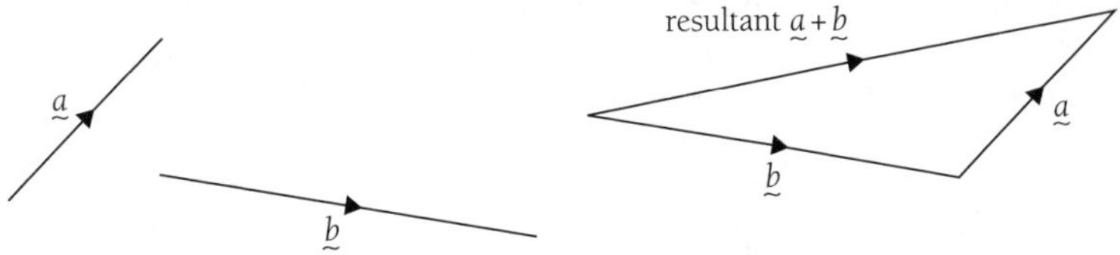

The addition of more than two vectors is done in the same manner.

Note: the vectors can be added in any order; the resultant is the same.

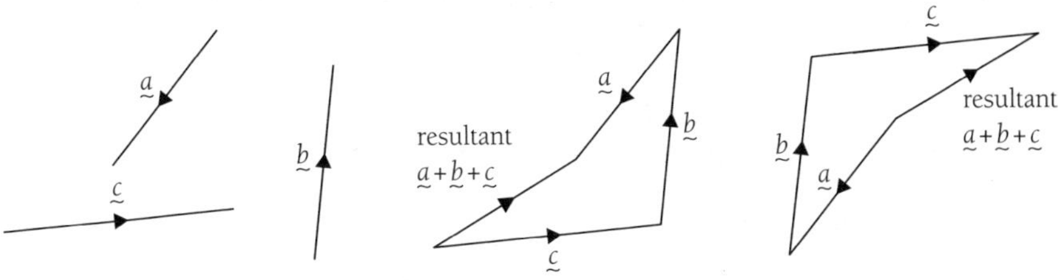

Example B

In the triangle ABC, $\underset{\sim}{a}$ is the vector joining points A to B and $\underset{\sim}{b}$ is the vector joining point A to C. M is the **midpoint** of BC.

Q. Find in terms of $\underset{\sim}{a}$ and $\underset{\sim}{b}$:

a. $\overrightarrow{BC}$

b. $\overrightarrow{AM}$

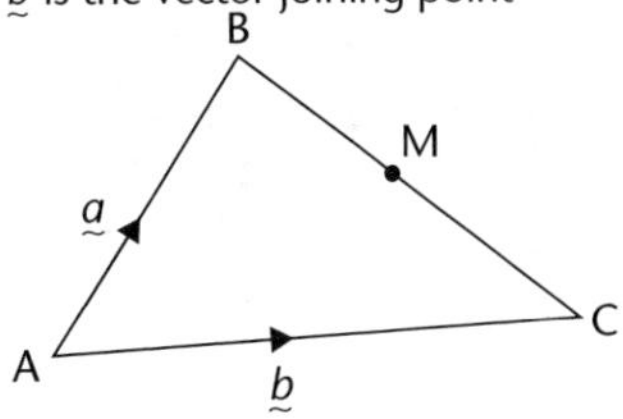

A. a. $\overrightarrow{BC} = \underset{\sim}{b} - \underset{\sim}{a} = \underset{\sim}{b} + (-\underset{\sim}{a})$

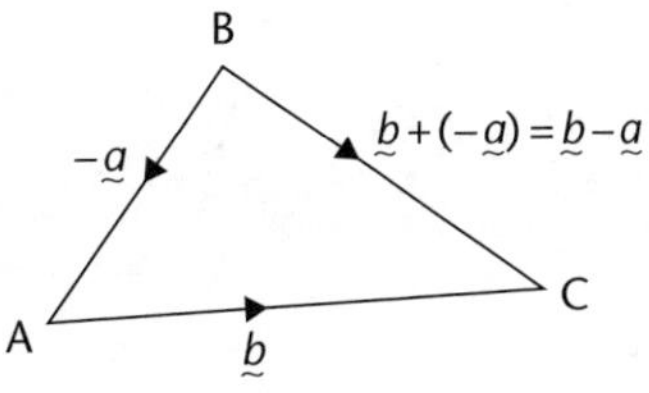

b. $\overrightarrow{AM} = \underset{\sim}{a} + \frac{1}{2}\overrightarrow{BC} = \underset{\sim}{a} + \frac{1}{2}(\underset{\sim}{b} - \underset{\sim}{a})$

$= \frac{1}{2}\underset{\sim}{a} + \frac{1}{2}\underset{\sim}{b}$

$= \frac{1}{2}(\underset{\sim}{a} + \underset{\sim}{b})$

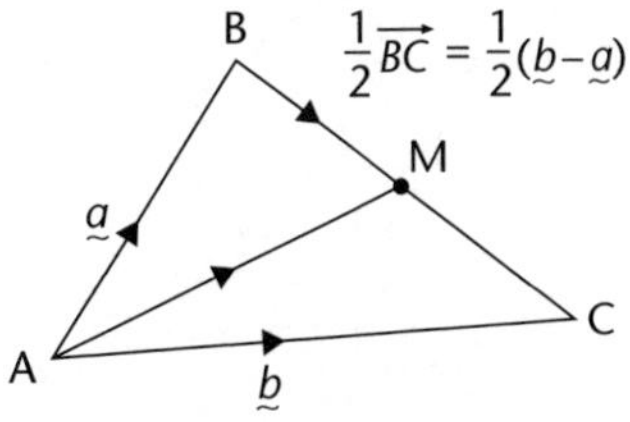

Example C

Q. If $\underset{\sim}{p}$ represents a force of 20 Newton in a direction S40°E and $\underset{\sim}{q}$ represents a force of 30 Newton in the direction N30°E:

a. Draw a scale diagram of these two vectors

b. Show the vector $\underset{\sim}{p} + \underset{\sim}{q}$ on a scale diagram

c. Use your scale diagram of $\underset{\sim}{p} + \underset{\sim}{q}$ to find the magnitude and direction of $\underset{\sim}{p} + \underset{\sim}{q}$.

A. a. Using a scale of 1 cm = 10 Newton

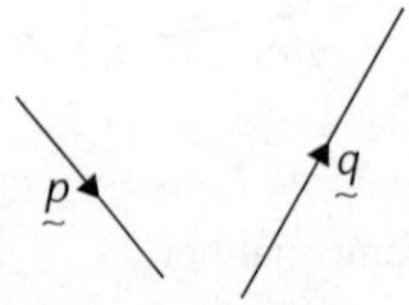

b.

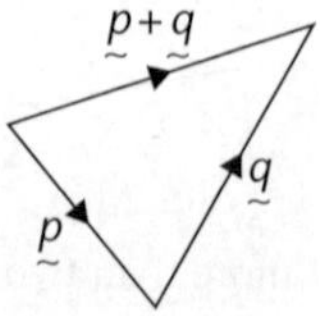

c. Measuring the length of $p + q$ gives approximately 3 cm, which equates to 30 Newton and the direction is N70°E

Unit 12.2 Activity 6A: Properties of vectors

Where necessary give your answer correct to two decimal places.

1. If $\underset{\sim}{a}$, $\underset{\sim}{b}$ and $\underset{\sim}{c}$ are the vectors given at right, show with diagrams:

a. $3\underset{\sim}{a}$ **b.** $-2\underset{\sim}{c}$

c. $\underset{\sim}{a} + \underset{\sim}{b}$ **d.** $2\underset{\sim}{c} - \underset{\sim}{a}$

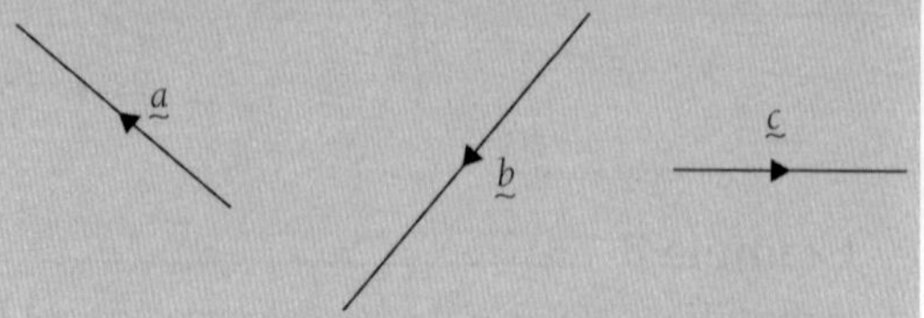

2. If $\underset{\sim}{x}$, $\underset{\sim}{y}$ and $\underset{\sim}{z}$ are the vectors given at right, show diagrammatically:

a. $\underset{\sim}{x} + \underset{\sim}{y}$ **b.** $\underset{\sim}{y} + \underset{\sim}{z}$

c. $\underset{\sim}{z} - \underset{\sim}{x}$ **d.** $\underset{\sim}{x} + \underset{\sim}{y} + \underset{\sim}{z}$

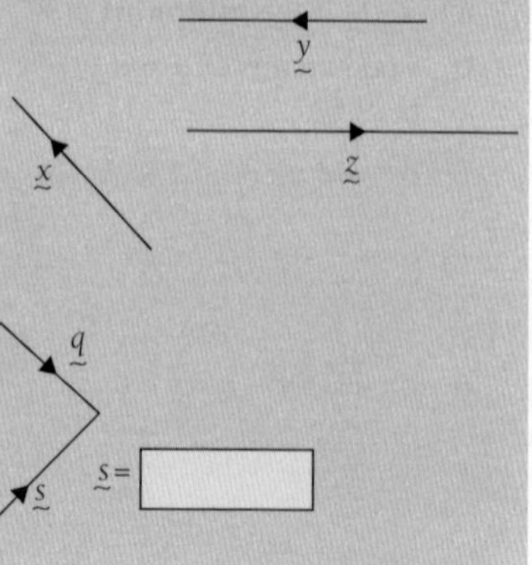

3. Write vector equations for each of the following:

a.

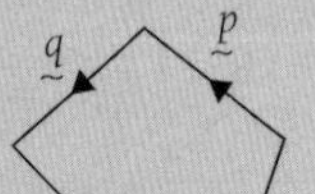

$\underset{\sim}{a} + \underset{\sim}{b} =$ ☐

b.

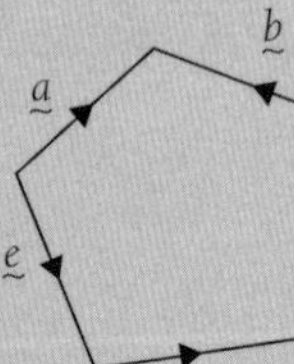

$\underset{\sim}{s} =$ ☐

c.

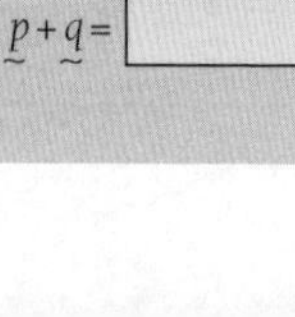

$\underset{\sim}{p} + \underset{\sim}{q} =$ ☐

d.

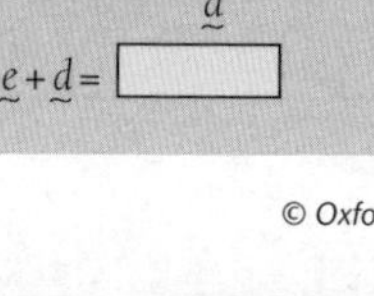

$\underset{\sim}{e} + \underset{\sim}{d} =$ ☐

4. **a.** Draw vectors for each of the following velocities. Use a scale of 1 cm = 10 km/hr.

i. $\underset{\sim}{a}$ = 35 km/hr in the direction due east

ii. $\underset{\sim}{b}$ = 40 km/hr in the direction north-east

iii. $\underset{\sim}{c}$ = 50 km/hr due south

iv. $\underset{\sim}{a} + \underset{\sim}{b}$

v. $\underset{\sim}{a} + \underset{\sim}{c}$

b. Use your scale to find the magnitude of:

i. $\underset{\sim}{a} + \underset{\sim}{b}$

ii. $\underset{\sim}{a} + \underset{\sim}{c}$

c. Use trigonometry to find the magnitude and direction of:

i. $\underset{\sim}{a} + \underset{\sim}{b}$

ii. $\underset{\sim}{a} + \underset{\sim}{c}$

5. A force of 10 kg wt acts at right angles to a force of 15 kg wt.

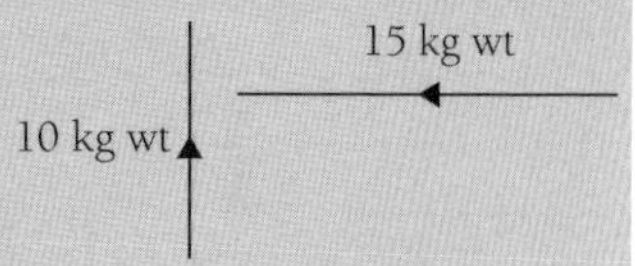

a. Drawing both vectors to scale, find the resultant of the sum of these vectors.

b. Define the resultant vector from part **a**. in terms of magnitude and direction.

c. Find the resultant of the subtraction of the 10 kg wt force from the 15 kg wt force, defining it in terms of magnitude and direction.

6. Mary is going to paddle across a river which is flowing at 2 m/s. She sets out from point A on one bank of the river, heading directly across the river and rowing at 1.5 m/s.

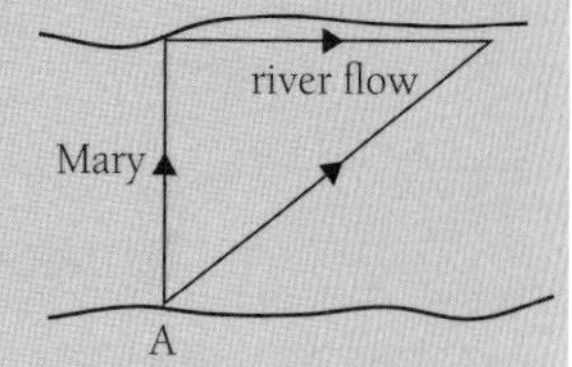

a. Draw and label a vector diagram with the appropriate values.

b. Calculate Mary's speed with respect to the bank.

c. Calculate the direction Mary ends up travelling.

7. An aeroplane heads due north at an air speed of 320 km/hr. It is carried off course by a steady wind of 40 km/hr blowing from the west.

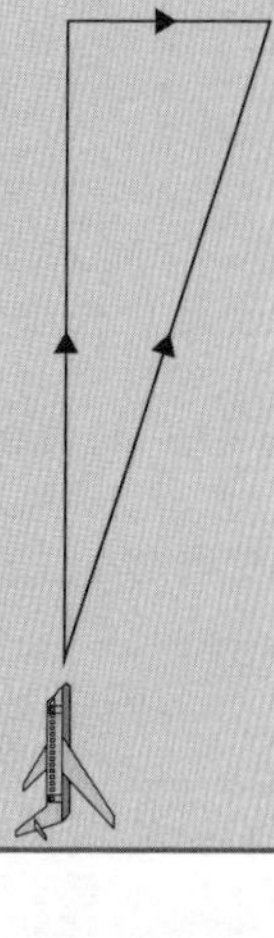

a. Copy the vector diagram and label each vector.

b. Calculate the actual speed of the aeroplane.

c. Calculate the direction in which the aeroplane is actually travelling. Give your answer as a compass bearing.

Unit 12.2 Trigonometry and Vectors

Topic 7: Geometric proofs using vectors

Topic 7 continues the coverage of vectors as summarised at the beginning of Topic 6 (p. 135), in particular applying vector geometry to problems.

Geometric proofs using vectors

Some properties of vectors that are useful for geometric proofs:

- the vectors $\underset{\sim}{a}$ and $k\underset{\sim}{a}$, where k is a scalar, are parallel.
- if $\underset{\sim}{a} = \underset{\sim}{b}$ then the $|\underset{\sim}{a}| = |\underset{\sim}{b}|$ and $\underset{\sim}{a}$ and $\underset{\sim}{b}$ are parallel.
- $\overrightarrow{PQ} + \overrightarrow{QR} = \overrightarrow{PR}$

  ↖ ↑

  same letter
- $\overrightarrow{PQ} = -\overrightarrow{QP}$
- If $\overrightarrow{OA} = k\overrightarrow{OB}$ then OA is parallel to OB and, because they have a point O in common, this means that the points O, A and B are **collinear** (on the same straight line).

Example A

Q. OABC is a quadrilateral and the points PQRS are the midpoints of the sides as shown in the diagram.

Let $\overrightarrow{OA} = \underset{\sim}{a}$, $\overrightarrow{AB} = \underset{\sim}{b}$ and $\overrightarrow{OC} = \underset{\sim}{c}$

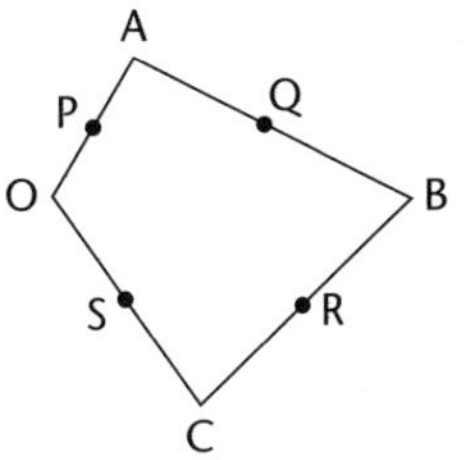

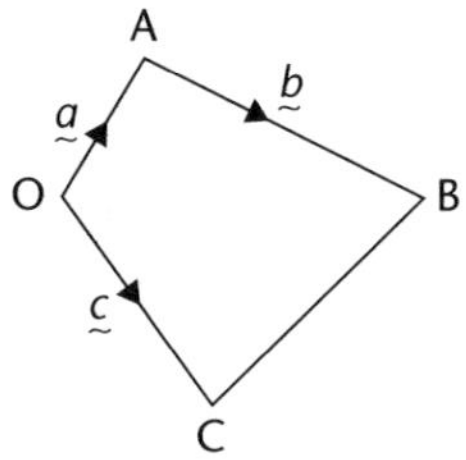

a. Find the vector $\overrightarrow{CB}$ in terms of vectors $\underset{\sim}{a}$, $\underset{\sim}{b}$ and $\underset{\sim}{c}$

b. Show that the quadrilateral PQRS is a parallelogram.

A. **a.** $\underset{\sim}{a} + \underset{\sim}{b} - \underset{\sim}{c} = \overrightarrow{CB}$

b. $\overrightarrow{PQ} = \frac{1}{2}\underset{\sim}{a} + \frac{1}{2}\underset{\sim}{b} = \frac{1}{2}(\underset{\sim}{a} + \underset{\sim}{b})$

$$\overrightarrow{SR} = \frac{1}{2}\underset{\sim}{c} + \frac{1}{2}\overrightarrow{CB}$$

$$= \frac{1}{2}\underset{\sim}{c} + \frac{1}{2}(\underset{\sim}{a} + \underset{\sim}{b} - \underset{\sim}{c})$$

$$= \frac{1}{2}\underset{\sim}{c} + \frac{1}{2}\underset{\sim}{a} + \frac{1}{2}\underset{\sim}{b} - \frac{1}{2}\underset{\sim}{c}$$

$$= \frac{1}{2}\underset{\sim}{a} + \frac{1}{2}\underset{\sim}{b}$$

$$= \frac{1}{2}(\underset{\sim}{a} + \underset{\sim}{b})$$

So $\overrightarrow{PQ} = \overrightarrow{SR}$ which means that $|\overrightarrow{PQ}| = |\overrightarrow{SR}|$ and that PQ and SR are parallel ... (1)

Similarly $\overrightarrow{PS} = \frac{1}{2}\underset{\sim}{c} - \frac{1}{2}\underset{\sim}{a} = \frac{1}{2}(\underset{\sim}{c} - \underset{\sim}{a})$

$$\overrightarrow{QR} = \frac{1}{2}\underset{\sim}{b} - \frac{1}{2}\overrightarrow{CB}$$

$$= \frac{1}{2}\underset{\sim}{b} - \frac{1}{2}(\underset{\sim}{a} + \underset{\sim}{b} - \underset{\sim}{c})$$

$$= \frac{1}{2}\underset{\sim}{b} - \frac{1}{2}\underset{\sim}{a} - \frac{1}{2}\underset{\sim}{b} + \frac{1}{2}\underset{\sim}{c}$$

$$= \frac{1}{2}\underset{\sim}{c} - \frac{1}{2}\underset{\sim}{a}$$

$$= \frac{1}{2}(\underset{\sim}{c} - \underset{\sim}{a})$$

So $\overrightarrow{QR} = \overrightarrow{PS}$ which means that $|\overrightarrow{PS}| = |\overrightarrow{QR}|$ and that PS and QR are parallel ... (2)

(1) and (2) together prove that PQRS is a parallelogram because opposite pairs of sides are equal in length and are parallel.

Unit 12.2 Activity 7A: Geometric proofs using vectors

Where necessary give your answer correct to two decimal places.

1. OABC is a trapezium.
AC is parallel to OB and $\overrightarrow{AC} = 2\overrightarrow{OB}$
If $\overrightarrow{OA} = \underset{\sim}{a}$ and $\overrightarrow{OB} = \underset{\sim}{b}$ express in terms of $\underset{\sim}{a}$ and $\underset{\sim}{b}$:

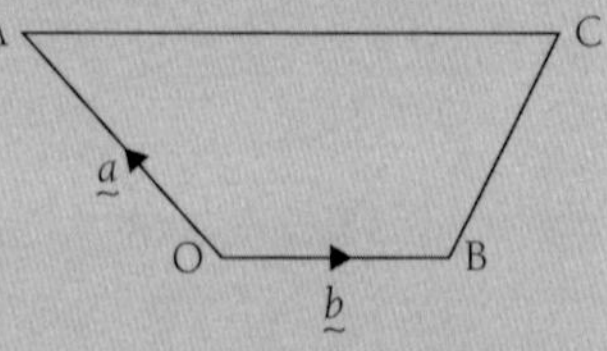

a. $\overrightarrow{AC}$
b. $\overrightarrow{OC}$
c. $\overrightarrow{BC}$

2. If ABC is any triangle and P and Q are the midpoints of AB and BC respectively, show that PQ is parallel to AC and half of its length (*hint*: let $\overrightarrow{AB} = \underset{\sim}{a}$ and $\overrightarrow{BC} = \underset{\sim}{b}$).

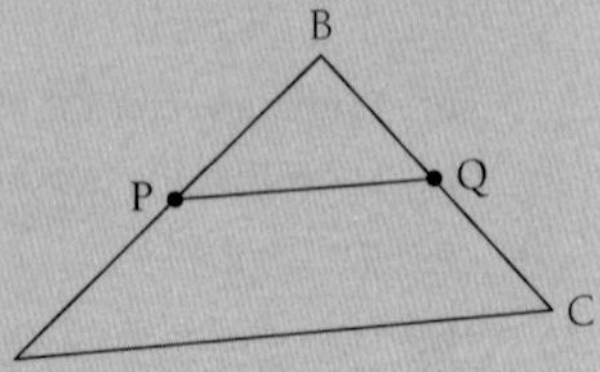

3. Simplify $\overrightarrow{CD} + \overrightarrow{DE} - \overrightarrow{AE} + \overrightarrow{AB} + \overrightarrow{CB}$.

4. In triangle ABC, where M is the midpoint of BC, show that:

$\overrightarrow{AM} = \frac{1}{2}(\overrightarrow{AB}) + (\overrightarrow{AC})$

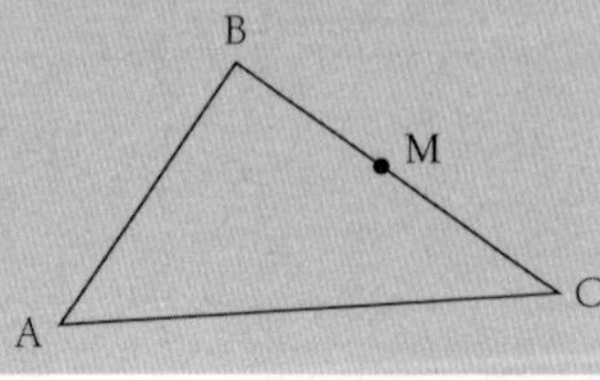

5. PQR is any triangle and M is the midpoint of PR. N is the point that divides QR in the ratio 1:2 and O is the midpoint of QM. If $\overrightarrow{PQ} = \underset{\sim}{a}$ and $\overrightarrow{PR} = \underset{\sim}{b}$ find in terms of $\underset{\sim}{a}$ and $\underset{\sim}{b}$:

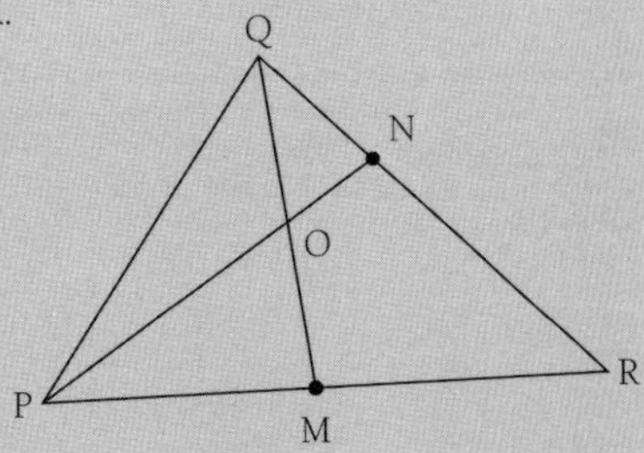

a. **i.** $\overrightarrow{QR}$
ii. $\overrightarrow{QN}$
iii. $\overrightarrow{PN}$
iv. $\overrightarrow{QM}$
v. $\overrightarrow{QO}$
vi. $\overrightarrow{PO}$

b. Comparing the vectors $\overrightarrow{PN}$ and $\overrightarrow{PO}$ what can be said about the points P, O and N?

6. In triangle ABC the midpoint of AC is the point N and the midpoint of the side BC is the point M. The point G divides BN in the ratio 1:2. If $\overrightarrow{AB} = \underset{\sim}{a}$ and $\overrightarrow{AC} = \underset{\sim}{b}$ find in terms of $\underset{\sim}{a}$ and $\underset{\sim}{b}$:

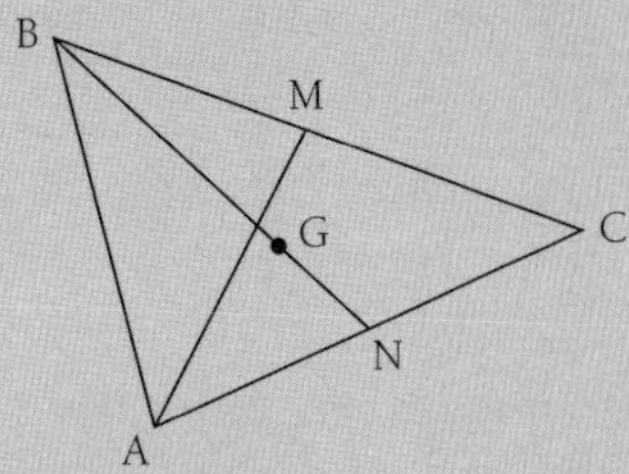

a. **i.** $\overrightarrow{BC}$
ii. $\overrightarrow{BN}$
iii. $\overrightarrow{BG}$
iv. $\overrightarrow{AG}$
v. $\overrightarrow{AM}$

b. Comparing the vectors $\overrightarrow{AG}$ and $\overrightarrow{AM}$ what can be said about the points A, G and N?

7. Show that the diagonals of a parallelogram bisect each other. (*Hint*: Assume the midpoints of the diagonals are different points, M and N, and then show that $\overrightarrow{OM} = \overrightarrow{ON}$)

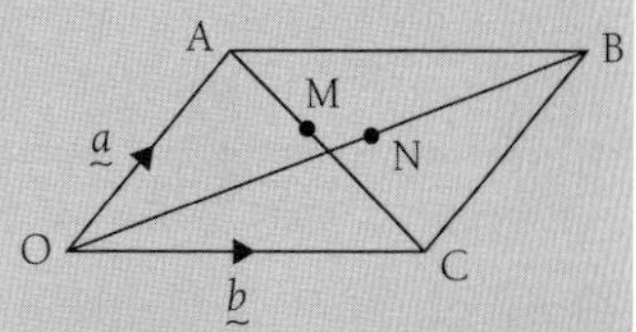

8. OABC is a parallelogram and M is the midpoint of side AB. H is the point that divides OM in the ratio 2:1 and G is the point that divides AC in the ratio 1:2.

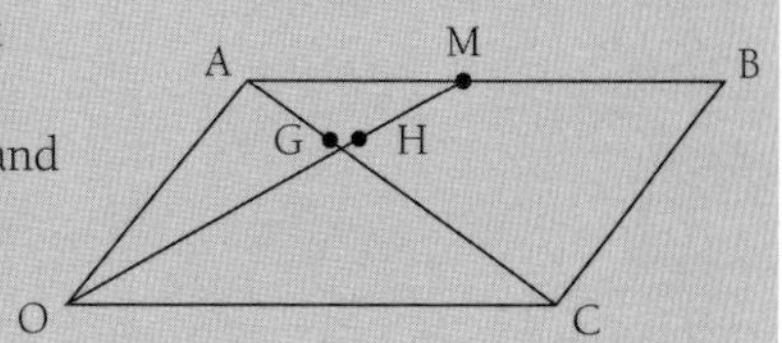

a. If $\overrightarrow{OA} = \underset{\sim}{a}$ and $\overrightarrow{OC} = \underset{\sim}{b}$, find in terms of $\underset{\sim}{a}$ and $\underset{\sim}{b}$:
i. $\overrightarrow{OH}$
ii. $\overrightarrow{OG}$

b. What can be said about the points G and H?

Unit 12.2 Trigonometry and Vectors

Topic 8: Further properties of vectors

Topic 8 continues the coverage of vectors as summarised at the beginning of Topic 6 (p. 135), looking at further properties of vectors: unit vectors and the angle between two vectors.

Summary

- **Unit vectors** are vectors that have a magnitude of one unit.
- The unit vector in the direction of the vector $\underset{\sim}{a}$ is denoted by $\hat{\underset{\sim}{a}}$ ('a hat').

$$\hat{\underset{\sim}{a}} = \frac{\underset{\sim}{a}}{|\underset{\sim}{a}|}$$

$\underset{\sim}{a} = |\underset{\sim}{a}| \times \hat{\underset{\sim}{a}}$ which means that we can write a vector as a product of a magnitude component $|\underset{\sim}{a}|$ and a direction component $\hat{\underset{\sim}{a}}$.

- The angle θ between two vectors $\underset{\sim}{a}$ and $\underset{\sim}{b}$ is defined as the angle between the vectors when they are in the positions as shown below:

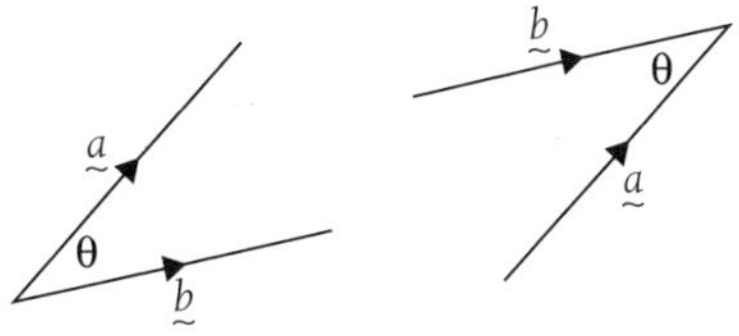

Example A

Q. The vectors $\underset{\sim}{a}$ and $\underset{\sim}{b}$ are given below:

a. Measure the vector $\underset{\sim}{a}$ and construct the vector $\hat{\underset{\sim}{a}}$, using 1 cm = 1 unit.

b. Show on a diagram the angle between the vectors $\underset{\sim}{a}$ and $\underset{\sim}{b}$.

c. Show on a diagram the angle between the vectors $\underset{\sim}{a} + \underset{\sim}{b}$ and $\underset{\sim}{b}$.

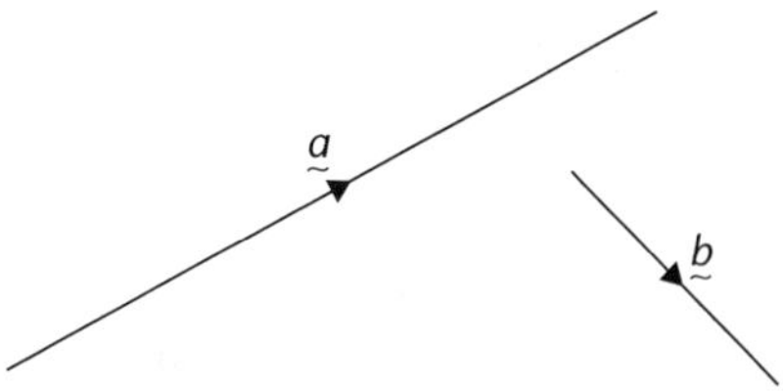

A. a. The vector $\underset{\sim}{a}$ measures 6 cm so $\hat{\underset{\sim}{a}}$ is 1 cm long in same direction as $\underset{\sim}{a}$.

b. The angle θ is the angle between the vectors $\underset{\sim}{a}$ and $\underset{\sim}{b}$.

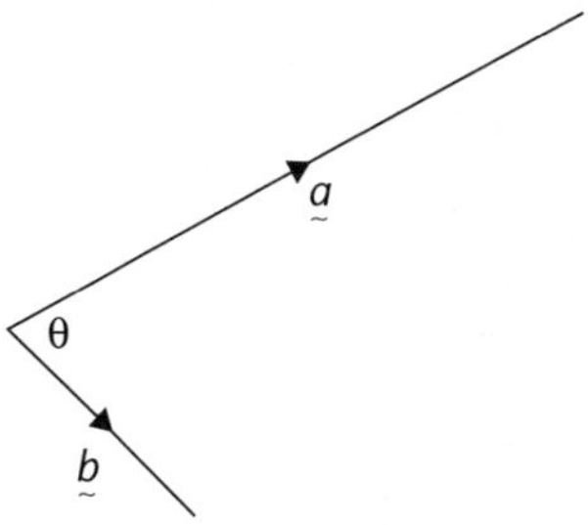

c. The angle α is the angle between the vectors $\underset{\sim}{a} + \underset{\sim}{b}$ and $\underset{\sim}{b}$.

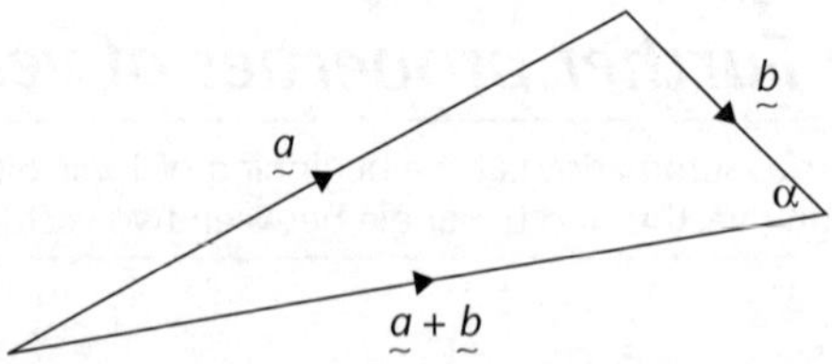

Example B

Q. If p represents a force of 20 Newton in a direction S40°E and q represents a force of 30 Newton in the direction N30°E, use trigonometry methods to find the magnitude and direction of $\underset{\sim}{p} + \underset{\sim}{q}$.

A. After calculating the angle in the magnitude triangle we can use the cosine rule to find the magnitude of $\underset{\sim}{p} + \underset{\sim}{q}$:

$|\underset{\sim}{p} + \underset{\sim}{q}|^2 = 20^2 + 30^2 - 2 \times 20 \times 30 \cos 70°$

$= 889.6$

$|\underset{\sim}{p} + \underset{\sim}{q}| = 29.82$ Newton

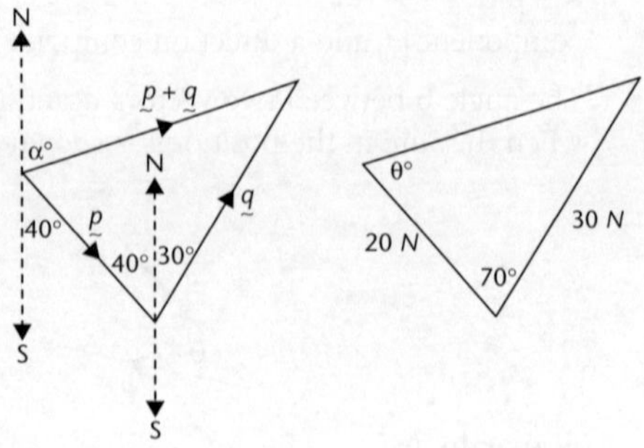

To find the direction (Nα°E) we need to first find the size of the angle θ using the sine rule:

$\dfrac{\sin\theta}{30} = \dfrac{\sin 70°}{29.82}$

$\sin\theta = 0.9454$

$\theta = 70.97°$

$180° - 40° - 70.97° = 69.03°$

The resultant $\underset{\sim}{p} + \underset{\sim}{q}$ has a magnitude of 29.82 Newton and a direction of N69.03°E

Example C

Q. Two forces act at an angle of 60°. If their resultant force is 15 kg wt and one of the forces is 10 kg wt, find the magnitude of the other force.

A. Drawing a diagram, approximately to scale:

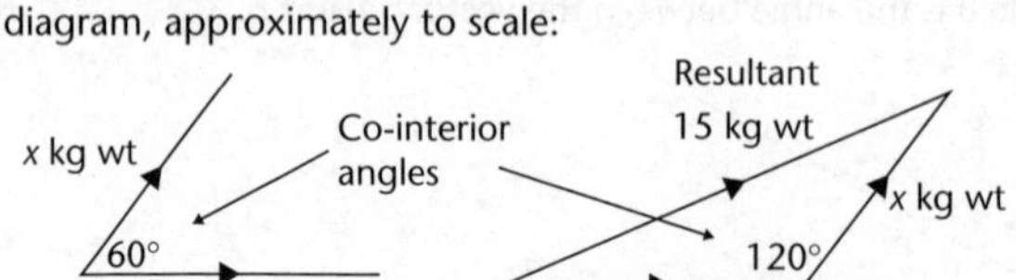

We can use the sine rule to find the value of x:

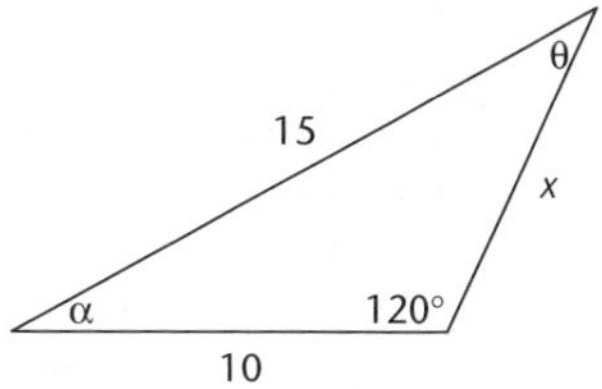

$$\frac{\sin\theta}{10} = \frac{\sin 120°}{15}$$

$\sin\theta = 0.5774$

$\theta = 35.264$

$\alpha = 180° - 35.264° - 120° = 24.74°$

$$\frac{x}{\sin 24.74°} = \frac{15}{\sin 120°}$$

$x = 7.25$

The other force has a magnitude of 7.25 kg wt.

Unit 12.2 Activity 8A: Further properties of vectors

Where necessary give your answer correct to two decimal places.

1. Two forces act at right angles. One of the forces has a magnitude of 5 kg wt and the resultant of the two forces has a magnitude of 6 kg wt. Find the magnitude of the second force and the angle between the resultant and the 5 kg wt force.
2. Two forces act at an angle of 60°. If their resultant force is 16 kg wt and one of the forces is 12 kg wt, find the magnitude of the other force.
3. The resultant of two forces is perpendicular to a force of magnitude 5 kg wt. If the other force is of magnitude 7 kg wt, find the angle between the two forces and the magnitude of the resultant.
4. A boat is rowed with a velocity of 6 km/hr directly towards the opposite bank of a river. The river flows at 2 km/hr.
   - **a.** Find the velocity of the boat relative to the bank of the river.
   - **b.** If the river is 100 metres wide, how far down the river below the point at which it was originally directed will the boat reach the opposite bank?
5. Once an aeroplane is airborne it can fly at a speed of 520 km/hr. On a particular day the destination of the aeroplane is due north of its departure point, however there is a wind of 50km/hr blowing in a north-westerly direction:
   - **a.** Calculate the direction in which the aeroplane needs to fly so that it reaches its destination.
   - **b.** What is the ground speed of the aeroplane during this flight?
6. Two tugboats are manoeuvring a ship as shown in the diagram. Find the resultant force on the ship and the angle this resultant makes with the smaller tug force.

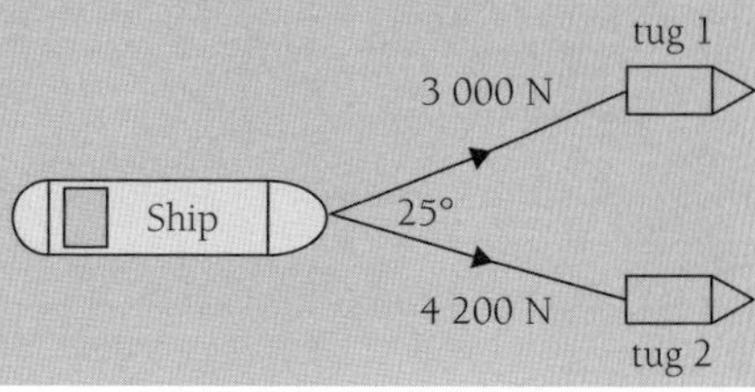

7. A ferry intends to travel from point A to point B across an inlet where a current of 6 knots is flowing due south. The bearing of point B from point A is 078°. If the ferry can travel at 18 knots, find:
   a. The bearing that the ferry should take so that it travels directly from A to B.
   b. The actual speed of the ferry.

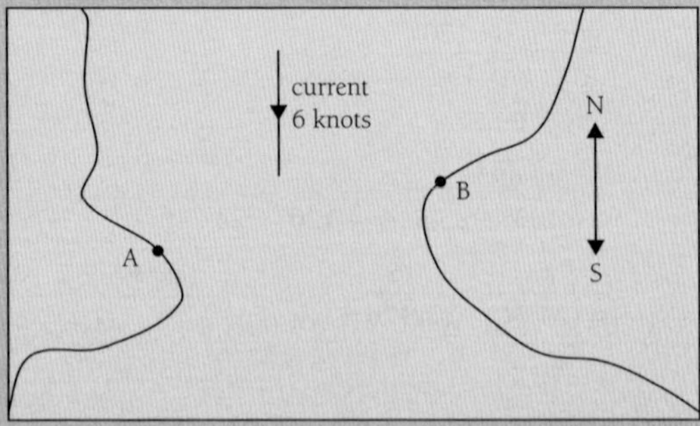

Unit 12.2 Trigonometry and Vectors
Topic 9: Rectangular components of vectors

Topic 9 continues the coverage of vectors as summarised at the beginning of Topic 6 (p. 135), looking at the geometry of vectors: position vector, magnitude, unit vector, parallel vectors, collinear points

Summary

Expressing vectors as a sum of vertical and horizontal components simplifies many calculations with vectors.

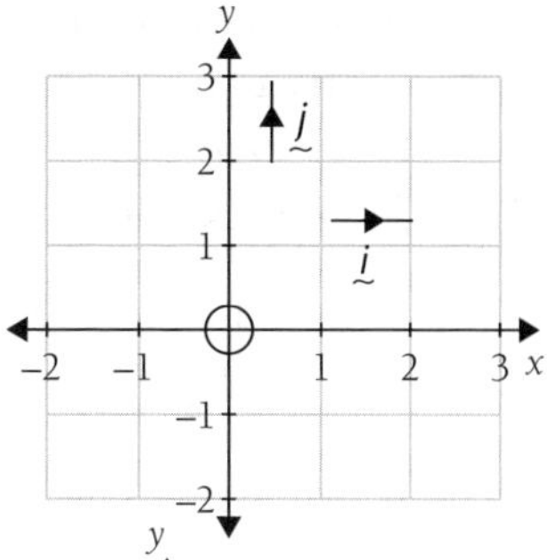

Using a set of Cartesian axes we define the unit vector in the positive direction of the x-axis as $\underset{\sim}{i}$, and we define the unit vector in the positive direction of the y-axis as $\underset{\sim}{j}$.

- The **position vector** of a point $P(x, y)$ is the vector $\overrightarrow{OP}$, where the point O is the origin.
  $\overrightarrow{OP} = x\underset{\sim}{i} + y\underset{\sim}{j}$

For example: The position vector of the point $Q(2, 3)$ is $\overrightarrow{OQ} = 2\underset{\sim}{i} + 3\underset{\sim}{j}$

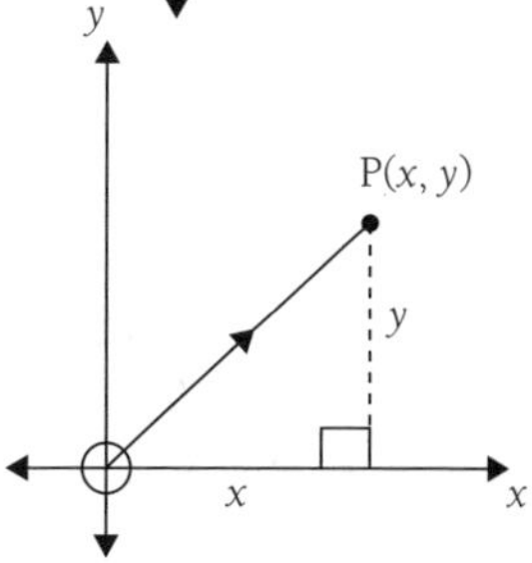

- The magnitude of the vector $\overrightarrow{OP} = x\underset{\sim}{i} + y\underset{\sim}{j}$ is
  $|\overrightarrow{OP}| = \sqrt{x^2 + y^2}$

This is verified by using Pythagoras' theorem in the right-angled triangle at right.

For example, the magnitude of the vector $\overrightarrow{OQ} = 2\underset{\sim}{i} + 3\underset{\sim}{j}$ is:
$|\overrightarrow{OQ}| = \sqrt{2^2 + 3^2} = \sqrt{13}$

- A unit vector in the direction of $\overrightarrow{OP}$ is
  $\hat{\overrightarrow{OP}} = \frac{1}{|\overrightarrow{OP}|}\overrightarrow{OP}$

For example, the unit vector in the direction of $\overrightarrow{OQ} = 2\underset{\sim}{i} + 3\underset{\sim}{j}$ is

$$\hat{\overrightarrow{OQ}} = \frac{1}{\sqrt{13}}(2\underset{\sim}{i} + 3\underset{\sim}{j}) = \frac{2}{\sqrt{13}}\underset{\sim}{i} + \frac{3}{\sqrt{13}}\underset{\sim}{j}$$

where $|\overrightarrow{OQ}| = \sqrt{13}$

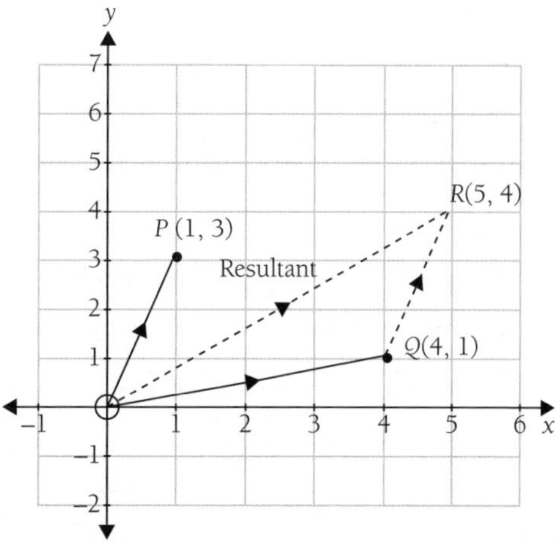

- Addition of vectors in component form:

$$\overrightarrow{OR} = \overrightarrow{OQ} + \overrightarrow{OP}$$
$$\overrightarrow{OR} = (4\underset{\sim}{i} + \underset{\sim}{j}) + (\underset{\sim}{i} + 3\underset{\sim}{j})$$
$$= (4 + 1)\underset{\sim}{i} + (1 + 3)\underset{\sim}{j}$$
$$= 5\underset{\sim}{i} + 4\underset{\sim}{j}$$

To add vectors in component form you add the $\underset{\sim}{i}$ component parts and the $\underset{\sim}{j}$ components parts.

Example A

Q. Find:

a. the position vectors $\overrightarrow{OP}$ and $\overrightarrow{OQ}$ of the points $P(3, -2)$ and $Q(4, 1)$

b. $\overrightarrow{OP} + \overrightarrow{OQ}$

c. the vector $\overrightarrow{PQ}$

d. the magnitude of $\overrightarrow{PQ}$

e. the unit vector in the direction of $\overrightarrow{PQ}$

A. **a.** The position vector of $(3, -2)$ is $\overrightarrow{OP} = 3\underset{\sim}{i} - 2\underset{\sim}{j}$ and the position vector of $(4, 1)$ is $\overrightarrow{OQ} = 4\underset{\sim}{i} + \underset{\sim}{j}$.

b. $\overrightarrow{OP} + \overrightarrow{OQ} = 3\underset{\sim}{i} - 2\underset{\sim}{j} + 4\underset{\sim}{i} + \underset{\sim}{j} = (3 + 4)\underset{\sim}{i} + (-2 + 1)\underset{\sim}{j} = 7\underset{\sim}{i} - \underset{\sim}{j}$

c. $\overrightarrow{PQ} = \overrightarrow{OQ} - \overrightarrow{OP}$
$= 4\underset{\sim}{i} + \underset{\sim}{j} - (3\underset{\sim}{i} - 2\underset{\sim}{j})$
$= (4 - 3)\underset{\sim}{i} + (1 + 2)\underset{\sim}{j}$
$= \underset{\sim}{i} + 3\underset{\sim}{j}$

d. $|\overrightarrow{PQ}| = \sqrt{1^2 + 3^2} = \sqrt{10}$

e. $\overrightarrow{PQ} = \frac{1}{\sqrt{10}}(\underset{\sim}{i} + 4\underset{\sim}{j}) = \frac{1}{\sqrt{10}}\underset{\sim}{i} + \frac{4}{\sqrt{10}}\underset{\sim}{j}$

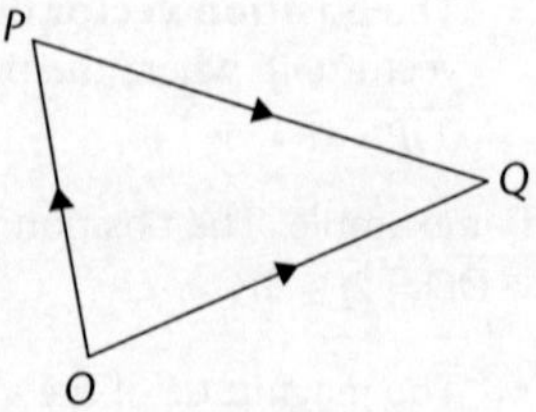

- Multiplication by a scalar
  $k(x\underset{\sim}{i} + y\underset{\sim}{j}) = (kx)\underset{\sim}{i} + (ky)\underset{\sim}{j}$ where k is a scalar
- The distance between two points
  For example: find the distance between the points P and Q.

  Solution:
  $\overrightarrow{PQ} = \overrightarrow{OQ} - \overrightarrow{OP}$
  $= 5\underset{\sim}{i} + 3\underset{\sim}{j} - (\underset{\sim}{i} + 6\underset{\sim}{j})$
  $= 4\underset{\sim}{i} - 3\underset{\sim}{j}$
  $|\overrightarrow{PQ}| = \sqrt{4^2 + -3^2} = \sqrt{25} = 5$

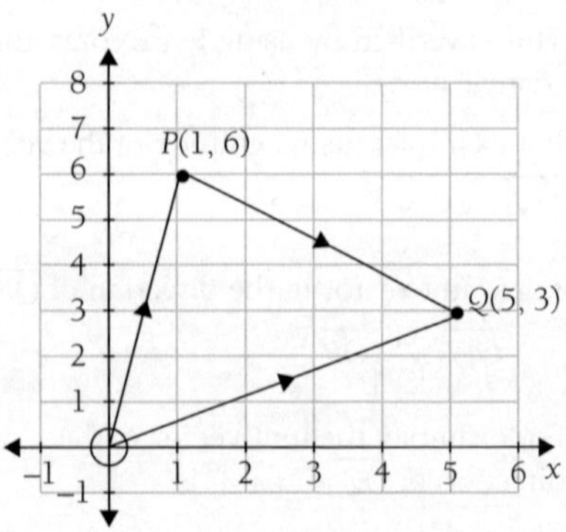

- Parallel vectors
  If vectors $\underset{\sim}{a}$ and $\underset{\sim}{b}$, in component form, are parallel, then $\underset{\sim}{a} = k\underset{\sim}{b}$ where k is a scalar.

Example B

Q. Show that $\underset{\sim}{a} = 4\underset{\sim}{i} - 3\underset{\sim}{j}$ is parallel to $\underset{\sim}{b} = 6\underset{\sim}{i} - 4.5\underset{\sim}{j}$.

A. The $\underset{\sim}{i}$ components are in the proportion $\frac{4}{6} = \frac{2}{3}$.

Are the $\underset{\sim}{j}$ components in the proportion $\frac{2}{3}$?

$\frac{3}{4.5} = \frac{2}{3}$ Yes. So we can write $\underset{\sim}{a} = \frac{2}{3}\underset{\sim}{b}$ and hence $\underset{\sim}{a}$ is parallel to $\underset{\sim}{b}$.

Example C

Q. If $\underset{\sim}{a} = 4\underset{\sim}{i} - \underset{\sim}{j}$, $\underset{\sim}{b} = -\underset{\sim}{i} + 2\underset{\sim}{j}$ and $\underset{\sim}{c} = 3\underset{\sim}{i} + 3\underset{\sim}{j}$ show that $\underset{\sim}{a} + 2\underset{\sim}{b} - \underset{\sim}{c}$ is parallel to the *x*-axis.

A. If a vector is parallel to the *x*- axis then it will have an $\underset{\sim}{i}$ component only.

$$\begin{aligned}\underset{\sim}{a} + 2\underset{\sim}{b} - \underset{\sim}{c} &= 4\underset{\sim}{i} - \underset{\sim}{j} + 2(-\underset{\sim}{i} + 2\underset{\sim}{j}) - (3\underset{\sim}{i} + 3\underset{\sim}{j})\\ &= (4 - 2 - 3)\underset{\sim}{i} + (-1 + 4 - 3)\underset{\sim}{j}\\ &= (-1)\underset{\sim}{i} + (0)\underset{\sim}{j}\\ &= -\underset{\sim}{i}\end{aligned}$$

$\underset{\sim}{a} + 2\underset{\sim}{b} - \underset{\sim}{c}$ has an $\underset{\sim}{i}$ component only, so $\underset{\sim}{a} + 2\underset{\sim}{b} - \underset{\sim}{c}$ is parallel to the *x*-axis.

- Collinear points (points that are in the same straight line)
  To show that the points *P*, *Q* and *R* are collinear, we need to show that $\overrightarrow{PQ} = k\overrightarrow{QR}$ where *k* is a scalar.

  This means that $\overrightarrow{PQ}$ is parallel to $\overrightarrow{QR}$ and if $\overrightarrow{PQ}$ is parallel to $\overrightarrow{QR}$, and they have the point *Q* in common, then *P*, *Q* and *R* are in the same straight line.

Example D

Q. The position vectors of the points *P*, *Q* and *R* are $-3\underset{\sim}{i} - 2\underset{\sim}{j}$, $-\underset{\sim}{j}$ and $6\underset{\sim}{i} + \underset{\sim}{j}$ respectively. Show that the points *P*, *Q* and *R* are collinear.

A. $\overrightarrow{OP} = -3\underset{\sim}{i} - 2\underset{\sim}{j}$

$\overrightarrow{OQ} = -\underset{\sim}{j}$

$\overrightarrow{OR} = 6\underset{\sim}{i} + \underset{\sim}{j}$

$$\begin{aligned}\overrightarrow{PQ} &= \overrightarrow{OQ} - \overrightarrow{OP}\\ &= -\underset{\sim}{j} - (-3\underset{\sim}{i} - 2\underset{\sim}{j})\\ &= 3\underset{\sim}{i} + \underset{\sim}{j}\end{aligned}$$

$$\begin{aligned}\overrightarrow{QR} &= \overrightarrow{OR} - \overrightarrow{OQ}\\ &= 6\underset{\sim}{i} + \underset{\sim}{j} - (-\underset{\sim}{j})\\ &= 6\underset{\sim}{i} + 2\underset{\sim}{j}\end{aligned}$$

So $\overrightarrow{QR} = 2\overrightarrow{PQ}$, which means that *PQ* and *QR* are parallel and since they have a point *Q* in common this means that points *P*, *Q* and *R* are collinear.

Unit 12.2 Activity 9A: Rectangular components of vectors

Where necessary give your answer correct to two decimal places.

1. For each of the following vectors, find:

a. The magnitude of the vector.

b. A unit vector in the direction of the vector.

i. $6\underset{\sim}{i} + 8\underset{\sim}{j}$

ii. $-3\underset{\sim}{i} + 5\underset{\sim}{j}$

iii. $5\underset{\sim}{i} - 12\underset{\sim}{j}$

iv. $\underset{\sim}{i} - 3\underset{\sim}{j}$

2. For the points $P(-2, 1)$ and $Q(3, 5)$ find:
  a. The position vectors $\overrightarrow{OP}$ and $\overrightarrow{OQ}$
  b. $\overrightarrow{PQ}$
  c. $|\overrightarrow{PQ}|$
  d. $\widehat{\overrightarrow{PQ}}$
  e. $\overrightarrow{OP} + \overrightarrow{OQ}$
  f. $3\overrightarrow{OQ} - \overrightarrow{OP}$

3. If vectors $\underset{\sim}{a} = \underset{\sim}{i} + 3\underset{\sim}{j}$, $\underset{\sim}{b} = 2\underset{\sim}{i} - 5\underset{\sim}{j}$ and $\underset{\sim}{c} = -2\underset{\sim}{i} - 3\underset{\sim}{j}$
  a. Find $\underset{\sim}{a} + \underset{\sim}{b} - \underset{\sim}{c}$
  b. Find $|\underset{\sim}{a} + \underset{\sim}{b}|$
  c. Find $\hat{\underset{\sim}{a}}$
  d. Show that $\underset{\sim}{a} + \underset{\sim}{c}$ is parallel to the x-axis.
  e. Show that $\underset{\sim}{b} + \underset{\sim}{c}$ is parallel to the y-axis.

4. The position vectors of the points P and Q are $\underset{\sim}{i} + 6\underset{\sim}{j}$ and $7\underset{\sim}{i} + 5\underset{\sim}{j}$ respectively. Show that the triangle OPQ is an isosceles triangle.

5. P, Q, R and S have the position vectors $-4\underset{\sim}{i} - 2\underset{\sim}{j}$, $-2\underset{\sim}{i} + \underset{\sim}{j}$, $2\underset{\sim}{i} - \underset{\sim}{j}$ and $-6\underset{\sim}{i} + 5\underset{\sim}{j}$ respectively. Show that PQ is parallel to RS.

6. The position vectors of the points P, Q and R are $-2\underset{\sim}{i} - 9\underset{\sim}{j}$, $-2\underset{\sim}{j}$ and $4\underset{\sim}{i} + 12\underset{\sim}{j}$ respectively. Show that the points P, Q and R are collinear.

7. If $\underset{\sim}{a} = 3\underset{\sim}{i} - 4\underset{\sim}{j}$, $b = 2\underset{\sim}{i} + \underset{\sim}{j}$ and $\underset{\sim}{c} = 4\underset{\sim}{i} + m\underset{\sim}{j}$, find the value of m such that $\underset{\sim}{a} + \underset{\sim}{b}$ is parallel to $\underset{\sim}{c}$.

8. Points P and Q have position vectors $-3\underset{\sim}{i} + 5\underset{\sim}{j}$, and $5\underset{\sim}{i} - \underset{\sim}{j}$. Find the position vector of the point M, the midpoint of PQ.

9. The point T divides SU in the ratio 1:2. If the position vectors of the points T and U are $\underset{\sim}{i} + 3\underset{\sim}{j}$ and $7\underset{\sim}{i} + 2\underset{\sim}{j}$ respectively, find the position vector of the point S.

Unit 12.2 Trigonometry and Vectors

Topic 10: Component vectors in three dimensions

Topic 10 continues the coverage of vectors as summarised at the beginning of Topic 6 (p. 135), looking at component vectors in three dimensions.

Vectors can be represented in three dimensions by introducing another axis at right angles to the x-y axes, as shown in the diagram at right.

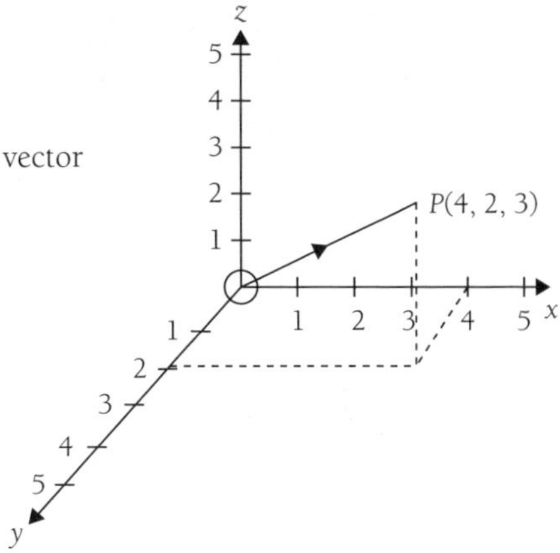

The unit vector in the direction of the z-axis is denoted by $\underset{\sim}{\hat{z}}$.

The position vector of the point P (4, 2, 3) is the vector $\overrightarrow{OP} = 4\underset{\sim}{i} + 2\underset{\sim}{j} + 3\underset{\sim}{k}$

Vector operations in three dimensions

- If $\overrightarrow{OP} = a\underset{\sim}{i} + b\underset{\sim}{j} + c\underset{\sim}{k}$ then

  $|\overrightarrow{OP}| = \sqrt{a^2 + b^2 + c^2}$

  and $\hat{\overrightarrow{OP}} = \dfrac{1}{\sqrt{a^2 + b^2 + c^2}}(a\underset{\sim}{i} + b\underset{\sim}{j} + c\underset{\sim}{k})$

- $\overrightarrow{PQ} = \overrightarrow{OQ} - \overrightarrow{OP}$
- If $\overrightarrow{OP} = 4\underset{\sim}{i} + 2\underset{\sim}{j} + 3\underset{\sim}{k}$ and $\overrightarrow{OQ} = -2\underset{\sim}{i} + 3\underset{\sim}{j} + 5\underset{\sim}{k}$ then

  $\overrightarrow{OP} + \overrightarrow{OQ} = (4 - 2)\underset{\sim}{i} + (2 + 3)\underset{\sim}{j} + (3 + 5)\underset{\sim}{k}$

  $= 2\underset{\sim}{i} + 5\underset{\sim}{j} + 8\underset{\sim}{k}$

Unit 12.2 Activity 10A: Component vectors in three dimensions

Where necessary give your answer correct to two decimal places.

1. For each of the following vectors find:

a. The magnitude of the vector.

b. A unit vector in the direction of the vector.

i. $\underset{\sim}{i} + 4\underset{\sim}{j} + 3\underset{\sim}{k}$

ii. $-2\underset{\sim}{i} + 2\underset{\sim}{j} - \underset{\sim}{k}$

iii. $3\underset{\sim}{i} - 4\underset{\sim}{j} - 5\underset{\sim}{k}$

iv. $7\underset{\sim}{i} - 3\underset{\sim}{j} + 4\underset{\sim}{k}$

2. For the points $P(-2, 1, 3)$ and $Q(3, 5, -1)$ find:

a. The position vectors $\overrightarrow{OP}$ and $\overrightarrow{OQ}$

b. $\overrightarrow{PQ}$

c. $|\overrightarrow{PQ}|$

d. $\hat{\overrightarrow{PQ}}$

e. $\overrightarrow{OP} + \overrightarrow{OQ}$

f. $\overrightarrow{OQ} - 2\overrightarrow{OP}$

3. If vectors $\underset{\sim}{a} = \underset{\sim}{i} + 3\underset{\sim}{j} + 2\underset{\sim}{k}$, $\underset{\sim}{b} = 2\underset{\sim}{i} - 5\underset{\sim}{j} + \underset{\sim}{k}$ and $\underset{\sim}{c} = -2\underset{\sim}{i} - 3\underset{\sim}{j} - 2\underset{\sim}{k}$

a. Find $\underset{\sim}{a} + \underset{\sim}{b} - \underset{\sim}{c}$

b. Find $|\underset{\sim}{a} + \underset{\sim}{b}|$

c. Find $\hat{\underset{\sim}{a}}$

d. Show that $\underset{\sim}{a} + \underset{\sim}{c}$ is parallel to the x-axis.

4. The position vectors of the points P and Q are $\underset{\sim}{i} + 6\underset{\sim}{j} - 2\underset{\sim}{k}$ and $7\underset{\sim}{i} + 5\underset{\sim}{j} - 4\underset{\sim}{k}$ respectively. Show that the triangle OPQ is an isosceles triangle.

5. Points S and T have position vectors $-3\underset{\sim}{i} + 5\underset{\sim}{j} + 3\underset{\sim}{k}$ and $5\underset{\sim}{i} - \underset{\sim}{j} - 2\underset{\sim}{k}$. Find the position vector of the point M, the midpoint of ST.

6. Show that the vector $\underset{\sim}{a} = 9\underset{\sim}{i} - 5\underset{\sim}{j} - 3\underset{\sim}{k}$ is parallel to the vector $\underset{\sim}{b} = -3\underset{\sim}{i} - \frac{5}{3}\underset{\sim}{j} - \underset{\sim}{k}$

7. If $\overrightarrow{OP} = 6\underset{\sim}{i} + 3j - 4\underset{\sim}{k}$, $\overrightarrow{OQ} = 3\underset{\sim}{i} + j + \underset{\sim}{k}$ and $\overrightarrow{OR} = 2\underset{\sim}{i} - j + 3\underset{\sim}{k}$ show that $|\overrightarrow{RP}| = 3|\overrightarrow{RQ}|$

8. If $\overrightarrow{OP} = 2\underset{\sim}{i} - \underset{\sim}{j} + 3\underset{\sim}{k}$, $\overrightarrow{OQ} = 5\underset{\sim}{i} - 3\underset{\sim}{j} + 2\underset{\sim}{k}$ and $\overrightarrow{OR} = -\underset{\sim}{i} + \underset{\sim}{j} + 4\underset{\sim}{k}$ show that P, Q and R are collinear.

9. If $\underset{\sim}{a} = 2\underset{\sim}{i} - 3\underset{\sim}{j} + \underset{\sim}{k}$, $\underset{\sim}{b} = 2\underset{\sim}{i} - 4\underset{\sim}{j} + 5\underset{\sim}{k}$ and $\underset{\sim}{c} = -\underset{\sim}{i} - 4\underset{\sim}{j} + 2\underset{\sim}{k}$ find the values of p and q such that $\underset{\sim}{a} + p\underset{\sim}{b} + q\underset{\sim}{c}$ is parallel to the x-axis.

Unit 12.2 Trigonometry and Vectors

Topic 11: Scalar product of vectors

Topic 11 concludes the coverage of vectors as summarised at the beginning of Topic 6 (p. 135), looking at the scalar product of vectors.

Definition

> The scalar product of vectors $\underset{\sim}{a}$ and $\underset{\sim}{b}$ is defined as $\underset{\sim}{a} \bullet \underset{\sim}{b} = |\underset{\sim}{a}|\,|\underset{\sim}{b}|\cos\theta$ where θ is the angle between the vectors $\underset{\sim}{a}$ and $\underset{\sim}{b}$

The scalar product is also called the dot product.

Important points about the scalar product

- The scalar product $\underset{\sim}{a} \bullet \underset{\sim}{b} = |\underset{\sim}{a}|\,|\underset{\sim}{b}|\cos\theta$ is a scalar quantity.
- The angle, θ, between two vectors, $\underset{\sim}{a}$ and $\underset{\sim}{b}$, is defined as the angle between the vectors when they are in the positions as shown below:

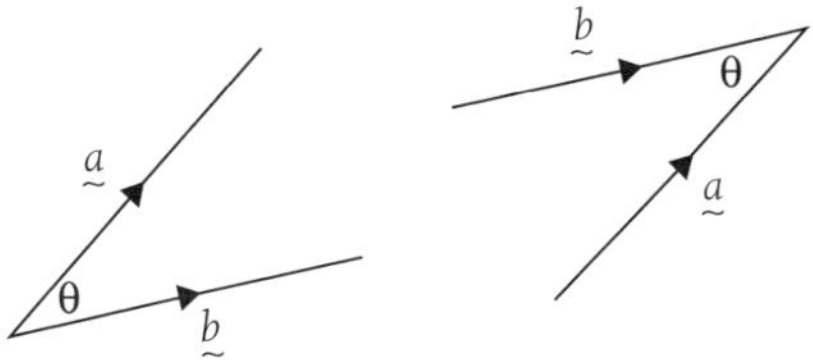

- $\underset{\sim}{a} \bullet \underset{\sim}{a} = |\underset{\sim}{a}|\,|\underset{\sim}{a}| \cos\theta = |\underset{\sim}{a}|^2 \times \cos 0^\circ = |\underset{\sim}{a}|^2$
- If $\underset{\sim}{a}$ is perpendicular to $\underset{\sim}{b}$ ($\underset{\sim}{a} \perp \underset{\sim}{b}$) then $\underset{\sim}{a} \bullet \underset{\sim}{b} = |\underset{\sim}{a}|\,|\underset{\sim}{b}| \cos 90^\circ = |\underset{\sim}{a}|\,|\underset{\sim}{b}| \times 0 = 0$
- If $\underset{\sim}{a} \bullet \underset{\sim}{b} = 0$ then one or more of the following is true: $|\underset{\sim}{a}| = 0, |\underset{\sim}{b}| = 0$ or $\underset{\sim}{a} \perp \underset{\sim}{b}$
- $\underset{\sim}{a} \bullet \underset{\sim}{b} = \underset{\sim}{b} \bullet \underset{\sim}{a}$
- $\underset{\sim}{a} \bullet (\underset{\sim}{b} + \underset{\sim}{c}) = \underset{\sim}{a} \bullet \underset{\sim}{b} + \underset{\sim}{a} \bullet \underset{\sim}{c}$
- $(k\underset{\sim}{a}) \bullet \underset{\sim}{b} = k(\underset{\sim}{a} \bullet \underset{\sim}{b})$ where k is a scalar.
- $(k\underset{\sim}{a}) \bullet (m\underset{\sim}{b}) = km(\underset{\sim}{a} \bullet \underset{\sim}{b})$ where k and m are scalars.
- $\hat{\underset{\sim}{a}} \bullet \hat{\underset{\sim}{a}} = |\hat{\underset{\sim}{a}}|\,|\hat{\underset{\sim}{a}}| \cos 0^\circ = 1 \times 1 \times 1 = 1$
- $\hat{\underset{\sim}{a}} \bullet \hat{\underset{\sim}{b}} = |\hat{\underset{\sim}{a}}|\,|\hat{\underset{\sim}{b}}| \cos\theta = 1 \times 1 \times \cos\theta = \cos\theta$

The scalar product for vectors in rectangular component form

In component form the unit vectors $\underset{\sim}{i}$, $\underset{\sim}{j}$ and $\underset{\sim}{k}$ have magnitudes of 1 and angles of 0° between themselves, so the scalar products $\underset{\sim}{i} \bullet \underset{\sim}{i} = 1 \times 1 \times \cos 0^\circ = 1$, and similarly $\underset{\sim}{j} \bullet \underset{\sim}{j} = 1$ and $\underset{\sim}{k} \bullet \underset{\sim}{k} = 1$.

The scalar products $\underset{\sim}{i} \bullet \underset{\sim}{j}$, $\underset{\sim}{j} \bullet \underset{\sim}{i}$, $\underset{\sim}{i} \bullet \underset{\sim}{k}$, $\underset{\sim}{k} \bullet \underset{\sim}{i}$, $\underset{\sim}{j} \bullet \underset{\sim}{k}$ and $\underset{\sim}{k} \bullet \underset{\sim}{j}$ are all equal to zero because the vectors $\underset{\sim}{i}$, $\underset{\sim}{j}$ and $\underset{\sim}{k}$ are all perpendicular ($\cos\theta = 0$) to each other.

Example A

Q. Find the scalar product of the vectors $\underset{\sim}{a} = 3\underset{\sim}{i} + 2\underset{\sim}{j}$ and $\underset{\sim}{b} = -2\underset{\sim}{i} + 4\underset{\sim}{j}$.

A. $\underset{\sim}{a} \bullet \underset{\sim}{b} = (3\underset{\sim}{i} + 2\underset{\sim}{j}) \bullet (-2\underset{\sim}{i} + 4\underset{\sim}{j})$

$= (3\underset{\sim}{i} \bullet -2\underset{\sim}{i}) + (3\underset{\sim}{i} \bullet 4\underset{\sim}{j}) + (2\underset{\sim}{j} \bullet -2\underset{\sim}{i}) + (2\underset{\sim}{j} \bullet 4\underset{\sim}{j})$

$= (3 \times -2)\underset{\sim}{i} \bullet \underset{\sim}{i} + (3 \times 4)\underset{\sim}{i} \bullet \underset{\sim}{j} + (2 \times -2)\underset{\sim}{j} \bullet \underset{\sim}{i} + (2 \times 4)\underset{\sim}{j} \bullet \underset{\sim}{j}$

$= 3 \times -2 + 0 + 0 + 2 \times 4$

$= -6 + 8$

$= 2$

In general:

> If $\underset{\sim}{a} = x_1\underset{\sim}{i} + y_1\underset{\sim}{j} + z_1\underset{\sim}{k}$ and $\underset{\sim}{b} = x_2\underset{\sim}{i} + y_2\underset{\sim}{j} + z_2\underset{\sim}{k}$ then $\underset{\sim}{a} \bullet \underset{\sim}{b} = x_1x_2 + y_1y_2 + z_1z_2$

Example B

Q. If $\underset{\sim}{a} = 3\underset{\sim}{i} + 2\underset{\sim}{j} - k$ and $b = -\underset{\sim}{i} + \underset{\sim}{j} - 3k$ find:

a. $\underset{\sim}{a} \bullet \underset{\sim}{b}$

b. The angle between vectors $\underset{\sim}{a}$ and $\underset{\sim}{b}$

A. a. $\underset{\sim}{a} \bullet \underset{\sim}{b} = 3 \times -1 + 2 \times 1 + -1 \times -3$

$= -3 + 2 + 3$

$= 2$

b. $\underset{\sim}{a} \bullet \underset{\sim}{b} = |\underset{\sim}{a}|\,|\underset{\sim}{b}| \cos \theta$ where θ is the angle between the vectors $\underset{\sim}{a}$ and $\underset{\sim}{b}$.

$\underset{\sim}{a} \bullet \underset{\sim}{b} = 2$, $|\underset{\sim}{a}| = \sqrt{3^2 + 2^2 + (-1)^2} = \sqrt{14}$ and $|\underset{\sim}{b}| = \sqrt{(-1)^2 + 1^2 + (-3)^2} = \sqrt{11}$

Substituting these values gives
$2 = \sqrt{14} \times \sqrt{11} \cos \theta$

so:

$$\cos \theta = \frac{2}{\sqrt{14}\ \sqrt{11}} = 0.16116\ldots$$

$\theta = 80.73°$

The angle between vectors $\underset{\sim}{a}$ and $\underset{\sim}{b}$ is 80.73° (to 2 dp).

Example C

Q. If the points P, Q and R have the position vectors $2\underset{\sim}{i} - \underset{\sim}{j} + 2\underset{\sim}{k}$, $-\underset{\sim}{i} + 3\underset{\sim}{j} - 3\underset{\sim}{k}$ and $\underset{\sim}{i} + 2\underset{\sim}{j} + \underset{\sim}{k}$, respectively, show that PQ is perpendicular to OR.

A. $\overrightarrow{OP} = 2\underset{\sim}{i} - \underset{\sim}{j} + 2\underset{\sim}{k}$,

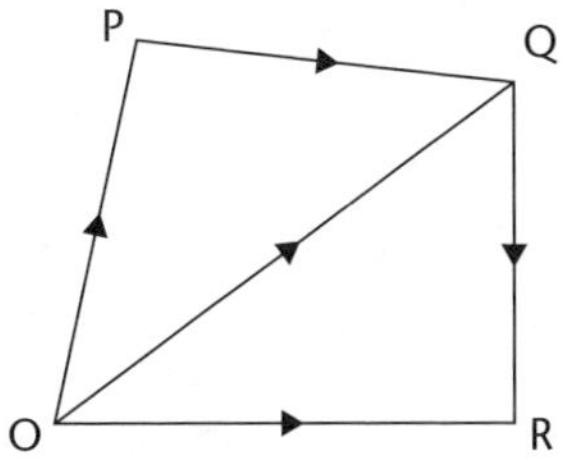

$\overrightarrow{OQ} = -\underset{\sim}{i} + 3\underset{\sim}{j} - 3\underset{\sim}{k}$

and $\overrightarrow{OR} = \underset{\sim}{i} + 2\underset{\sim}{j} + \underset{\sim}{k}$

$\overrightarrow{PQ} = \overrightarrow{OQ} - \overrightarrow{OR}$

$= (-\underset{\sim}{i} + 3\underset{\sim}{j} - 3\underset{\sim}{k}) - (2\underset{\sim}{i} - \underset{\sim}{j} + 2\underset{\sim}{k})$

$= -3\underset{\sim}{i} + 4\underset{\sim}{j} - 5\underset{\sim}{k}$

$\overrightarrow{PQ} \bullet \overrightarrow{OR} = -3 \times 1 + 4 \times 2 - 5 \times 1 = 0$

Since neither the magnitude of PQ nor the magnitude of OR is zero this means the PQ is perpendicular to OR.

Unit 12.2 Activity 11A: Scalar product of vectors

Where necessary give your answer correct to two decimal places.

1. Find the scalar product for each of the following pairs of vectors:

a.

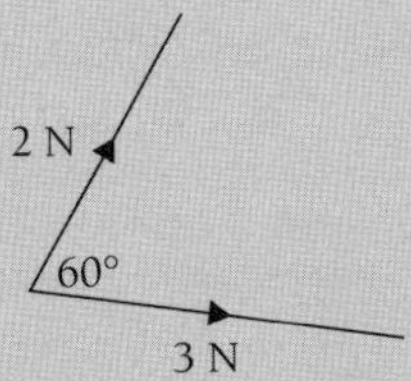

b.

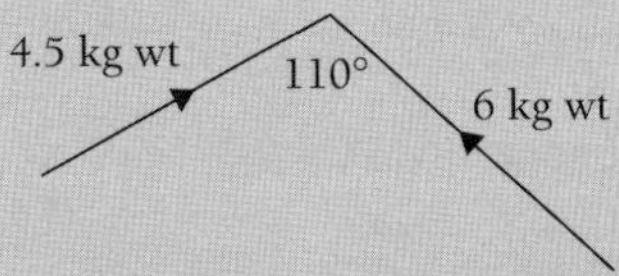

c.

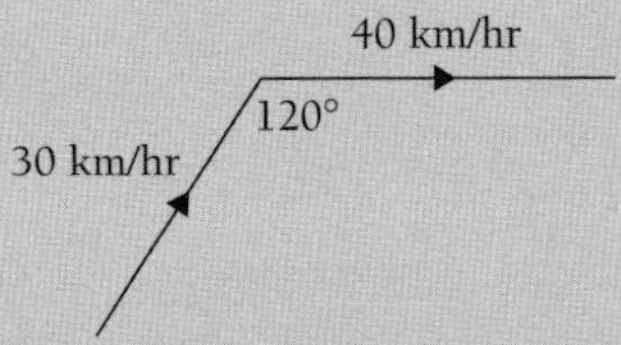

d.

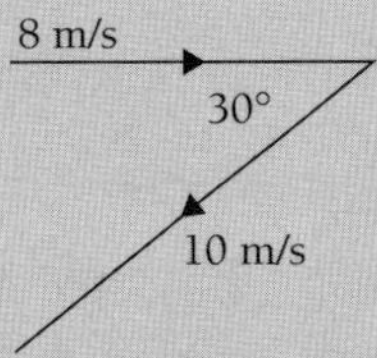

e.

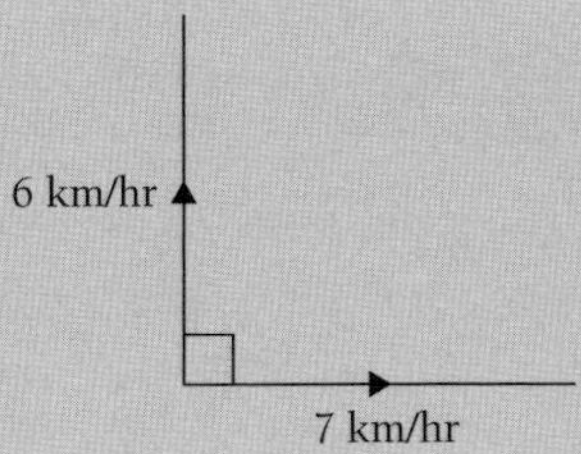

f.

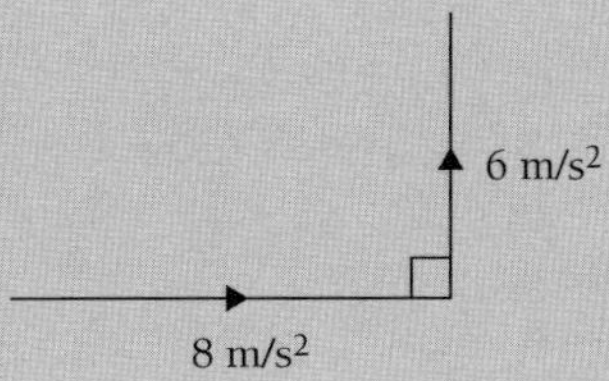

2. For the vectors given on the diagram at right, find:

a. $\underset{\sim}{a} \bullet \underset{\sim}{b}$
b. $\underset{\sim}{a} \bullet \underset{\sim}{c}$
c. $\underset{\sim}{a} \bullet \underset{\sim}{d}$
d. $\underset{\sim}{a} \bullet \underset{\sim}{e}$
e. $\underset{\sim}{c} \bullet \underset{\sim}{b}$
f. $\underset{\sim}{b} \bullet \underset{\sim}{d}$
g. $\underset{\sim}{b} \bullet \underset{\sim}{e}$
h. $\underset{\sim}{c} \bullet \underset{\sim}{d}$
i. $\underset{\sim}{c} \bullet (\underset{\sim}{a} + \underset{\sim}{b})$

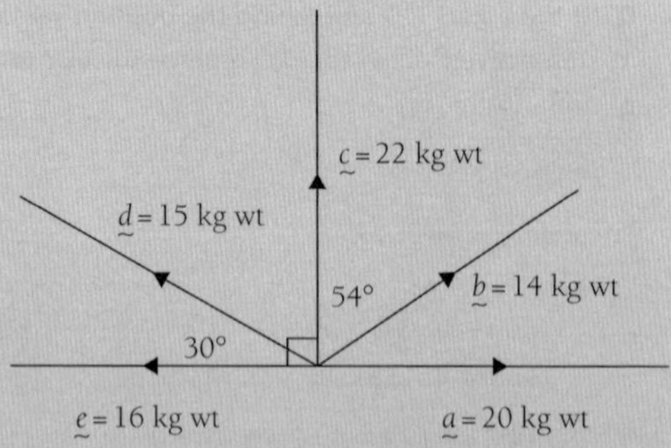

3. If $\underset{\sim}{a} = 2\underset{\sim}{i} + 3\underset{\sim}{j}$ and $\underset{\sim}{b} = 4\underset{\sim}{i} - \underset{\sim}{j}$ find:

a. $\underset{\sim}{a} \bullet \underset{\sim}{b}$
b. The angle between vectors $\underset{\sim}{a}$ and $\underset{\sim}{b}$

4. If the position vectors of the points *P* and *Q* are $2\underset{\sim}{i} + 3\underset{\sim}{j}$ and $7\underset{\sim}{i} - 5\underset{\sim}{j}$ respectively, find the angle between these position vectors.

5. If $\underset{\sim}{p} = \underset{\sim}{i} + 3\underset{\sim}{j} - \underset{\sim}{k}$ and $\underset{\sim}{q} = 3\underset{\sim}{i} - \underset{\sim}{j} + 2\underset{\sim}{k}$ find:

a. $\underset{\sim}{p} \bullet \underset{\sim}{q}$
b. The angle between vectors $\underset{\sim}{p}$ and $\underset{\sim}{q}$

6. Find the value of *t* such that vectors $\underset{\sim}{a} = 2\underset{\sim}{i} - 5\underset{\sim}{k} + \underset{\sim}{k}$ and $\underset{\sim}{b} = -3\underset{\sim}{i} + t\underset{\sim}{j} - 7\underset{\sim}{k}$ are perpendicular.

7. If the points *P*, *Q* and *R* have the position vectors $3\underset{\sim}{i} - 2\underset{\sim}{j} + 4\underset{\sim}{k}$, $5\underset{\sim}{i} + 3\underset{\sim}{j} - 2\underset{\sim}{k}$ and $6\underset{\sim}{i} + 5\underset{\sim}{j}$ respectively,

a. Show that the triangle *PQR* is a right-angled triangle.
b. Find the size of the other angles in triangle *PQR*.

8. If the vectors $\underset{\sim}{a}$, $\underset{\sim}{b}$ and $\underset{\sim}{c}$ are vectors whose magnitudes are not zero, state the conclusions that can be made for the following equations:

a. $\underset{\sim}{a} \bullet \underset{\sim}{b} = 0$
b. $\underset{\sim}{a} \bullet \underset{\sim}{b} = |\underset{\sim}{a}| \times |\underset{\sim}{b}|$
c. $\underset{\sim}{a} \bullet \underset{\sim}{b} = \underset{\sim}{a} \bullet \underset{\sim}{c}$
d. $\underset{\sim}{c} \bullet \underset{\sim}{a} - \underset{\sim}{c} \bullet \underset{\sim}{b} = 0$

9. Use the vectors in the given triangle to establish the cosine rule for finding $|\underset{\sim}{c}|$.

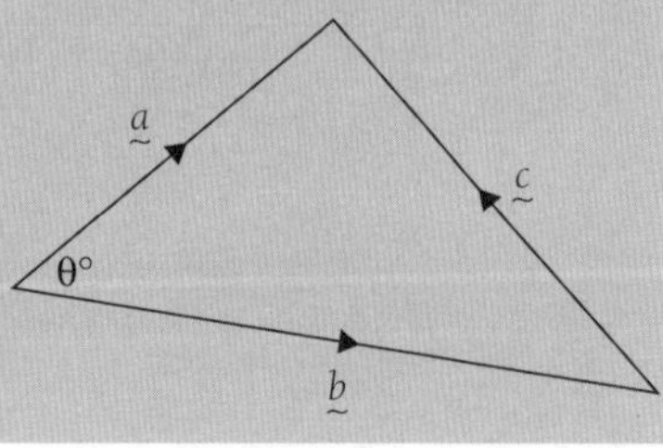

10. Use the given vectors on the rhombus ABCD to prove that the diagonals of a rhombus intersect at right angles.

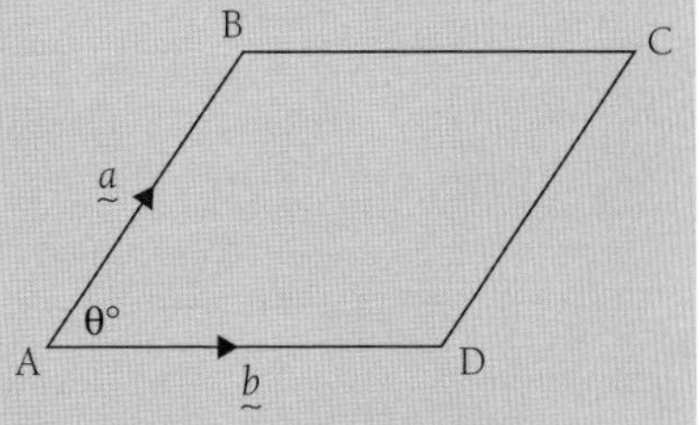

11. Use the given vectors in the diagram to show that the angle in a semicircle (∠PQR) is a right angle.

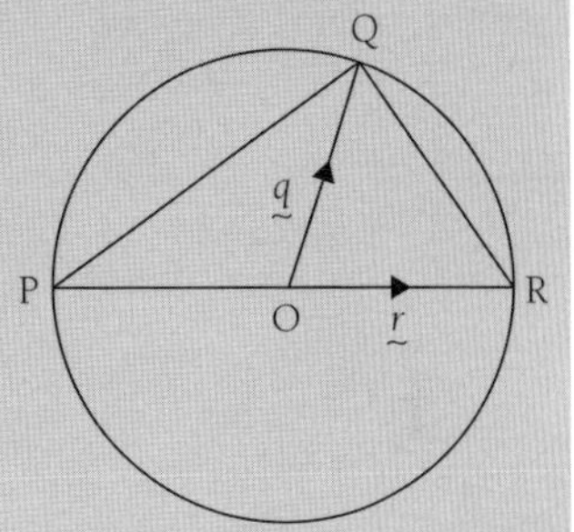

12. Use the vectors given in the diagram to prove that the line that bisects the angle at the apex of an isosceles triangle (point B) is the **perpendicular bisector** of the third side (AC) of the triangle.

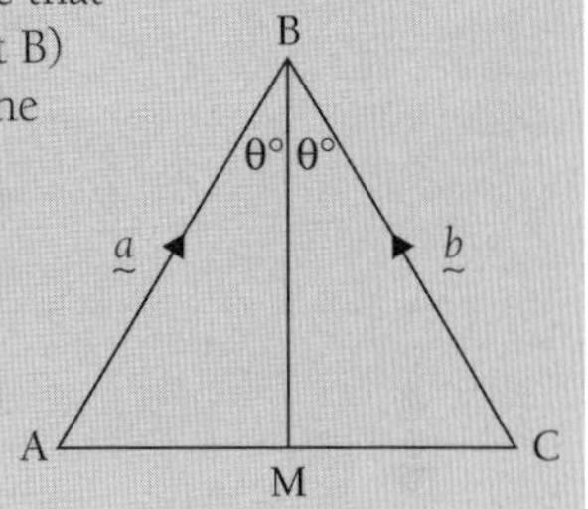

Unit 12.3 Calculus

Topic 1: The gradient function

Unit 12.3 focuses on calculus and the application of differentiation, which is used in many different areas, such as the flow of water for example, or the velocity of a vehicle. The objective in Topic 1 is to:

- Demonstrate an understanding of the relationship between derivatives and gradients.
- Evaluate the derivative at a point to find the gradient.

Introduction

In Grade 11, straight lines were studied in Unit 11.2 Topic 9 'The gradient of a straight line'. That Topic included an important feature of straight lines, their **gradient** or **slope**.

This feature was used in problems where there was a constant **rate of change** of one variable when compared with another variable. In such cases it was found that the graph obtained was a straight line and the rate of change was equal to the gradient of the line.

It is important to have a good understanding of the Grade 11 content and the concepts of gradients and rates of change in the context of a straight line, before considering the gradient of a curve, as discussed in this Topic.

Rate of change of any variable

In most situations the graph of one variable against the other is not a straight line. In these cases the rate of change of one variable when compared with the other is not constant, but varies according to the position on the curve.

Example A

Q. Find the rate of change of y with respect to x on the graph $y = x^2$ when $x = 2$.

A. The graph is that of $y = x^2$.

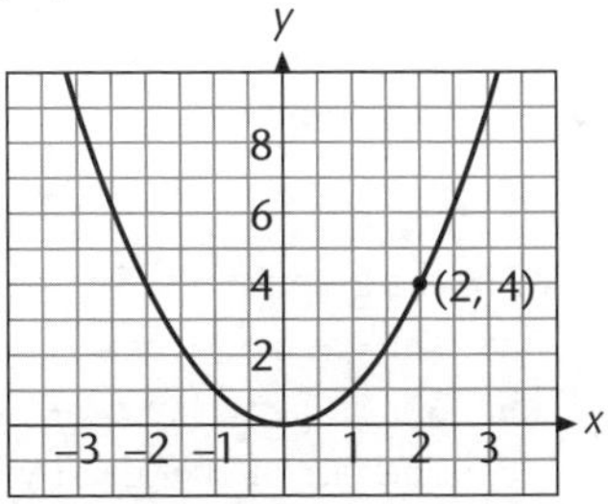

If a magnifying glass with sufficiently powerful magnification was used to examine the graph of $y = x^2$ at the point (2, 4), the graph would appear straight in the neighbourhood of that point.

The gradient or slope of the graph at the point (2, 4) is the gradient of this straight line. This is also the rate of change of y with respect to x when $x = 2$.

If the student lacks a graphical calculator or other device which enables the graph to be 'blown up' at a point so as to appear straight, then the slope may be found using the technique given below.

1. Select a pair of values very close to the x-value in question (in this case 2) one above and one below, eg $x = 1.99$ and $x = 2.01$.
2. Calculate the corresponding y-values (in this case using the rule $y = x^2$).

x	1.99	2.01
y	3.9601	4.0401

3. Calculate the slope of the line joining these two points, using $m = \frac{y_2 - y_1}{x_2 - x_1}$.
Here, $\boldsymbol{m} = \frac{4.0401 - 3.9601}{2.01 - 1.99} = 4$.

Example B

Q. For the function $y = 2x - 4$ calculate:

1. The gradient when $x = 1$.
2. The slope when $x = 3$.
3. The rate of change of y with respect to x, when $x = 2$.

A. 1. The student should recognise that the equation $y = 2x - 4$ is that of a straight line having a constant gradient of 2. Thus at $x = 1$, the gradient is 2 (since gradient = 2 for all values of x).

2. Similarly, as slope and gradient have the same meaning, the slope when $x = 3$ is 2 (slope is constant).

3. The rate of change of y with respect to x is the same as the gradient and will also equal 2.

Thus the answer to all three questions is 2.

[These values can also be found by calculation as in Example A.]

Unit 12.3 Activity 1A: Calculating gradients

1. For the graph of $y = x^2$, calculate the gradient when:
   a. $x = 1$ b. $x = -3$

2. For the graph of $y = x^2$, calculate the slope when:
   a. $x = 0$ b. $x = -2$

3. For the graph of $y = x^2$, calculate the rate of change of y with respect to x when:
   a. $x = 2$ b. $x = -1$

4. For the graph of $y = 3x - 1$, calculate:
   a. The gradient when $x = 2$.
   b. The slope when $x = 4$.
   c. The rate of change of y with respect to x, when $x = 5$.

5. For the graph of $y = 4 - 2x$, calculate:
   a. The gradient when $x = 1$.
   b. The slope when $x = 3$.
   c. The rate of change of y with respect to x, when $x = -2$.

6. For the graph of $y = 4$, calculate:
   a. The gradient when $x = 2$.
   b. The slope when $x = -2$.
   c. The rate of change of y with respect to x, when $x = -1$.

7. For the graph of $y = x^3$, calculate:
   a. The slope when $x = 2$.
   b. The gradient when $x = 3$.
   c. The rate of change of y with respect to x, when $x = -1$.

8. For the graph of $y = 2x^2 + x$, calculate:
   a. The slope when $x = 1$.
   b. The gradient when $x = 2$.
   c. The rate of change of y with respect to x, when $x = -3$.

The practical effects of variable gradients

'Steepness' is a term commonly used by people to describe gradient. The greater the gradient, the greater the 'steepness'. Situations involving variable gradients are often encountered in daily life.

Example C

The driver of a car approaches a hill as shown below:

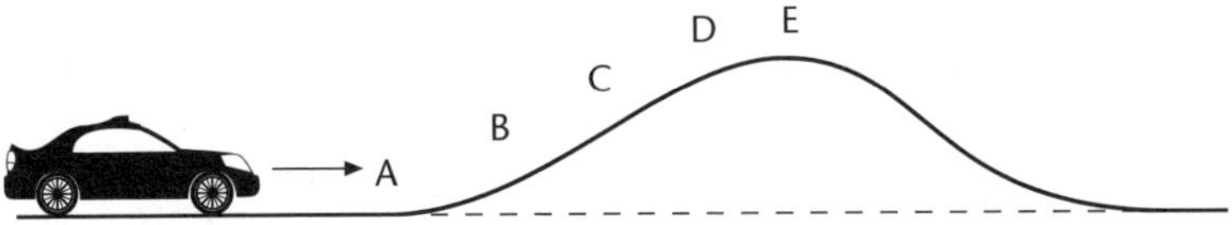

At point A the driver, going at a *constant* speed and never accelerating, is on the 'flat'. As the car moves from A to B to C the hill gets 'steeper and steeper'; without acceleration the driver will notice an increasing slowing effect. As the driver proceeds from C to D to E the hill becomes less and less steep. The gradient is decreasing and the 'slowing' effect will decrease. Once the car passes the top of the hill at E it will start to speed up again as it goes down the hill. In mathematical terms, C, the point of greatest steepness, is called a **point of inflection**.

Unit 12.3 Activity 1B: Variable gradients in practical contexts

1. a. Describe the passage of a car that is going at a constant speed as it begins to travel up the hill. Throughout the journey the driver never accelerates.

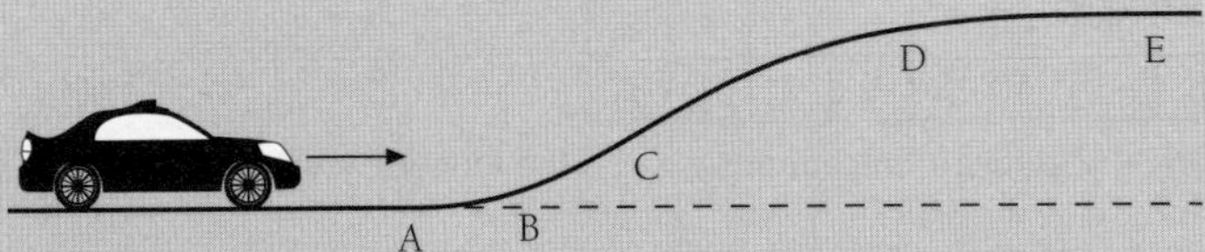

   b. Which point is the point of inflection?

2. Repeat question 1 for the hill below.

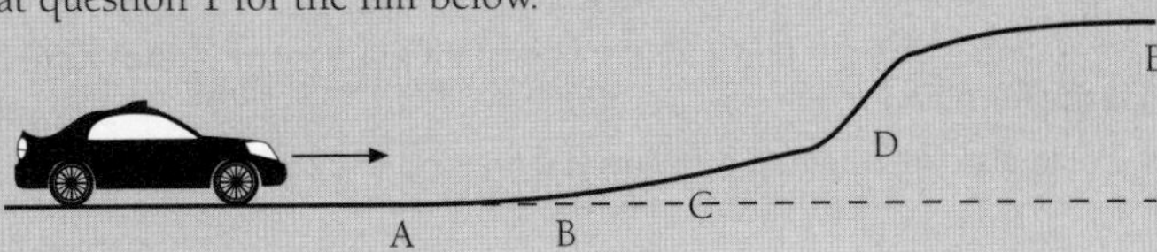

3. Draw a graph that would describe the following trip:
I am proceeding along a flat road. I notice a slight slope at A which increases as I pass from A to B to C. After I pass through C the hill I have come up starts to flatten out as I pass through D and the road after E is once again flat.

4. Draw a graph that describes this child's path down an amusement park slide:
It started gently but quickly got steeper. Then the slide seemed to be getting flatter and flatter till it was actually flat then it suddenly got steeper and steeper again til I landed in the water below.

5. Draw a graph that might describe this company's profits:
We started out making a profit of about K1 000 a day in January. This slowly increased over 6 months when we were making K1 500 a day. Our profits rapidly increased till the end of August. After that the rate of increase of profit decreased, until the beginning of December when the profit was constant at K3 000 a day until the end of the month.

Gradient function

The function that arises from the calculation of the gradient, slope or rate of change at every point of a function's graph is called the **gradient function** for that particular graph.

At points on the graph where there is a tangent, the gradient function gives the slope of the tangent to the curve at that point.

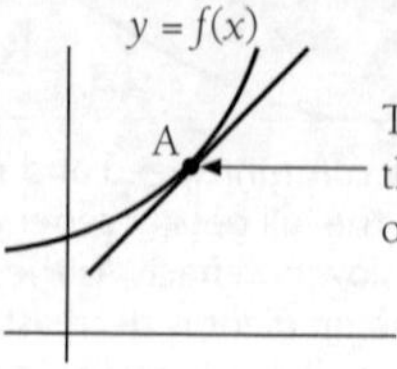

The gradient of the function $f(x)$ at the point A is equal to the gradient of this **tangent** at the point A.

Example D

Q. Find a formula for the gradient function of the graph $y = x^2$.

A. Using a graphical calculator or the method of Example A, the student will obtain the following values:

x	−3	−2	−1	0	1	2	3
y	9	4	1	0	1	4	9
gradient	−6	−4	−2	0	2	4	6

Quite clearly the gradient function for $y = x^2$ is the function $y = 2x$ (for example, when $x = -3$, grad $= 2 \times -3 = -6$).

Example E

Q. Find a formula for the gradient function of $y = 3x + 1$.

A. $y = 3x + 1$ represents a straight line and has a constant gradient of 3, thus its gradient function has the formula $y = 3$.

(This result could also be obtained by using the method in Example D.)

Unit 12.3 Activity 1C: The gradient function

1. For the function $y = x^3$:

a. Copy and fill in this table of values.

x	-3	-2	-1	0	1	2	3
y	-27			0		8	
gradient	-27			0		12	

b. Obtain a formula for the gradient function.

2. For the function $y = x^4$:

a. Copy and fill in a table of values similar to that in question **1**.

b. Find the formula for the gradient function.

3. Repeat the above method to find formulae for the gradient functions of the following:

a. $y = x^5$ **b.** $y = 2x^2$ **c.** $y = x^2 - x$

d. $y = (x + 2)^2$ **e.** $y = 2x + 1$ **f.** $y = 7$

Sketching graphs of gradient functions from the graphs of functions

Example F

Q. 1. Sketch the graph of the gradient function for the function $y = f(t)$ whose graph is shown here. y is measured in thousands and t in months.

2. If t represents time measured in months from January 1 and $f(t)$ represents the number of immigrants arriving in a city, interpret the gradient function.

A. 1. Examination of this graph shows that:

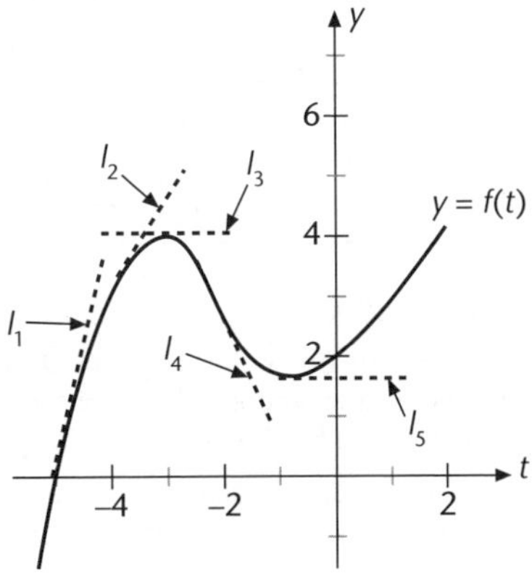

- From –5 to –3, the gradient decreases.

  The gradients of the graph at the points (–5, 0), (–4, 3), (–3, 4) are equal to the gradients of the lines l_1, l_2, l_3, which are the tangents at those points.

  These gradients are decreasing because the tangents are becoming less steep.

- Once the graph has passed through (–3, 4) the gradients become negative, reaching their lowest value at the point (–2, 2.5).

  This lowest value is shown by the gradient of the line l_4 which is not a tangent but which has a slope with the same value as the gradient function at that point. (–2, 2.5) is a point of inflection.

- After (–2, 2.5), the gradients gradually increase, taking the value 0 at the point (–1, 1.7), and increasing from there on.

Using this information the following sketch of the gradient function is drawn:

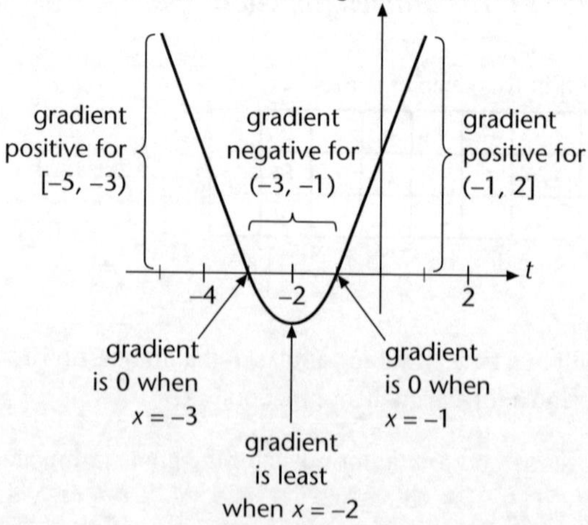

Note: The interval notation [–5, –3) means $-5 \le x < -3$; (–3, –1) means $-3 < x < 1$; (–1, 2] means $-1 < x \le 2$. (A curved bracket *excludes* an endpoint, a square bracket *includes* an endpoint.)

2. From five to three months before 1 January there was an increasing number of immigrants arriving in the city. The rate of increase steadily decreased until October ($t = -3$) when the actual number of immigrants began to decrease, reaching its greatest rate of decrease near the beginning of November ($t = -2$). The number continued to decrease until the beginning of December ($t = -1$), when the number of immigants began increasing again with the rate of increase steadily increasing until at least 1 February ($t = 2$).

Note: If you find it hard initially to sketch freehand the graph of a gradient function then you can get a good sketch as follows:

1. Make a large neat copy of the graph given by plotting points on fine-mesh graph paper.

2. Calculate approximate values for the gradients at those points, then plot those values on another graph.

Example G

A large copy on fine-mesh paper of the graph in Example F allows values of $y = f(t)$ to be read off more easily. A good sketch of the gradient function could then be made by calculating the gradients at –5, –4.5, –4, –3.5, –3, –2.5, –2, . . . , 2, 2.5, 3 and plotting them.

(A less accurate but still reasonable sketch could be obtained by calculating the gradients at –5, –4, –3, –2, –1, 0, 1, 2, 3.)

The gradient at any point is calculated using the gradient formula (see next example).

Example H

Q. Calculate the gradient at the point (–4, 3) for $y = f(t)$ of Example F.

A. Choose a pair of points close to $x = -4$ (here $x = -3.9$ and $x = -4.1$ are chosen).

Let $(-3.9, f(-3.9))$, $(-4.1, f(-4.1))$ be the points corresponding to –3.9 and –4.1, which are read off your graph drawn on fine-mesh paper.

The gradient at (–4, 3) can be estimated to be $\dfrac{f(-3.9)-f(-4.1)}{-3.9-(-4.1)}$.

[gradient = $\dfrac{y_2-y_1}{x_2-x_1}$].

Unit 12.3 Activity 1D: Sketching graphs of gradient functions

1. a. Sketch the gradient function for the graph shown.

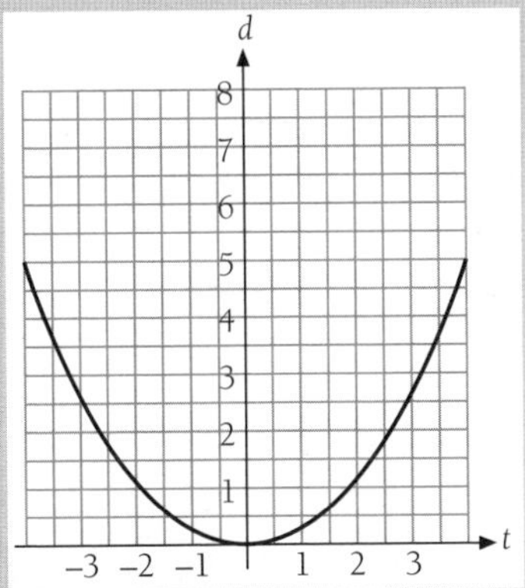

b. If the graph represents the distance *d* from a house in km, with *t* being time measured in hours, interpret the gradient function.

2. a. Sketch the gradient function for the graph shown.

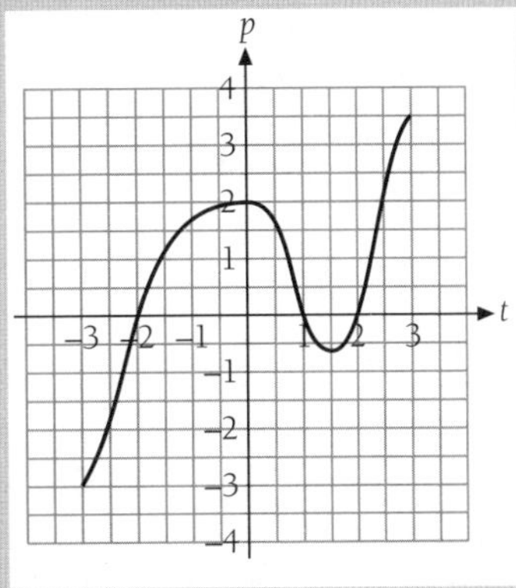

b. The graph represents the profit *p* of a business in units of K1 000 per day and *t* represents time in months measured from 1 March. Interpret the gradient function.

3. a. Sketch the graph of a curve passing through these points:
(–3, 0), (–2, 5), (–1, 8), (0, 9), (1, 8), (2, 5), (3, 0)

b. Sketch the gradient function for the graph drawn.

4. a. Sketch the graph of a curve passing through these points:
(–2, –6), (–1, 0), (–0.5, 0.4), (0, 0), (0.5, –0.4), (1, 0), (2, 6)

b. Sketch the gradient function of the above graph.

5. a. Sketch the graph of a curve passing through the following points:
(–2, 0.25), (–1, 0.5), (0, 1), (1, 2), (2, 4), (3, 8)

b. Sketch the gradient function.

Unit 12.3 Calculus

Topic 2: Derivatives

Topic 2 builds on the information in the previous Topic and looks at differentiation, which is used in many different areas, such as the flow of water for example, or the velocity of a vehicle. In this Topic, we:

- Find the derivatives of expressions given in expanded form that have natural number exponents.
- Use $\frac{dy}{dx}$ and $f'(x)$ notations.
- Evaluate the derivative at a point to find the gradient.

Introduction

Derived functions or **derivatives** are a part of mathematics called **calculus**.

The derivative of a function f at point x in the function's domain is defined to be the value of the gradient function at x.

Example A

It was shown in Example D of the previous Topic that the gradient function of $f(x) = x^2$ is the function $y = 2x$.

By using identical methods it can be shown that the derivative or gradient function of x^3 is $3x^2$ and of x^4 is $4x^3$.

Generally: The derivative of x^n is nx^{n-1}, for all real n

Example B

The derivative of x^5 is $5x^4$ and the derivative of x^{43} is $43x^{42}$.

Note:

1. The derivative of x is 1.

Proof: x can be written as x^1. The derivative of x^1 is $1x^0 = 1$.

2. The derivative of 1 is 0.

Proof: 1 can be written as x^0. The derivative of x^0 is $0x^{-1} = 0$.

Differentiation

The process of *finding the derivative* is called **differentiating**. Functions that *can be differentiated* are said to be **differentiable**. Substitution is used to find the value of the derivative for a particular value in the domain of the function.

Example C

Q. Differentiate x^7 and find the value of the derivative when $x = 2$.

A. Differentiating x^7 gives the derivative $7x^6$.

When $x = 2$, the value of the derivative is $7 \times 2^6 = 448$. [substituting $x = 2$]

$f'(x)$, $\frac{dy}{dx}$ or y' all mean the derivative of the function, $y = f(x)$.
$f'(a)$ means the derivative of the function $y = f(x)$ at $x = a$.

Example D

1. $f(x) = x^{24}$ $\quad \therefore f'(x) = 24x^{23}$

2. $y = x^{11}$ $\quad \therefore \frac{dy}{dx} = 11x^{10}$

3. $u = r^{13}$ $\quad \therefore \frac{du}{dr} = 13r^{12}$

4. $g = h^9$ $\quad \therefore g' = 9h^8$

When a term has a coefficient a, the term is differentiated using the rule:

$$\text{If } y = ax^n \text{ then } \frac{dy}{dx} = nax^{n-1}$$

Example E

1. $y = 5x^3$

$\therefore \frac{dy}{dx} = 5 \times 3 \times x^2$

$= 15x^2$

2. $g(x) = 6x^4$

$\therefore g'(x) = 6 \times 4 \times x^3$

$= 24x^3$

If an expression contains several terms, the expression is differentiated by differentiating each term separately. Brackets may need to be removed first.

Example F

Q. Differentiate: **1.** $y = x^{10} + x^9 - x^7 + 2x$ **2.** $u = 8t^6 + 7t^2 + 5$

A. 1. $\frac{dy}{dx} = 10x^9 + 9x^8 - 7x^6 + 2$

2. $\frac{du}{dt} = 48t^5 + 14t + 0$

$\therefore \frac{du}{dt} = 48t^5 + 14t$

Example G

Q. Differentiate: $(x^2 + 1)^2$

A. $f(x) = (x^2 + 1)^2$

$= x^4 + 2x^2 + 1$ [expanding]

$\therefore f'(x) = 4x^3 + 4x$ [differentiating each term]

Good setting out helps to avoid making mistakes. Always take care *not* to write incorrect mathematical statements.

Example H

Q. Differentiate x^7.

A. $f(x) = x^7$ or $y = x^7$

$\therefore f'(x) = 7x^6$ $\quad \therefore \frac{dy}{dx} = 7x^6$

Note: Writing $x^7 \rightarrow 7x^6$ is acceptable but $x^7 = 7x^6$ is not, because $x^7 \neq 7x^6$.

Unit 12.3 Activity 2A: Differentiation

1. Differentiate:

a. x^5 **b.** x^{22} **c.** x^{48} **d.** $3x^7$

e. $5x^9$ **f.** $10x^2$ **g.** $\frac{x^2}{3}$ **h.** $8x^3 - 3x^2 + 6$

i. $\frac{x^6}{2}$ **j.** $\frac{4x^2}{3}$ **k.** 9 **l.** $5x + 7$

m. $2x^2 + 3x$ **n.** $7x^{10} - \frac{8x^{12}}{5}$ **o.** $(x + 1)^2$ **p.** $(2x + 1)^2$

q. $(3x + 4)$ **r.** $(x^2 + x)^2$ **s.** $(2x - 3)^2$ **t.** $x(x + 3)$

u. $(x + 1)(x + 4)$ **v.** $(x^2 + 1)(x + 3)$ **w.** $x^3(x^2 + 3x + 2)$ **x.** $(x^2 + 1)(x^2 + 2x + 2)$

y. $x(x + 2) + (x + 1)(x + 3)$ **z.** $x^2(x^2 - 3) + \frac{1}{2}(x^2 + 3x)$

2. **a.** $y = 4x^3$. Find $\frac{dy}{dx}$. **b.** $y = 6x^2$. Find $\frac{dy}{dx}$.

c. $f(x) = 5x^2 + 3x - 2$. Find $f'(x)$. **d.** $h(t) = 3t^3 + 4t^2 - 3t + 7$. Find $h'(t)$.

e. $g(u) = 2(u^2 + 3u) + 3(u + 4)$. Find $g'(u)$.

f. $w = 3x^2 + 2t$. Differentiate w with respect to x where t is a constant.

g. $w = 3x^2 + 2t$. Differentiate w with respect to t where x is a constant.

h. $y = 4x^5 + \frac{x}{7}$. Find y'. **i.** $p = 2r(r + 1)(r + 3)$. Find $\frac{dp}{dr}$.

j. $u = (2t + \frac{3}{t})(3t^2 + 4t)$. Find u'.

3. **a.** $f(x) = x^2 + 3x$

i. Find $f'(2)$. **ii.** Find $f'(3.5)$.

iii. Solve the equation $f'(x) = 4$.

b. $g(x) = x^2 - 2x + 3$

i. Find $g(2)$. **ii.** Find $g'(2)$.

iii. Find $g'(3)$. **iv.** Solve $g'(x) = 5$.

c. $h(x) = 2x^3 - 3x^2 + 5x - 11$

i. Find the derivative of h when x is 5.

ii. Find $h'(x)$ when x is 2.

iii. Solve $h'(x) = 5$.

d. $y = 3(x + 1)^2$

i. Find y when $x = -2$. **ii.** Find $\frac{dy}{dx}$ when $x = \frac{1}{2}$.

iii. Evaluate $\left.\frac{dy}{dx}\right|_{-1}$
This means 'find the value of $\frac{dy}{dx}$ when $x = -1$'.

iv. Solve $\frac{dy}{dx} = 5$.

e. $y = \frac{1}{3}x^3 - x^2 - 8x + 2$

i. What is y when $x = 2$?
ii. Find y' when $x = 1$.
iii. Solve $\frac{dy}{dx} = 0$.
iv. Solve $y' = 7$.

Differentiation with negative indices

Expressions involving negative **indices** are differentiated in the same way as expressions with positive indices, taking care with the sign.

Example I

Q. Differentiate the following:

1. $f(x) = x^{-2}$ **2.** $\frac{1}{x^3}$ **3.** $\frac{4}{x^5}$ **4.** $\frac{3}{5x^6}$

A. 1. $f(x) = x^{-2}$

$f'(x) = -2x^{-3}$ $\left(\text{or } \frac{-2}{x^3}\right)$

2. $\frac{1}{x^3} = x^{-3}$

$(x^{-3})' = -3x^{-4}$ $\left(\text{or } \frac{-3}{x^4}\right)$

3. $\frac{4}{x^5} = 4x^{-5}$

$(4x^{-5})' = -20x^{-6}$ $\left(\text{or } \frac{-20}{x^6}\right)$

4. $\frac{3}{5x^6} = \frac{3}{5}x^{-6}$

$\left(\frac{3}{5}x^{-6}\right)' = \frac{-18}{5}x^{-7}$ $\left(\text{or } \frac{-18}{5x^7}\right)$

Unit 12.3 Activity 2B: Differentiation with negative indices

1. Differentiate each of the following:

a. x^{-4} **b.** $\frac{1}{x^5}$ **c.** $\frac{1}{u^7}$ **d.** $\frac{3}{u^4}$ **e.** $\frac{5}{x^6}$

f. $\frac{2}{3y^5}$ **g.** $\frac{3}{4x^2}$ **h.** $\frac{1}{5x^4}$ **i.** $\frac{3}{7x^2}$ **j.** $\frac{2}{x^3} + \frac{3}{x^4}$

2. For each of the following, find $\frac{dy}{dx}$:

a. $y = x^{-5}$ **b.** $y = \frac{1}{x^4}$ **c.** $y = \frac{5}{x^5}$ **d.** $y = \frac{1}{2x^8}$ **e.** $y = \frac{2}{3x^9}$

3. For each of the following, find $f'(x)$:

a. $f(x) = x^{-7}$ **b.** $f(x) = 2(1 + x^{-2})$ **c.** $f(x) = x^3 + \frac{1}{x^2}$

d. $f(x) = \frac{x^2+1}{x^2}$ **e.** $f(x) = (1 + x^{-1})(1 + x^{-2})$

4. Find the derivatives of the following:

a. $\frac{1}{x^5} - x^2$ **b.** $\frac{3}{x} + x^3$ **c.** $(x + \frac{1}{x})(1 + \frac{1}{x})$

d. $(x + \frac{1}{x})^2$ **e.** $\frac{2}{x^2}(3 - 2x + x^2)$

Unit 12.3 Calculus

Topic 3: Tangents to curves

Topic 3 builds on the information in the previous two Topics and examines tangents to a curve. As advised at the beginning of Topic 1 in this Unit, it is important to have a good understanding of the Grade 11 content (Unit 11.2 'Graphs and Functions') as a basis before studying Unit 12.3 'Calculus'.

The particular objective of Topic 3 is to:

- Evaluate the derivative at a point to find the gradient.
- Find the equation of a tangent to a curve.

Gradients of tangents

The following result from Topic 2 'Derivatives' is extremely important:

The gradient of the tangent to the curve $y = f(x)$ at $(a, f(a))$ is $f'(a)$.

Example A

Q. Find the gradient of the tangent to the curve, $y = x^2$, when $x = 2$.

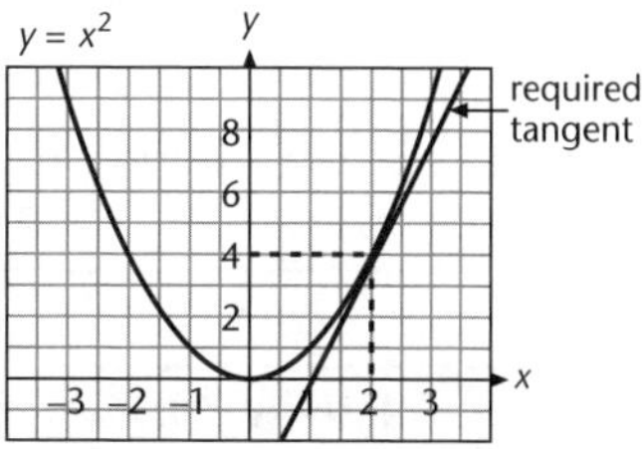

A. The gradient of the tangent at x is $\frac{dy}{dx} = 2x$ [differentiating]

$\therefore$ the gradient when $x = 2$ is $2 \times 2 = 4$ [substituting $x = 2$ into $2x$]

Only the x-coordinate of a point is required in order to evaluate the gradient of the tangent at a point (x, y).

Example B

Q. Find the gradient of the tangent to the curve $y = x^3 - 5x^2 + 10x - 12$ at the point $(1, -6)$.

A. The gradient of the tangent is $y' = 3x^2 - 10x + 10$ [differentiating]

At the point $(1, -6)$, $x = 1$ so $y' = 3(1)^2 - 10(1) + 10$

$= 3$

The gradient of the tangent at $(1, -6)$ is 3.

Example C

Q. Find the point on the curve $y = x^2$ where the gradient of the tangent is 3.

A. Gradient of the tangent is $2x$. [differentiating x^2]

$\therefore$ point with tangent having gradient 3 satisfies $2x = 3$.

$\therefore\ x = 1.5$ [dividing]

at $x = 1.5$, $y = (1.5)^2$ [since $y = x^2$]

$= 2.25$

The point on $y = x^2$ where the tangent has gradient 3 is (1.5, 2.25).

Example D

Q. Find the x-coordinates of the points on the curve $y = x^3 - 3x^2 - 19x + 10$ where the tangent has gradient 5.

A. The gradient of the tangent is $3x^2 - 6x - 19$ [differentiating $x^3 - 3x^2 - 19x + 10$]

Setting the gradient equal to 5 gives:

$3x^2 - 6x - 19 = 5$

$3x^2 - 6x - 24 = 0$ [rearranging]

$x^2 - 2x - 8 = 0$ [dividing by 3]

$(x + 2)(x - 4) = 0$ [factorising]

$x = -2$, or 4 [solving for x]

Unit 12.3 Activity 3A: Gradients of tangents

1. Find the gradient of the tangent to each of the following curves at the given point:

a. $y = x^2 + x$ when $x = 2$ **b.** $y = x^2 - 2x$ when $x = 1$

c. $y = x^2 + x$ when $x = -2$ **d.** $y = x^3 - x^2$ when $x = 1$

2. Find the gradient of the tangent to each of the following curves at the point indicated:

a. $y = x^2$ at $(-2, 4)$ **b.** $y = 2x^2$ at $(-1, 2)$

c. $y = \frac{x^2}{2}$ at $(3, 4.5)$ **d.** $y = \frac{x^3}{2} + 1$ at $(1, 1.5)$

3. Find the point or points on each of the following graphs where the tangents have the gradient indicated:

a. $y = x^2$, gradient is 3 **b.** $y = \frac{1}{3}x^3$, gradient is 4

c. $y = 2x^3$, gradient is -1 **d.** $y = 2x + 3$, gradient is 2

Equations of tangents and normals

The equation of the tangent at a given point

> The equation of the tangent to a curve $y = f(x)$ at a point $p(x_1, y_1)$ can be found using the equation $y - y_1 = m(x - x_1)$, where $m = f'(x_1)$

Example E

Q. Find the equation of the tangent to the curve $f(x) = 2x^3 - x^2$ at the point (–1, –3).

A. The gradient of the curve at the point (–1, –3) is given by $f'(-1)$:

$f'(x) = 6x^2 - 2x$

so $f'(-1) = 6(-1)^2 - 2(-1) = 6 + 2 = 8$; the gradient ($m$) at (–1, –3) is 8

Substituting in $y - y_1 = m(x - x_1)$:

$y - (-3) = 8(x - (-1))$

$y + 3 = 8(x + 1)$

$y = 8x + 5$

The equation of the tangent at (–1, –3) is $y = 8x + 5$

The equation of the normal at a given point

The **normal** to a curve $y = f(x)$ at a point P is a straight line that is **perpendicular** (at right angles) to the tangent at point P.

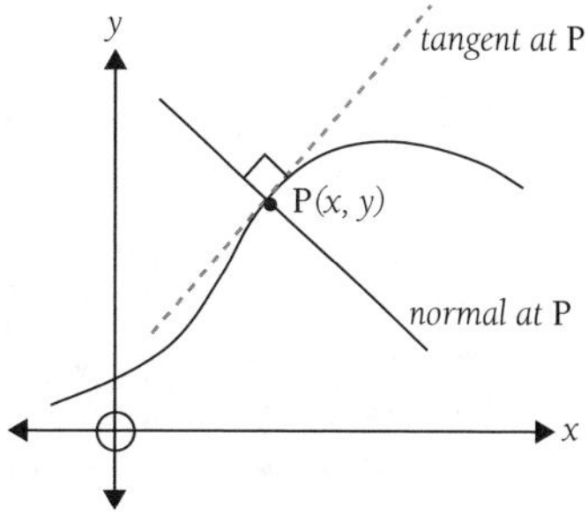

> The equation of the normal to a curve $y = f(x)$ at a point $P(x_1, y_1)$ can be found using the equation $y - y_1 = m(x - x_1)$ where $m = \dfrac{-1}{f'(x_1)}$

Example F

Q. Find the equation of the normal to the curve $f(x) = 2x^3 - x^2$ at the point (–1, –3).

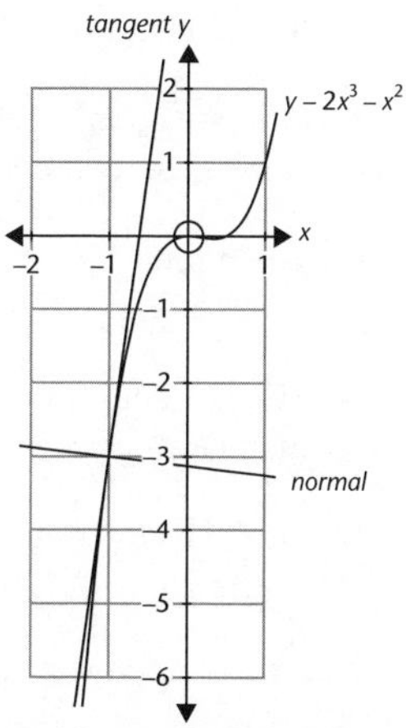

A. The gradient of the curve at the point (–1, –3) is given by $f'(-1)$:

$f'(x) = 6x^2 - 2x$

so $f'(-1) = 6(-1)^2 - 2(-1) = 6 + 2 = 8$

The gradient of the tangent at the point (–1, –3) is 8 so the gradient of the normal at the point (–1, –3) is $-\dfrac{1}{8}$

The equation of the normal is found by substituting in

$y - y_1 = m(x - x_1)$

$y - (-3) = -\dfrac{1}{8}(x - -1)$

$y + 3 = -\dfrac{1}{8}(x + 1)$

$8y + 24 = -x - 1$

$x + 8y + 25 = 0$

The equation of the normal at (–1, –3) is $x + 8y + 25 = 0$

Unit 12.3 Activity 3B: Equations of tangents

1. Find the equation of the tangent to the curve at the given points:

a. $y = x^2 - 3x + 2$ at $(-1, 6)$ **b.** $y = 3x^2 - 4x - 5$ at $(2, -1)$

c. $y = x^3 + x^2$ at $(-1, 0)$ **d.** $y = 4x^2 - 5$ at $(\frac{1}{2}, -4)$

2. For each of the following find the equation of the normal to the curve at the given point:

a. $y = 2x^2 - 5$ at the point $(2, 3)$ **b.** $y = 4x^2 - 3x + 2$ at the point $(-1, 9)$

c. $y = x^3 - 2x^2 + 3x - 1$ at the point $(2, 5)$

3. Find the equation of the tangent and the normal to the curve $y = 2x^3 + 2x - 7$ at the point where $x = -1$.

4. Find the equation of the tangent to the curve $y = x^2 - 5x + 3$ at the point where the gradient is 1.

5. Find the equation of the normal to the curve $y = x^2 - 3x + 1$ at the point where the gradient is 2.

6. Find the equations of the tangents to the curve $y = x^3 - 6x$ at the points where the gradient is 6.

7. The normal to the curve $y = 3x^2 - 2x + 1$ at the point $(1, 2)$ intersects with the curve at another point. Find the coordinates of this other point of intersection. (–5/12, 113/48).

8. **a.** Find the equation of the tangent to the curve $y = x^3 - 9x^2 + 20x - 8$ at the point $(1, 4)$.

b. At what points of the curve is the tangent parallel to the line $4x + y - 3 = 0$?

9. Find the coordinates of the point on the curve $y = 2x^2 - 5x$ at which the tangent to the curve is perpendicular to the tangent at the point $(1, -3)$.

10. The normal to the curve $y = x^2 - 4x$ at the point $(3, -3)$ cuts the x-axis at A and the y-axis at B. Find the equation of the normal and the coordinates of A and B.

11. The tangent to the curve $y = 2x^2 + ax + b$ at the point $(-2, 11)$ is perpendicular to the line $x - 2y + 7 = 0$. Find the values of a and b.

12. Find the equations of the normals to the parabola $y = \frac{1}{4}x^2$ at the points $(-2, 1)$ and $(-4, 4)$. Show that the point of intersection of these two normals lies on the parabola.

Unit 12.3 Calculus

Topic 4: Increasing and decreasing functions: turning points

In topic 4 we continue with straightforward derivatives and integrals:

- Locate turning points where $f'(x) = 0$.
- Determine the nature of turning points.

Increasing and decreasing functions

A function $y = f(x)$ is **increasing** on a set of values if, as the x-values increase, the function's y-values increase.

A function $y = f(x)$ is **decreasing** on a set of values if, as the x-values increase, the function's y-values decrease.

Example A

The graph of $y = f(x)$ is shown. At the points A (where $x = 1$) and B (where $x = 8$), the curve is neither increasing nor decreasing. The values $x = 1$ and $x = 8$ are called **stationary values** of the function.

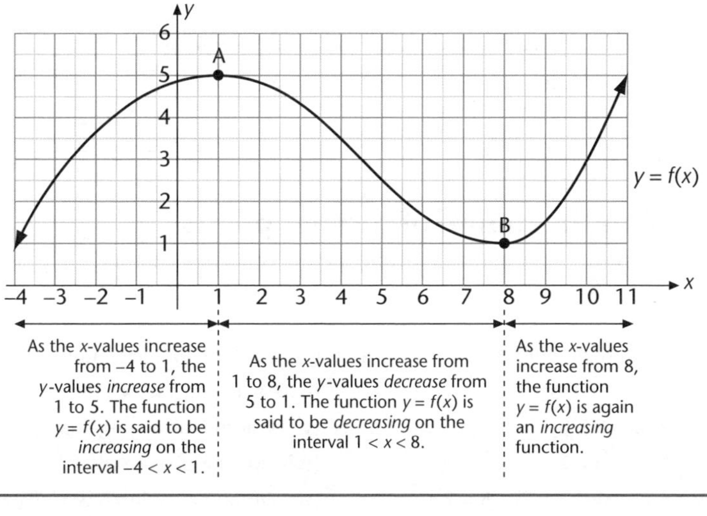

Turning points

A point at which a function changes from increasing to decreasing or vice versa is called a **turning point** (abbreviation **TP**). Turning points are stationary values of the function.

If the function changes from increasing to decreasing, the turning point is called a **local maximum**. In Example A, the point $A(1, 5)$ is a local maximum because the y-value at $x = 1$ is the maximum value of the function for all x-values close to 1.

Note: When the x-value exceeds about 11, the value of the function becomes greater than 5.

Local minimums are turning points where the function changes from being a decreasing function to being an increasing function. In Example A, the point B (8, 1) is a local minimum.

Tangents to graphs

In the diagram below, several tangents to the curve $y = g(x)$ are drawn.

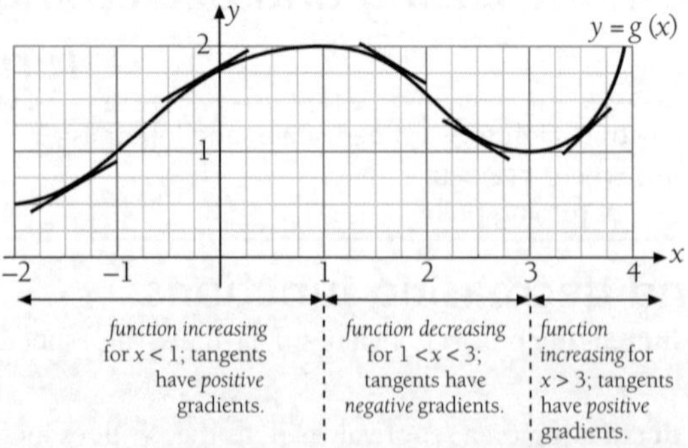

Any tangent drawn at a point on the graph of a function which is increasing is 'tilted up' and hence has a *positive* gradient. Thus at points on the graph of a function which is increasing, the derivative will be positive. Similarly when the function is decreasing the derivative is negative. This can be written:

> If $y = f(x)$ is an increasing function then $f'(x) > 0$.
>
> Similarly, if $y = f(x)$ is decreasing then $f'(x) < 0$.

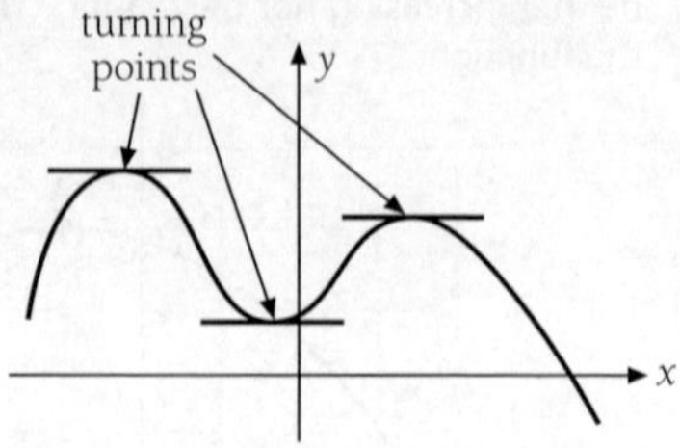

At turning points, tangents are horizontal and have a gradient of 0.

> At turning points, $f'(x) = 0$.

Determining the nature of turning points

A graph will show clearly whether a turning point is a local maximum or minimum.

Example B

Q. Find the turning point(s) of $y = 2x^2 - 8x + 4$ and determine their nature.

A. $y' = 4x - 8$ [differentiating $y = 2x^2 - 8x + 4$]

At TP, $4x - 8 = 0$ [$y' = 0$ at the turning point(s)]

$\therefore\ x = 2$ [solving for x]

At $x = 2$, $y = 2 \times 2^2 - 8 \times 2 + 4$ [substituting $x = 2$ in $y = 2x^2 - 8x + 4$]

$\therefore\ y = -4$

$\therefore$ coordinates of the TP are $(2, -4)$.

$y = 2x^2 - 8x + 4$ is a **quadratic** function and its graph is a **parabola**.

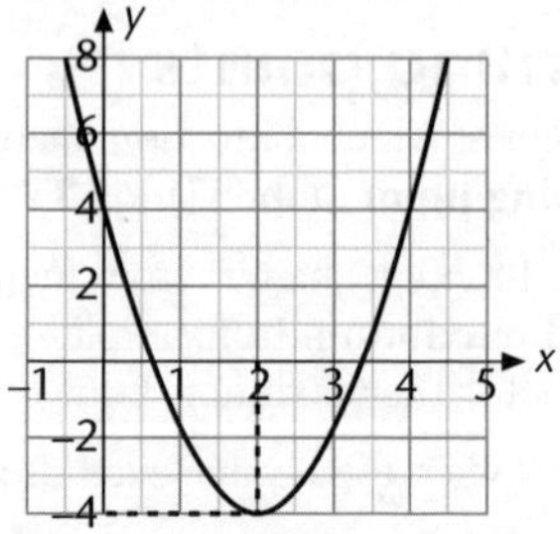

The **y-intercept** is (0, 4) and from the diagram, (2, −4) is a local minimum.

Example C

Q. **1.** Find the turning points of $y = \frac{1}{3}x^3 - x^2 - 8x + 3$ and determine their nature.

2. For what set of values is this function decreasing?

A.

$$y' = x^2 - 2x - 8 \qquad \text{[differentiating } y\text{]}$$

$$y' = 0 \text{ at any TP}$$

$$\therefore\ x^2 - 2x - 8 = 0$$

$$\therefore\ (x - 4)(x + 2) = 0 \qquad \text{[factorising]}$$

$$x = -2 \text{ or } 4$$

when $x = -2$, $y = \frac{1}{3}(-2)^3 - (-2)^2 - 8(-2) + 3$ [substituting $x = -2$ in y]

$$= -\frac{8}{3} - 4 + 16 + 3$$

$$= 12\frac{1}{3} \qquad \therefore \text{ TP is } (-2, 12\tfrac{1}{3})$$

when $x = 4$, $y = \frac{1}{3} \times 4^3 - 4^2 - 8 \times 4 + 3$ [substituting $x = 4$ in y]

$$= -23\frac{2}{3} \qquad \therefore \text{ TP is } (4, -23\tfrac{2}{3})$$

Substituting $x = 0$ in $y = \frac{1}{3}x^3 - x^2 - 8x + 3$ shows that the function has a y-intercept of (0, 3).

A sketch of the graph shows that $(-2, 12\frac{1}{3})$ is a local maximum and $(4, -23\frac{2}{3})$ is a local minimum.

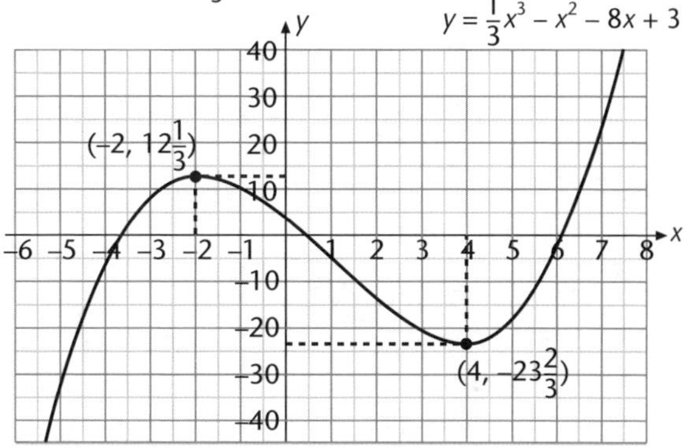

2. The function is *decreasing* for all values in the interval $-2 < x < 4$.

Another technique for determining the nature of turning points involves evaluating the gradient function in the neighbourhood of the turning point.

Example D

Q. Find the turning points of $y = \frac{1}{3}x^3 - x^2 - 8x + 3$ and determine their nature.

A. From Example C, the turning points occur at $x = 2$ and $x = 4$. The table shows the values of the gradient function in the neighbourhoods of these two x-values.

x	−3	−2	−1	3	4	5
$f'(x)$	+	0	−	−	0	+

[for example $f'(-3) = (-3)^2 - 2 \times (-3) - 8 = 7$ (positive)]

Between $x = -3$ and$x = -1$, the graph changes from increasing (positive gradient) to decreasing (negative gradient), so $x = -2$ is a maximum.

Between $x = 3$ and $x = 5$, the graph changes from decreasing (negative gradient) to increasing (positive gradient), so $x = 4$ is a minimum.

So $(-2, 12\frac{1}{3})$ is a maximum, and $(4, -23\frac{2}{3})$ is a **minimum turning point** for the curve $y = \frac{1}{3}x^3 - x^2 - 8x + 3$.

Note: The nature of the turning points can also be determined by examining the values of the function in the neighbourhood of the turning points. For example, the table below shows that there is a minimum value at $x = 4$.

x	3	4	5
$f(x)$	-21	$-23\frac{2}{3}$	$-20\frac{1}{3}$

Unit 12.3 Activity 4A: Turning points

For each of the following functions, find the coordinates of the turning points and determine their nature. State the values for which the functions are increasing and those for which the functions are decreasing.

1. $y = x^2 - 4x + 1$

2. $y = 6 - 8x - x^2$

3. $y = 5 + 3x - 2x^2$

4. $y = 4 - 3x - 2x^2 - \frac{1}{3}x^3$

5. $y = 8 - 6x - \frac{5}{2}x^2 - \frac{1}{3}x^3$

6. $y = 2x^3 + 12x^2 + 18x - 1$

7. $y = x^3 - 12x + 2$

8. $y = 27x - x^3 + 2$

9. $y = -3x^3 - 3x^2$

10. a. State whether the curve $y = 5 - 4x^2 - x^3$ is increasing, decreasing or stationary when:

i. $x = -2\frac{2}{3}$ **ii.** $x = 2$ **iii.** $x = -1\frac{1}{2}$ **iv.** $x = 0$

b. Find the nature of any turning points.

11. f is a real number function such that $f(x) = (x + 1)(x - 1)^2 = x^3 - x^2 - x + 1$.

a. Sketch a graph of f, marking the intercepts on the axes.

b. Differentiate to find $f'(x)$.

c. Find the values of x that give turning points of f, stating the nature of these turning points.

d. For what set of values of x is f a decreasing function?

e. Find $f(0)$.

f. Find $f'(0)$.

g. Find the equation of the tangent to the graph of f, at $x = 0$.

12. $y = Ax^2 + Bx + C$ is a quadratic function. Its derivative, y', has a value of -2 when $x = 1$ and a value of 2 when $x = 2$. In addition, the point $(1, -1)$ lies on the graph of this function. Find A, B and C.

Alternative tests for turning points

The following are alternative tests for distinguishing maximum and minimum points of a graph and **stationary points of inflection**. In particular, these tests should be applied to the point $x = a$ if it is found that $f'(a) = f'(a) = 0$ (ie **test failure** occurs). The sign of the gradient is considered in the neighbourhood of the stationary point.

Test for a maximum point

If A is a maximum point at $x = a$, then the gradient of the graph changes from positive (+) to negative (–) as x increases, as shown in the figure. The test for a maximum point is made by using a numerical value for x just less than a and one for x just greater than a, and substituting each value into the derivative $f'(x)$.

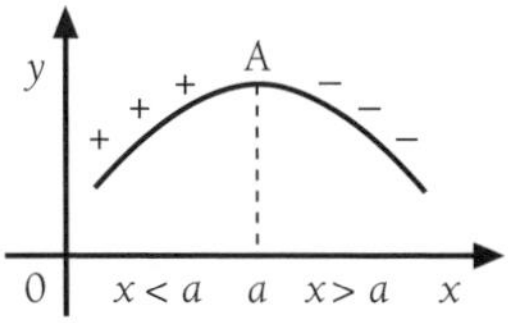

Sign of gradient

These values for the gradients are entered in the table alongside to verify the maximum point at $x = a$.

x	$x < a$	$x = a$	$x > a$
$f'(x)$	$f'(x) > 0$	$f'(a) = 0$	$f'(x) < 0$
Slope		maximum	

Test for a minimum point

If B is a minimum point at $x = b$, then the gradient of the graph changes from negative (–) to positive (+) as x increases, as shown in the figure. The test for a minimum point is made using a numerical value for x just less than b, and one for x just greater than b, and substituting each value into the derivative $f'(x)$.

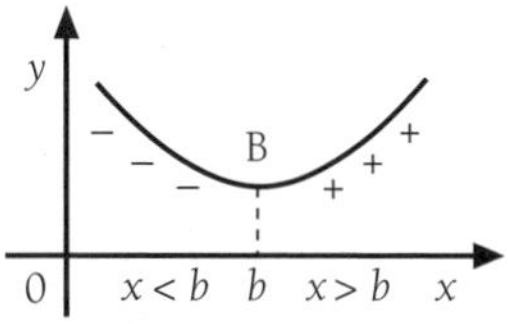

Sign of gradient

These values for the gradients are entered in the table alongside to verify the minimum point at $x = b$.

x	$x < b$	$x = b$	$x > b$
$f'(x)$	$f'(x) < 0$	$f'(b) = 0$	$f'(x) > 0$
Slope		minimum	

Test for a stationary point of inflection

If C is a stationary point of inflection at $x = c$, then the gradient of the graph stays positive (or negative) on both sides of the stationary point, as x increases.

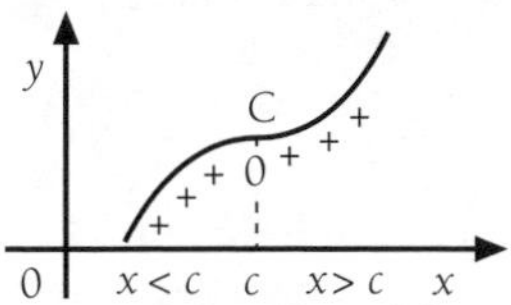

Sign of gradient (+)

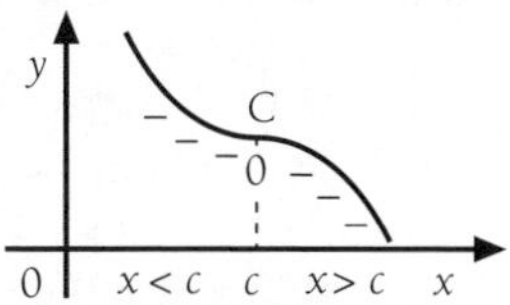

Sign of gradient (–)

x	$x < c$	$x = c$	$x > c$
$f'(x)$	$f'(x) > 0$	$f'(c) = 0$	$f'(x) > 0$
Slope		stationary point of inflection	

x	$x < c$	$x = c$	$x > c$
$f'(x)$	$f'(x) < 0$	$f'(c) = 0$	$f'(x) < 0$
Slope		stationary point of inflection	

Polynomials

A **polynomial graph** will have equation $y = a_n x^n + a_{n-1} x^{n-1} + \dots + a_3 x^3 + a_2 x^2 + a_1 x + a_0$.

It is useful to have some idea of the graphs of polynomials. The following table shows examples that *could* represent some polynomials.

Degree	Graph	Description
2 (quadratic)	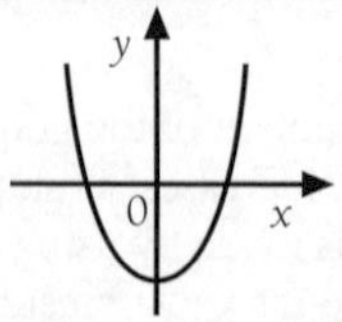	At most, one turning point. No points of inflection.
3 (cubic)	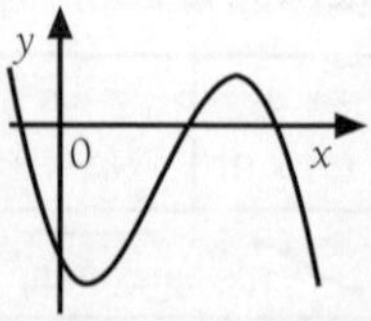	At most, two turning points. One point of inflection.
4 (**quartic**)	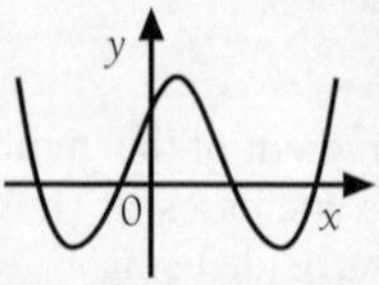	At most, three turning points. Two points of inflection.

Unit 12.3 Activity 4B: Curve sketching

1. Sketch a continuous curve $y = f(x)$ with these properties:

  a. $f(3) = 2, f'(3) = 0$, $f(x)$ is **concave down** for all x.

  b. $f'(3) = f'(5) = 0$, $f(5) < 0$ and $f(3) = 0$.

  c. $f(3) = f'(3) = f''(3) = 0$, $f(x)$ is **concave up** for $x > 3$ and $f(x)$ is concave down for $x < 3$.

2. Find the coordinates of the turning points of $f(x) = 5 + 12x - x^3$.

3. $f(x) = x^4 - 8x^2 + 16$ has three turning points. Find their coordinates and identify which are maximum points and which are minimum points. Show full working.

4. The equation of a curve is $y = x(x - 2)(x + 3)$.

  a. State the **x-intercepts** and the y-intercept.

  b. Using calculus, find the x-coordinates of the maximum and minimum points and distinguish between them using suitable tests.

  c. Sketch the curve showing the main features.

5. $y = 2x(x - 2)(x + 2)$ has two turning points.

  a. Find their coordinates and identify which is which, using suitable tests.

  b. Identify the x-intercepts and y-intercept.

  c. Sketch the curve showing its properties.

6. The equation of a curve is $f(x) = 2x(2 + x)(3 - x)$.

a. Find the x- and y-intercepts of the graph and explain how you obtained your answers.

b. Find the coordinates of the maximum and minimum points.

c. Sketch the curve.

7. Find and identify the stationary points of the function $f(x) = 2x^5 - 5x^4 + 20$.

8. Find the coordinates of the turning points for $y = x(x - 4)^4$.

9. The figure shows the graph of $f(x) = x^2(3 - x)$. The dotted line AB is a tangent to the curve at B, and $y = 0$ is a tangent to the curve at the origin. Calculate the coordinates of A and B (do not do a scale drawing).

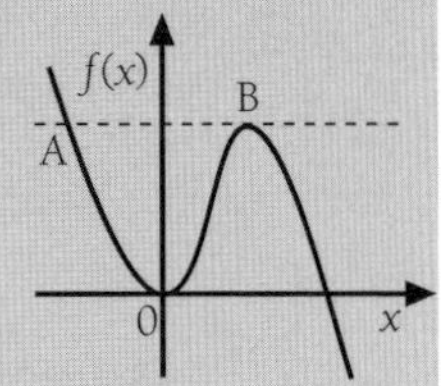

10. The equation of a polynomial curve of degree 4 is $y = 3x^4 - 4x^3$.

a. Find the coordinates of all the stationary points and distinguish between them. Justify your answers.

b. Explain why the x-intercepts are 0 and $\frac{4}{3}$.

c. Find the non-stationary point of inflection. Justify your answer.

d. Sketch the curve showing the main features.

e. Write down the values of x where the curve is concave down.

11. Find the coordinates of the stationary points for the function $f(x) = x^4 - 9x^2$ and show tests for distinguishing the maximum, minimum and points of inflection. Find all inflection points, the intercepts, and hence sketch the curve.

12. a. Find and identify the stationary points of the function $f(x) = x^5 - x$. If the curve crosses the x-axis at $x = a$, $x = b$, $x = c$, find a, b and c. Hence, sketch the curve.

b. From the graph of $f(x) = x^5 - x$ find all the values of x for which:

i. $f(x)$ is decreasing. **ii.** $f(x)$ is concave up.

13. Investigate the turning points for the curve $y = 2x^4 + px^2$.

a. For what values of p does this curve have three turning points?

b. If $p = -1$ find the turning points and sketch the curve.

Unit 12.3 Calculus

Topic 5: Optimisation (maxima and minima problems)

In Topic 5 we look at finding maximum or minimum values of a variable quantity:

- Apply calculus techniques to solve optimisation problems.

Maxima and minima

Optimisation involves finding the maximum or minimum values of a variable quantity. If the quantity varies according to a particular formula, turning points can be located to find these optimal values.

Example A

Q. The distance s of a boat from a port is given by $s = 36t - 0.1t^2$, where s is in metres and t is in minutes. Find the greatest distance of the ship from the port.

A. s is a **quadratic function** of t. The following sketch of the graph shows an inverted parabola.

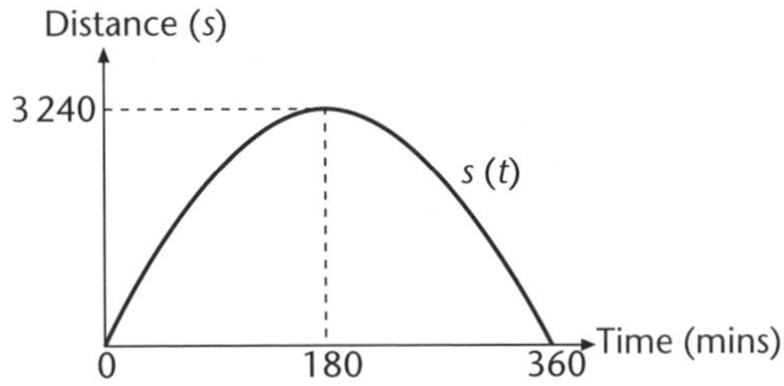

The maximum corresponds to the greatest distance from the port.

Differentiating s gives $s' = 36 - 0.2t$

At the maximum distance, $s' = 0$

$\therefore\ 36 - 0.2t = 0$

$t = 180$ minutes [solving]

$\therefore$ maximum distance $= 36 \times 180 - 0.1 \times 180^2 = 3\,240$ m [substituting in s]

In many problems, the formula for the quantity to be optimised is not given directly. The information given in the problem must be used to develop this formula. Considering a few simple cases first can help with this.

Example B

Q. Daniel has 6 m of wire. He bends it to give three sides of a rectangular shape so that two of these sides have the same length. Find the length of each side so that the shape has the largest possible area.

A. Some possible lengths and the resulting areas are shown:

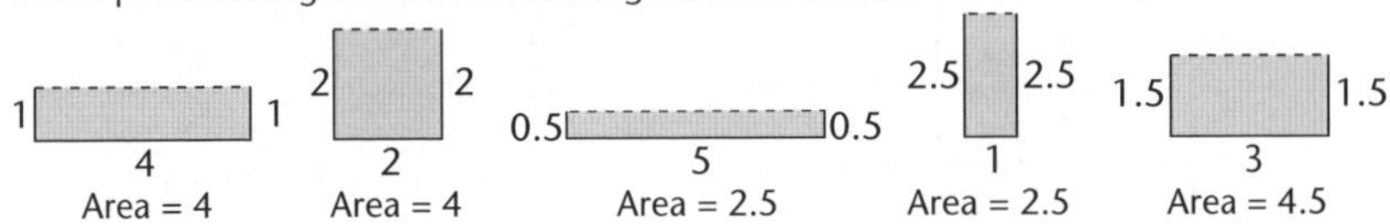

The sketches suggest a maximum area of 4.5 m^2 when the equal two sides are 1.5 m and the third side is 3 m.

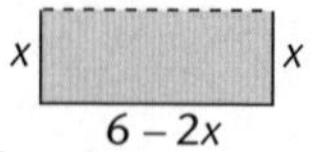

This can be confirmed using calculus:

Let the two equal sides have length x, and the area of the shape be A.

$\therefore$ Length of the third side $= 6 - 2x$ [since total length = 6]

$$\therefore \text{ area, } A = x(6 - 2x) \quad \text{[area is length} \times \text{breadth]}$$

$$= 6x - 2x^2$$

$$\therefore A' = 6 - 4x \quad \text{[differentiating } A\text{]}$$

Area, A, is a maximum when $A' = 0$

$$\therefore 6 - 4x = 0$$

$$\therefore x = 1\frac{1}{2}$$

$$\therefore \text{ maximum area } A = 1\frac{1}{2}\left(6 - 2 \times 1\frac{1}{2}\right) \quad \text{[substituting } x = 1\frac{1}{2} \text{ into } A = x(6 - 2x)\text{]}$$

$$= 4\frac{1}{2} \text{ m}^2$$

Use of spreadsheets

Example C

Area of rectangle problem (Example B) done using a **spreadsheet**:

This pair of worksheets and charts shows how a maximum or minimum problem can easily be tackled using a spreadsheet:

x	$6 - 2x$	Area
0	6	0
0.5	5	2.5
1	4	4
1.5	3	4.5
2	2	4
2.5	1	2.5
3	0	0

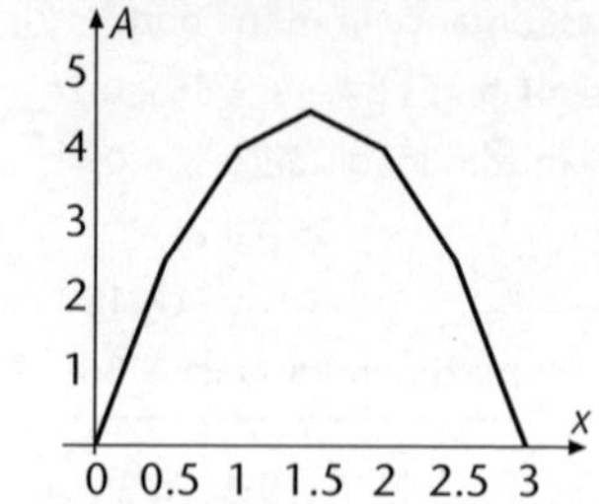

x	$6 - 2x$	Area
1.2	3.6	4.32
1.3	3.4	4.42
1.4	3.2	4.48
1.5	3	4.5
1.6	2.8	4.48
1.7	2.6	4.42
1.8	2.4	4.32

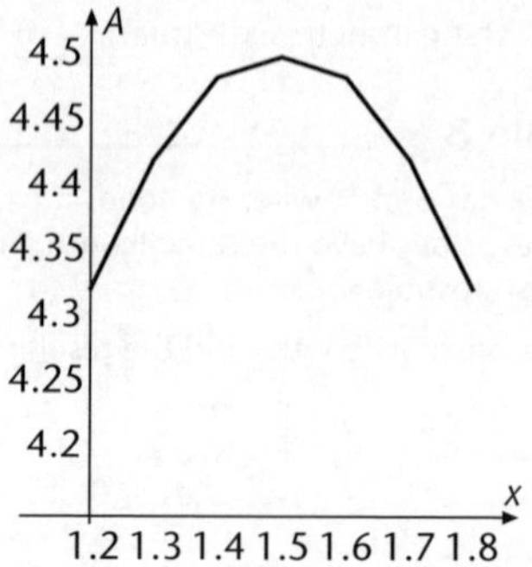

This set of diagrams was created in about four minutes using the Excel spreadsheet. The first worksheet shows there is a maximum area near (or at) 1.5. This suggests the next set of x-values to investigate (values around 1.5) in the search for that value of x which gives the maximum area.

In this case it is easy to see the answer is 1.5. In other cases, though, the value may be some decimal whose accuracy can be made as precise as is necessary by doing repeated spreadsheets.

The formula to be optimised must be expressed in terms of one variable. Use substitution where necessary for this.

Example D

Q. Find the minimum positive value of $x + y$ if $xy = 9$.

A. Let $V = x + y$

but, $y = \frac{9}{x}$ [making y the **subject** of $xy = 9$]

$\therefore V = x + \frac{9}{x}$ [substituting into $V = x + y$]

$= x + 9x^{-1}$ [since $\frac{1}{x} = x^{-1}$]

For a minimum value, $V' = 0$

$\therefore 1 - 9x^{-2} = 0$ [differentiating V and putting $V' = 0$]

$\therefore 1 - \frac{9}{x^2} = 0$ [since $x^{-2} = \frac{1}{x^2}$]

$\therefore x^2 - 9 = 0$ [multiplying by x^2]

$\therefore x = 3$ [solving for a positive value of x]

and $y = 3$ [substituting into $y = \frac{9}{x}$]

$\therefore$ The minimum value of $x + y$ is 6. [substituting $x = y = 3$]

Unit 12.3 Activity 5A: Maxima and minima

1. The distance D (in metres) of a boat from a port is given by $D = 48t - 0.08t^2$ where t is in minutes. Find the maximum distance of the boat from the port.

2. The volume of water (in litres) in a tank is given by $V = 360t - 6t^2$, where t is measured in minutes.
   a. Find the maximum volume of water in the tank.
   b. At what time is the maximum volume reached?

3. The speed of a boat in kilometres per hour is given by $12t^2 - 60t + 90$ over a period of 6 hours (t is in hours). For this six-hour period find:
   a. the maximum speed.
   b. the minimum speed.

4. The depth, d, of water at a pier t hours after sunrise is given by $d = t^3 - 12t^2 + 95t$, where d is measured in centimetres from a mark on the pier.
   a. When is high tide and when is low tide?
   b. At these times what is the depth of water from the mark on the pier?

5. What is the maximum possible area bounded by a fence of length 100 m if the fence forms a rectangle with a riverbank (which is not fenced), as shown in the diagram?

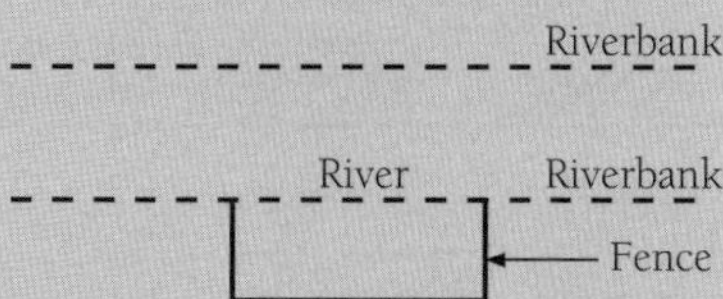

6. What is the maximum possible area of a trapezium made by bending a piece of wire 10 cm long, as shown in the diagram (4th side, not wire, shown dotted)? [Note the two right-angled bends.]

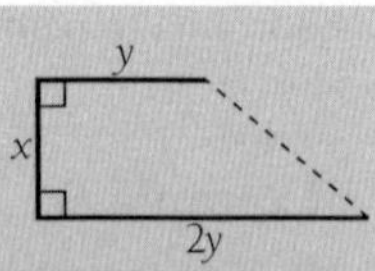

7. A villager is about to build two large gardens alongside a river, one for pigs and the other for vegetables, as shown in the diagram. He has only 240 metres of fencing available, but he wants to enclose the largest possible area.

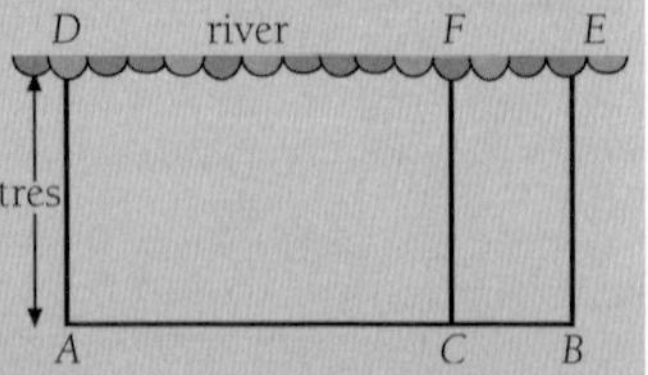

a. Suppose the width of the strip of land is x metres. Show that the length AB is $(240 - 3x)$ metres and write down an expression for the total area, A m^2, of the two gardens.

b. Find a value of x for which this area is a maximum, and find the maximum area.

8. A farmer constructs a rectangular enclosure by means of a fence of total length 150 m and two gates of length 3 m. These are situated as shown in the following diagram. Find the maximum area enclosed.

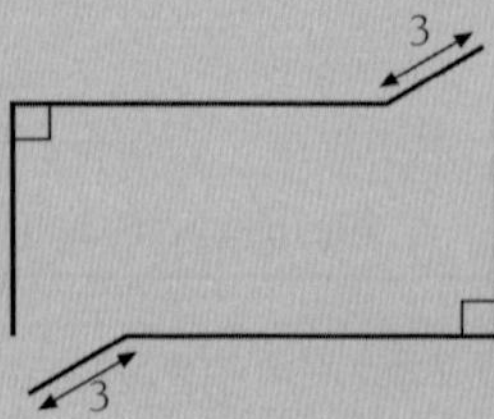

9. What is the maximum value of XY if $X + 2Y = 10$?

10. What is the maximum value of UV if $3U + 2V = 12$?

11. A rectangle 10 cm by 8 cm has four squares of side length x cm removed from each corner, as shown in the diagram.

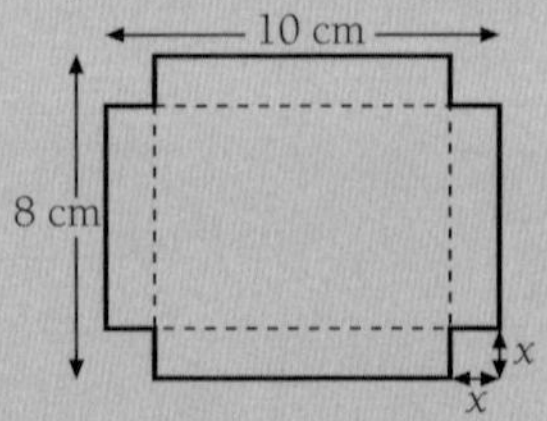

The sides are folded up to make a box as shown.

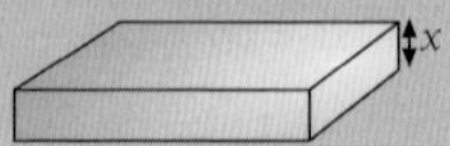

a. Find a formula for the volume of the box in terms of x.

b. Find the value of x which gives the maximum volume.

12. A piece of string of length L is cut into two pieces. One is formed into a square and the other into a circle. What are the dimensions of each shape in order that the combined area is a minimum?

Unit 12.3 Calculus

Topic 6: Rates of change

Many practical problems involve rates of change and we can use differentiation to solve them. In Topic 6 we:

- Solve rates of change problems (such as kinematics).

Rates of change

When two variables are related by a formula, such as $y = f(x)$, the rate at which the variable y changes when compared to the variable x is written as $\frac{dy}{dx}$ or $f'(x)$.

Many practical situations involve rates of change, and differentiation is often used to solve such problems, as shown below.

Example A

The quantity of goods stored in a warehouse is $q(t)$ at time t, where q is measured in kilograms and t is measured in days. The rate of change at q at time t can be shown to be $q'(t)$, the derivative of q with respect to t.

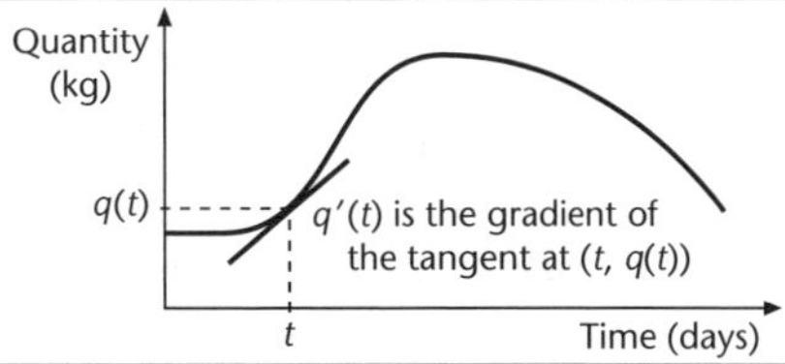

The following example deals with a similar situation, with the relationship between q and t given by a formula.

Example B

Q. The quantity of butter stored in a warehouse during a year is given by $q(t) = 100 + 30t - 2t^2$ where t is the time in months and q is the quantity of butter measured in tonnes.

1. How much butter is there at the beginning and end of the year?
2. Write an expression for the rate of change of butter stock after t months.
3. At what time was there zero rate of change?

A. **1.** At the beginning of the year, $t = 0$ months. The quantity of butter is:

$q(0) = 100 + 30 \times 0 - 2 \times 0^2$ [substituting $t = 0$ into $q(t) = 100 + 30t - 2t^2$]

$= 100$ tonnes

At the end of the year, $t = 12$ and the quantity of stock is:

$q(12) = 100 + 30 \times 12 - 2 \times 12^2$ [substituting $t = 12$]

$= 172$ tonnes

2. The rate of change is $q'(t) = 30 - 4t$ tonnes per month [differentiating]

3. When the rate of change is zero, $q'(t) = 0$

$\therefore\ 30 - 4t = 0$

$\therefore\ t = 7\frac{1}{2}$ months [solving]

Velocity – a rate of change

When an object is moving, the distance it travels and the time taken can be used to provide information about its velocity. Velocity is the *rate* at which the distance from a reference point is changing with respect to time. Since this distance can be increasing or decreasing, velocity can be positive or negative.

Note: The 'speedo' in a car displays the speed of the car, which is always positive (or zero). Speed and velocity are often treated as the same, when in fact they are different.

Example C

Q. The distance of a car from a reference point, A, is given by $s(t) = 80t + 5$, where t is in hours and s in km.

1. How far is the car from A after 3 hours?

2. Give a formula for the velocity of the car at time t.

3. The car had enough fuel for a $4\frac{1}{2}$ hour journey. What distance was it from A when it had to refuel?

A. 1. After 3 hours the distance from A is:

$s = 80 \times 3 + 5$ [substituting $t = 3$ in $s(t)$]

$= 245$ km

2. Velocity is found by differentiating the formula for distance with respect to time.

$\therefore$ velocity = 80 km h^{-1} [since $\frac{ds}{dt} = 80$]

This means that the velocity is constant.

3. After $4\frac{1}{2}$ hours the car would be a distance from A of:

$80 \times 4\frac{1}{2} + 5 = 365$ km [by substitution]

Acceleration

The acceleration of a moving object is equal to the rate of change of the velocity with respect to time.

Example D

Q. Find an expression for the acceleration of the car in Example C.

A. The acceleration is found by differentiating the formula for velocity with respect to time.

$\therefore$ acceleration = 0 [as the derivative of 80 is 0]

This means that the velocity is constant.

Unit 12.3 Activity 6A: Rates of change

1. The weight of water in a tank over a year varies with time according to the equation $w = 0.007t^2 + 0.15t + 50$, where w is the weight in kilograms of water in the tank and t is the time in days.
   a. What was the initial weight of water?
   b. What was the final weight of water?
   c. What was the rate of change of w after 64 days?
   d. What was the rate of change of w after 150 days?

2. $s = 0.01t^3 + 0.2t^2 + 5t + 2$ gives the distance in metres of a car from a point at time t in seconds. This formula applies for one minute. Find the velocity:
   a. Initially.
   b. When t is 10 seconds.
   c. When t is 30 seconds.
   d. When t is 1 minute.

3. The table following shows the amount of copper being mined on various days during a 60-day period. A is the amount of copper mined in a day in tonnes, t is the number of the day being considered.

t	1	21	41	51	60
A	1.4	2.6	3	2.9	2.639

   a. Show that the formula, $A(t) = 1.319 + 0.082t - 0.001t^2$ models the situation described.
   b. How much copper was mined on the 45th day?
   c. Write an expression for the rate of change of the amount of copper mined on day t.
   d. Find the rate of change of the amount of copper being mined on the 11th day.
   e. By drawing a graph or otherwise, find which day had the greatest copper production.

4. The length of the side of a cube is $(5 + 0.1t)$ cm where t is the time measured in minutes. This formula holds true for 10 minutes.
   a. What is the length of the cube's side after 10 minutes?
   b. What is the rate of change of the length of the cube's side after t minutes?
   c. What is the area of a face of the cube after t minutes?
   d. What is the area of a face of the cube after 4 minutes?
   e. What is the total surface area of the cube after 5 minutes?
   f. What is the volume of the cube after t minutes?
   g. What is the volume of the cube after 5 minutes?
   h. What is the rate of change of the volume after t minutes?

5. A rectangle has the length of its base given by $3 + 0.1t$ metres and its height given by $1 + 0.2t$ metres, where t is measured in hours.
   a. What is the initial length of the base of the rectangle?
   b. At what rate is the length of the base increasing?
   c. At what rate is the height increasing?
   d. At what time does the length of the base equal the length of the height?
   e. What is the area of the rectangle at time t?
   f. What is the area of the rectangle after 3 hours?
   g. What is the rate at which the area changes after t hours?
   h. At what time does the rectangle increase in area by 5 m^2 per hour?

6. The length of the base of a rectangle is given by $7 - 0.1t$ metres and its height by $2 + 0.2t$ metres where t is measured in hours.
   a. What are the rates of change of the length of its base and height?
   b. Explain what the values obtained in **a.** mean.
   c. What is the time when base and height are the same?
   d. When does the rectangle cease to exist?
   e. What is the area of the rectangle at time t?
   f. What is the rate of change of the area of the rectangle?
   g. What is the initial rate of change of the area of the rectangle?
   h. When is the area increasing by 0.6 m^2 per hour?
   i. When is the area decreasing by 0.3 m^2 per hour?
   j. By sketching a graph or otherwise, determine the maximum area.

The chain rule and related rates of change

The **chain rule** is a formula that relates one derivative to two (or more) other derivatives. The rule relating one derivative to two other derivatives is:

$$\frac{dy}{dx} = \frac{dy}{du} \times \frac{du}{dx}$$

The chain rule that relates one derivative to three other derivatives is:

$$\frac{dy}{dx} = \frac{dy}{du} \times \frac{du}{dw} \times \frac{dw}{dx}$$

where (of course) any variables can be used instead of x, y, u and w.

Note: The chain rule can be checked by considering each derivative to be a fraction made up of two **differentials**, then 'cancelling' the differentials, as demonstrated below.

$$\frac{dy}{dx} = \frac{dy}{\cancel{du}} \times \frac{\cancel{du}}{\cancel{dw}} \times \frac{\cancel{dw}}{dx} \qquad \left[\text{cancelling } \frac{du}{du} \text{ and } \frac{dw}{dw}\right]$$

$$= \frac{dy}{dx}$$

Problems involving **related rates of change** and the chain rule can be solved using the series of steps outlined here:

1. Introduce and define the variables, where necessary.
2. State any derivatives that are given.
3. State the derivative that you are asked to find.
4. Write down the chain rule in order to identify the other derivative(s) that will be needed to complete the chain rule.
5. Find the other derivative(s) required by the chain rule.
6. Substitute all the derivatives found into the chain rule so that the required derivative can be found.

Note: Find the rate of change of y with respect to x, means find $\frac{dy}{dx}$. However, if a rate of change problem does not specify the other variable, it can always be assumed to be with respect to time, ie 'find the rate of change of y' means find $\frac{dy}{dt}$, the rate of change of y with respect to time.

Example D

Q. The radius of a circular oil slick is increasing at the rate of 0.05 metres per second in calm waters. At what rate is the area (A) of the oil slick increasing when its radius is 20 metres?

A. *Introduce the variables*

After time, t seconds, let the radius of the oil slick be r metres and the area be A m^2.

Given

$$\frac{dr}{dt} = 0.05$$

To find

$\frac{dA}{dt}$ when $r = 20$ m

Chain rule

When more than one derivative is involved, expect to use the chain rule.

To identify the **third derivative**, write out the chain rule with a gap for the unknown.

$$\frac{dA}{dt} = \frac{dr}{dt} \times \left(\quad\right)$$

The derivative that goes in the bracket is $\frac{dA}{dr}$

[since $\frac{dr}{dt} \times \frac{dA}{dr} = \frac{dA}{dt}$]

Find the other derivative

The equation connecting A and r that can be differentiated to obtain $\frac{dA}{dr}$ is the formula for the area of a circle $A = \pi r^2$.

So $\frac{dA}{dr} = 2\pi r$ [differentiating]

Use the chain rule $\frac{dA}{dt} = \frac{dr}{dt} \times \frac{dA}{dr}$

$= 0.05 \times 2\pi r$ [substitute for $\frac{dr}{dt}$ and $\frac{dA}{dr}$]

$= 6.28$ [$r = 20$, $\pi = 3.14$ to 2 dp]

$\frac{dA}{dt} = 6.28$ m²/sec (2 dp)

Note: A is measured in m², t is measured in seconds, so $\frac{dA}{dt}$ is measured in m²/sec.

Example E

Q. A kite is 20 metres above the ground and is moving horizontally at a constant rate of 0.5 metres per second as the string is let out. At what rate is the angle of inclination of the string to the horizontal ground changing when 50 metres of string has been let out?

A. *Introduce the variables* After time, t seconds, let θ radians be the angle the string makes with the horizontal ground, and let the projection of the string on the ground be x metres. (K is the position of the kite.)

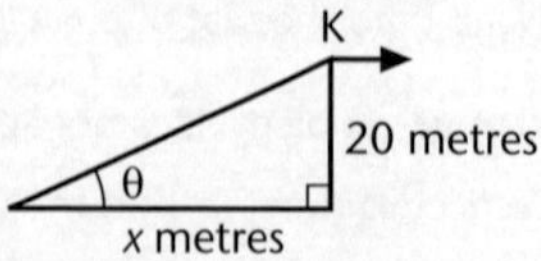

Given The horizontal speed of the kite is $\frac{dx}{dt} = 0.5$ ms⁻¹

To find $\frac{d\theta}{dt}$ when the length of string is 50 metres

Chain rule $\frac{d\theta}{dt} = \frac{dx}{dt} \times \left(\quad \right)$

The derivative that goes in the bracket is $\frac{d\theta}{dx}$ since

$\frac{d\theta}{dt} = \frac{dx}{dt} \times \frac{d\theta}{dx}$ (find $\frac{dx}{d\theta}$ then use the **reciprocal rule**

to obtain $\frac{d\theta}{dx}$).

Find the other derivative The equation connecting θ and x is

$\tan\theta = \frac{20}{x}$ [$\tan\theta = \frac{\text{opp}}{\text{adj}}$]

$x = \frac{20}{\tan\theta}$ [rearranging]

$x = 20\cot\theta$ [since $\cot\theta = \frac{1}{\tan\theta}$]

$\frac{dx}{d\theta} = 20 \times (-\text{cosec}^2\,\theta)$ [derivative of $\cot\theta$ is $-\text{cosec}^2\,\theta$]

$= \frac{-20}{\sin^2\theta}$ [since $\text{cosec}\,\theta = \frac{1}{\sin\theta}$]

$\frac{d\theta}{dx} = -\frac{\sin^2\theta}{20}$ [inverting both sides]

Use the chain rule $\left(\frac{d\theta}{dt} = \frac{dx}{dt} \times \frac{d\theta}{dx}\right)$

$$\frac{d\theta}{dt} = 0.5 \times -\frac{\sin^2\theta}{20} \qquad \left[\text{substitute } \frac{dx}{dt} = 0.5\right]$$

$$= -\frac{\sin^2\theta}{40} \quad \ldots (1) \qquad [\text{simplifying}]$$

Need to find θ when the length of string let out is 50 metres.

$\sin\theta = \frac{20}{50}$ $\qquad$ $[\sin\theta = \frac{\text{opp}}{\text{hyp}}]$

Substitute for sin θ in (1)

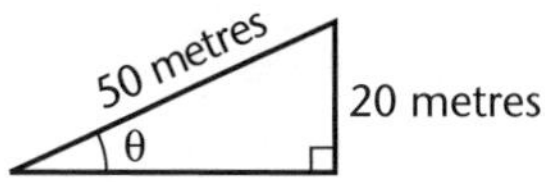

$$\frac{d\theta}{dt} = \frac{-\left(\frac{20}{50}\right)^2}{40}$$

$= -0.004$ radians/second

Note: The negative sign indicates clockwise change in angle size (ie reducing angle).

Example F

Q. The height of the lamp on a lamp-post above a wall is 4 m, and a man who is 2 metres tall walks away from the lamp-post at a constant speed of 1.5 m s^{-1}. Find the rate of change of the distance of the top of his shadow from the wall when he is 10 metres from the lamp-post.

A. *Introduce the variables* The lamp-post is OL, the man is AH and AT is the shadow. Suppose in time *t* seconds the man has walked *x* metres from O to A, and his velocity is $\frac{dx}{dt}$.

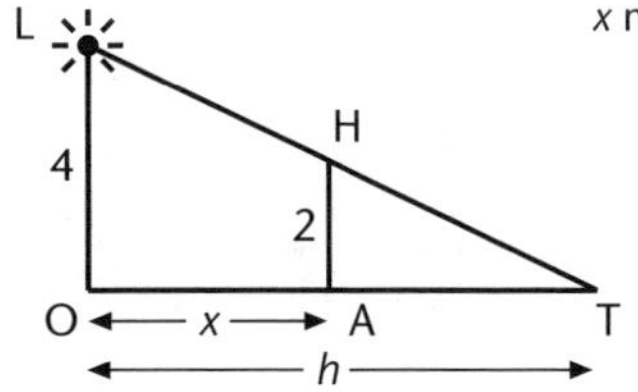

Given $\frac{dx}{dt} = 1.5$ m s^{-1}

To find $\frac{dh}{dt}$ when $x = 10$ metres

Chain rule $\frac{dh}{dt} = \frac{dx}{dt} \times \left(\quad\right)$

The derivative that goes in the bracket is $\frac{dh}{dx}$

Find the other derivative The relationship between *h* and *x* is found by using the similar triangles LOT and HAT.

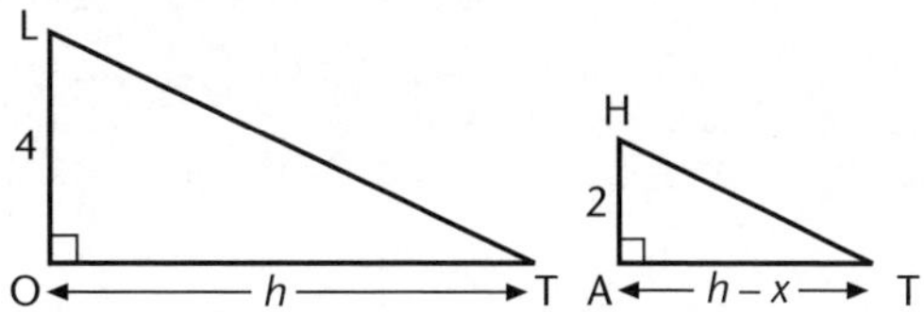

From the diagram:

$$\frac{4}{2} = \frac{h}{h - x}$$ [rule for similar triangles]

$$4h - 4x = 2h$$ [cross-multiplying]

$$2h = 4x$$

$$h = 2x$$

$$\frac{dh}{dx} = 2, \text{ which is a constant}$$ [differentiating]

Use the chain rule

$$\frac{dh}{dt} = \frac{dx}{dt} \times \frac{dh}{dx}$$

$$= 1.5 \times 2$$ [substitute $\frac{dx}{dt} = 1.5$ and $\frac{dh}{dx} = 2$]

$$= 3$$

Since $\frac{dh}{dt}$ is a constant, there is no need to substitute $x = 10$, so the rate of change of the top of the shadow is 3 m s^{-1}.

Unit 12.3 Activity 6B: Chain rule

1. The radius of an expanding circular ink blot is given by $r = 5 + t$, where the radius is measured in mm and the time t in seconds.

- **a.** What was the original radius of the ink blot?
- **b.** Find the radius of the ink blot when $t = 10$ seconds.
- **c.** At what rate is the radius of the ink blot increasing when $t = 10$ seconds?
- **d.** At what rate is the area of the ink blot increasing when $t = 10$ seconds?

2. The volume of a spherical bubble is increasing at the rate of 6 cm^3 per second. Find the rate at which the radius of the bubble is increasing when the radius is 4 cm.

3. The side of a square is increasing at the rate of 3 cm s^{-1}. When the length of a side is 6 cm, find the rate of increase of:

- **a.** The perimeter.
- **b.** The area.

4. A wine glass is shaped so that the volume of liquid it contains is $\frac{1}{2}\pi h^2$, where h cm is the depth of liquid in the glass. Suppose wine is poured into the glass at a rate of 7 cm^3 per second. At what rate is the depth of wine increasing when the depth of the wine is 6 cm?

5. A spherical balloon is being inflated, and its radius r cm after t seconds is given by $r = 3t$. Find the rate of increase of the surface area of the balloon and the rate of increase of the volume of the balloon after 5 seconds.
(*Hint:* Surface area of a sphere is $4\pi r^2$, and volume is $\frac{4}{3}\pi r^3$.)

6. A cube of ice melts at the rate of 12 cm^3 per second. Find the rate at which the sides are decreasing when the length of a side is 20 cm.

7. Oil is flowing into a vertical cylindrical tank of radius 6 cm at the rate of 8 cm^3/minute. How fast is the level of oil in the tank rising? (Volume of a cylinder is $V = \pi r^2 h$.)

8. A ladder 4 metres long is leaning against a wall. If the base of the ladder (which rests on horizontal ground) is being pulled away from the wall at a speed of 3 metres per second, how fast is the top of the ladder sliding down the wall at the instant when the base of the ladder is 2 metres from the wall?

9. Patricia loves swimming. She enters the sea at S, the foot of the wharf which is 12 metres high and strikes out at right angles to the wharf at 3 km per hour. She has a rope, PT, one end of which is fastened around her waist for safety. The rope passes through Tom's hands, which are level with the top of the wharf. Find the speed at which the rope is passing through Tom's hands when Patricia has swum 5 metres.

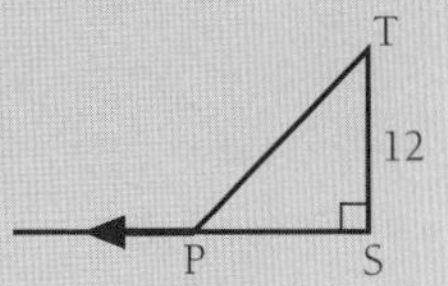

10. A ladder 5 metres long rests in a vertical plane with one end A on a horizontal floor and the other end B against a vertical wall. The end A is pulled away from the wall at a constant speed of 2 m s^{-1}. If θ is the inclination of the ladder to the horizontal find $\frac{d\theta}{dt}$, when A is 3 metres from the wall.

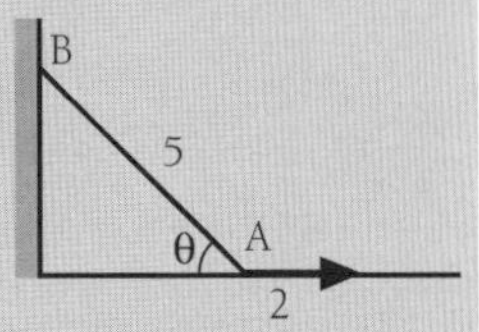

Unit 12.3 Calculus

Topic 7: Derivatives of trigonometric functions

In Topic 2 we looked at derivatives. Topic 7 deals with the derivatives of trigonometric, exponential and logarithmic functions. If necessary, revise the trigonometric identities, exponential laws, and laws of logarithms before working through this Topic.

Trigonometric derivatives

Example A

From the above investigation it can be seen that the derivative of sin x is cos x. Using the chain rule it can be verified that the derivative of cos x is –sin x.

Method

$y = \cos x$

$y = \sin\left(\frac{\pi}{2} - x\right)$ [using the identity for sin (A – B)]

Let $u = \frac{\pi}{2} - x$ and $y = \sin u$

$\frac{du}{dx} = -1$ and $\frac{dy}{du} = \cos u$ [differentiating]

$\frac{dy}{dx} = \frac{dy}{du} \cdot \frac{du}{dx}$ [chain rule]

$= \cos u \,.\, -1$

$= -\cos u$

$= -\cos\left(\frac{\pi}{2} - x\right)$ [substituting $u = \frac{\pi}{2} - x$]

$= -\sin x$ [using the identity for cos (A – B)]

The derivatives of sin x and cos x are stated below. Note that for trigonometric derivatives, unless stated otherwise, it is always assumed that the angle is in radians.

$$\frac{d}{dx}(\sin x) = \cos x \qquad \frac{d}{dx}(\cos x) = -\sin x$$

To find the derivatives of cos ($f(x)$) and sin ($f(x)$) the chain rule is used.

Example B

Q. Find the derivative of $y = \sin(f(x))$.

A. Let $u = f(x)$ and so $y = \sin u$.

$\therefore \frac{du}{dx} = f'(x)$ and $\frac{dy}{du} = \cos u$ [differentiating]

$\frac{dy}{dx} = \frac{dy}{du} \cdot \frac{du}{dx}$ [using the chain rule]

$= (\cos u)\, f'(x)$

$= f'(x) \cos f(x)$ [since $u = f(x)$]

The derivative of $y = \cos f(x)$ can be found in a similar way.

$$\frac{d}{dx}(\sin f(x)) = f'(x) \cos f(x) \qquad \frac{d}{dx}(\cos f(x)) = -f'(x) \sin f(x)$$

Example C

Q. Find $\frac{dy}{dx}$ for $y = \sin 3x$.

A. $\frac{dy}{dx} = 3 \cos 3x$ [since $\frac{d}{dx}(\sin f(x)) = f'(x) \cos f(x)$]

Example D

Q. Find $\frac{dy}{dx}$ for $y = -3 \cos (1 + 2x^3)$.

A. $\frac{dy}{dx} = -3 \,.\, 6x^2 \,.\, -\sin (1 + 2x^3)$ [since $\frac{d}{dx}(\cos (f(x)) = -f'(x) \sin f(x)$]

$= 18x^2 \sin (1 + 2x^3)$

When an equation is expressed in the form $(f(x))^n$ the **bracket rule** can be used before differentiating.

Example E

Q. Find $\frac{dy}{dx}$ for $y = 4 \sin^3 (2x)$.

A. $y = 4 (\sin 2x)^3$ [rewriting in the form $(f(x))^n$]

$\frac{dy}{dx} = 4 \,.\, 3 \,.\, (2 \cos 2x)(\sin 2x)^2$ [using the bracket rule]

$= 24 \cos 2x \sin^2 (2x)$

Unit 12.3 Activity 7A: Differentiating trigonometric functions

1. Find the derivatives of each of the following (some derivatives will require the use of the bracket rule, and some will require the use of trigonometric identities):

a. $y = \cos 3x$ **b.** $y = 2 \sin 4x$ **c.** $y = \cos x + \sin(x + \pi)$
d. $y = 5 \cos (1 - 2x)$ **e.** $y = 3 \sin (2x - \pi)$ **f.** $y = \cos^2 x$
g. $s = \sin^3 (x - \pi)$ **h.** $x = \frac{1}{\cos^2 t}$ **i.** $y = \cos 2x - 4 \sin 3x$
j. $y = \sqrt{\cos 2t}$ **k.** $y = \sin (2x + \frac{1}{6}\pi)$ **l.** $y = \cos (nx + \pi)$
m. $y = \sin (\frac{1}{4}\pi - \frac{1}{2}x)$ **n.** $y = \frac{3}{\sqrt{\sin 2t}}$ **o.** $y = (1 + \sin x)^4$
p. $y = \sin (\cos x)$ **q.** $y = \cos x \tan x$ **r.** $y = (\sin x + \cos x)^2$
s. $y = \sin^2 (2\theta) + \cos^2 (2\theta)$

2. a. If $y = 5 \cos (1 - 2x)$, find $y'(12)$. *Hint:* Use answer to **1. d.**
b. If $y = \sqrt{\cos 2t}$, find $y'\left(\frac{\pi}{6}\right)$. *Hint:* Use answer to **1. j.**

3. There is a chairlift to the top of a hill and a ski slope down the other side. The hill is modelled with a curve whose equation is $y = 200 \sin \frac{x}{200}$.

a. Find the height of the hill, h, and the intercept A.
b. What is the gradient of the slope when $x = 150\pi$?

4. The equation of the curve shown is $y = \sin x$. The tangent to the curve at P meets the x-axis in the point Q.

Find the coordinates of Q.

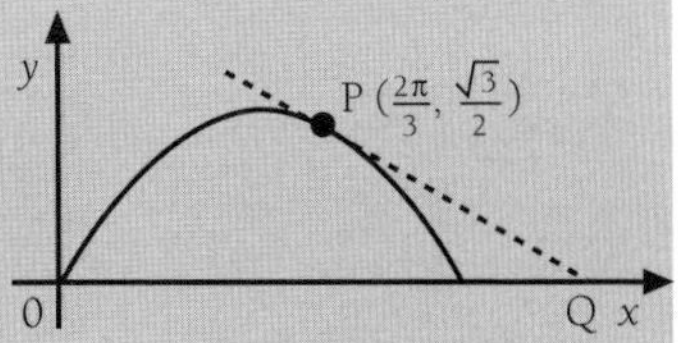

5. The diagram shows the cross-section of a hillside. The equation of the hill is $y = 100 \cos \dfrac{\pi x}{400}$. An observation tower BC is to be erected at the top of the hill so that the bottom of the hill A is just visible from the point C.

a. Verify that the intercept at A is $x = 200$, and the intercept at B is $y = 100$.

b. Find the height of the tower.

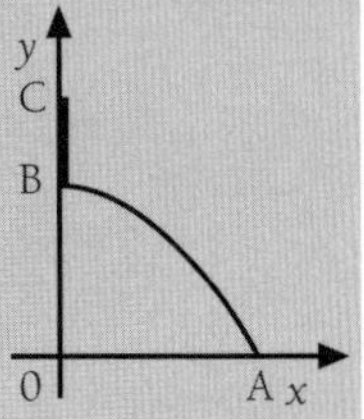

Investigation

The diagram shows the cross-section of a hill, with a lake at the foot of the hill. The equation of the whole curve is $y = 2 \sin 2x + 1$.

For each axis the scale is 1 unit represents 100 metres.

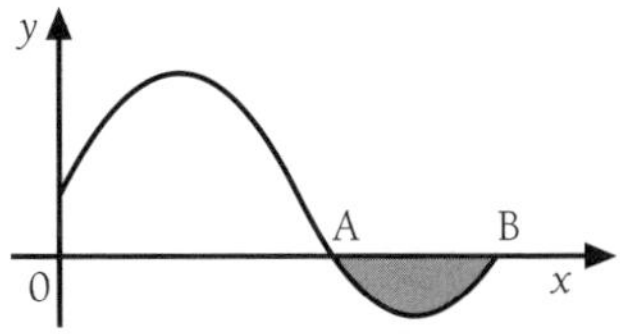

1. Find the height of the hill.

2. Verify that the distance OA is 183.26 metres (2 dp).

3. Find the length of the lake.

4. If a vertical fence post was erected at A, what is the acute angle between the fence post and the hillside?

Unit 12.3 Calculus

Topic 8: The product rule and the quotient rule for differentiation

To find the derivative of two functions that are multiplied together, we can use the product rule. The quotient rule is useful when one function is divided by another.

The product rule

The **product rule** is used to find the derivative of two functions which are multiplied together.

The formula is easier to remember (or learn) in the following form:

$$\text{If } p(x) = f \, . \, g \text{ then } p'(x) = f \, . \, g' + g \, . \, f'$$

This product rule formula is frequently expressed in **Leibnitz** form:

$$\text{If } y = u.v \text{ where } u \text{ and } v \text{ are functions of } x \text{ then } \frac{dy}{dx} = u \, . \, \frac{dv}{dx} + v \, . \, \frac{du}{dx}$$

The derivative of the original product in the investigation can now be found.

Example A

Q. If $y = 2x \sin x$, find $\frac{dy}{dx}$.

A. Let $f(x) = 2x$ and $g(x) = \sin x$

Then $f'(x) = 2$ and $g'(x) = \cos x$

$\frac{dy}{dx} = f \, . \, g' + g \, . \, f'$ [product rule for derivative of $f \, . \, g$]

$= 2x \, . \, \cos x + (\sin x) \, . \, 2$ [substituting]

$= 2(x \cos x + \sin x)$

The quotient rule

The **quotient rule** is useful when one function is divided by another:

$$\text{If } q(x) = \frac{f}{g}, \, g \neq 0 \text{ then } q'(x) = \frac{g \, . \, f' - f \, . \, g'}{g^2}$$

Alternatively:

$$\text{If } y = \frac{u}{v}, \, v \neq 0 \text{ then } \frac{dy}{dx} = \frac{v\frac{du}{dx} - u\frac{dv}{dx}}{v^2}$$

Example B

Q. If $s = \frac{2u+1}{2u-1}$ find $\frac{ds}{du}$, providing $2u - 1 \neq 0$; ie $u \neq \frac{1}{2}$.

A. $f(u) = 2u + 1$ and $g(u) = 2u - 1$ [where $s = \frac{f}{g}$]

$f'(u) = 2$ $g'(u) = 2$

$$\frac{ds}{du} = \frac{(2u-1).\,2 - (2u+1).\,2}{(2u-1)^2} \quad \text{[by quotient rule]}$$

$$= \frac{4u - 2 - 4u - 2}{(2u-1)^2}$$

$$= \frac{-4}{(2u-1)^2}$$

Further examples demonstrate how more complex functions can be differentiated:

Example C

Q. Find derivatives of the following:

1. $y = \frac{x}{\sqrt{1-2x}}$, providing $x < \frac{1}{2}$ **2.** $y = 2x \cos 4x$

A. (Using Leibnitz notation)

1. Let $u = x$ and $v = \sqrt{1-2x}$, so $y = \frac{u}{v}$ and

$$\frac{du}{dx} = 1 \text{ and } \frac{dv}{dx} = \frac{1}{2}(-2)(1-2x)^{\frac{-1}{2}} \quad \text{[using the bracket rule]}$$

$$= \frac{-1}{\sqrt{1-2x}}$$

$$\frac{dy}{dx} = \frac{\sqrt{1-2x} - x\left(\frac{-1}{\sqrt{1-2x}}\right)}{\left(\sqrt{1-2x}\right)^2} \quad \text{[using the quotient rule]}$$

$$= \frac{\sqrt{1-2x} + \frac{x}{\sqrt{1-2x}}}{(1-2x)} \quad \text{[simplifying]}$$

$$= \frac{(1-2x) + x}{(1-2x)\sqrt{1-2x}}$$ [multiplying numerator and denominator by $\sqrt{1-2x}$]

$$= \frac{1-x}{(1-2x)^{\frac{3}{2}}} \quad \text{[simplifying]}$$

2. Let $u = 2x$ and $v = \cos 4x$, so $y = uv$ and

$\frac{du}{dx} = 2$ and $\frac{dv}{dx} = -4 \sin 4x$

Therefore $\frac{dy}{dx} = 2x(-4 \sin 4x) + (\cos 4x) .\, 2$ [using the product rule]

$= -8x \sin 4x + 2 \cos 4x$

More trigonometric derivatives

The quotient rule can be used to find the derivatives of other trigonometric functions.

Example D

Q. Find the derivative of $y = \tan x$.

A. $y = \tan x$

$= \dfrac{\sin x}{\cos x}$

$= \dfrac{u}{v}$, where $u = \sin x$ and $v = \cos x$

Thus $\dfrac{du}{dx} = \cos x$ and $\dfrac{dv}{dx} = -\sin x$ [differentiating]

$\dfrac{dy}{dx} = \dfrac{v\dfrac{du}{dx} - u\dfrac{dv}{dx}}{v^2}$ [using the quotient rule]

$= \dfrac{\cos x \, . \, \cos x - \sin x \, . \, (-\sin x)}{\cos^2 x}$ [substituting from above]

$= \dfrac{\cos^2 x + \sin^2 x}{\cos^2 x}$ [simplifying]

$= \dfrac{1}{\cos^2 x}$ [using $\sin^2 x + \cos^2 x = 1$]

$= \sec^2 x$ [$\dfrac{1}{\cos x} = \sec x$]

This gives the derivative of tan x as:

$$\frac{d}{dx}(\tan x) = \sec^2 x, \text{ providing } \cos x \neq 0$$

Using the chain rule, it can be shown that:

$$\frac{d}{dx}(\tan f(x)) = f'(x) \sec^2 f(x), \text{ providing } \cos f(x) \neq 0$$

Example E

Q. If $y = 3 \tan (2x - \pi)$, find $\dfrac{dy}{dx}$.

A. $y = 3 \tan (2x - \pi)$

$\dfrac{dy}{dx} = 3 \, . \, 2 \, . \, \sec^2 (2x - \pi)$ [$f(x) = 2x - \pi$, $f'(x) = 2$]

$= 6 \sec^2 (2x - \pi)$

Example F

Q. Find the derivative of $y = \text{cosec } x$.

A. $y = \dfrac{1}{\sin x}$ [by definition of cosec x]

$= \dfrac{u}{v}$, where $u = 1$ and $v = \sin x$, so $\dfrac{du}{dx} = 0$ and $\dfrac{dv}{dx} = \cos x$

$\dfrac{dy}{dx} = \dfrac{\sin x \,.\, 0 - 1 \,.\, \cos x}{\sin^2 x}$ [using the quotient rule]

$= \dfrac{-\cos x}{\sin x \,.\, \sin x}$ [simplifying]

$= \dfrac{-1}{\sin x} \,.\, \dfrac{\cos x}{\sin x}$

$= -\text{cosec } x \cot x$ $[\dfrac{1}{\sin x} = \text{cosec } x$ and $\dfrac{\cos x}{\sin x} = \cot x]$

$$\frac{d}{dx} \text{cosec } x = -\text{cosec } x \cot x, \text{ providing } \sin x \neq 0$$

More generally:

$$\frac{d}{dx} \text{cosec } f(x) = -f'(x) \text{ cosec } f(x) \cot f(x), \text{ providing } \sin f(x) \neq 0$$

In a similar way, the following results can be obtained:

y	$\frac{dy}{dx}$	Example y	Example $\frac{dy}{dx}$
$\sec x$	$\sec x \tan x$		
$\sec f(x)$	$f'(x) \sec f(x) \tan f(x)$	$3 \sec 2x$	$6 \sec 2x \tan 2x$
$\cot x$	$-\text{cosec}^2 x$		
$\cot f(x)$	$-f'(x) \text{ cosec}^2 f(x)$	$5 \cot 3x^2$	$-30x \text{ cosec}^2 3x^2$

Unit 12.3 Activity 8A: Differentiation rules and more trigonometric derivatives

1. Find the derivatives of each of the following:

a. $5 \tan 2x$ **b.** $\sec(2x - \pi)$ **c.** $\text{cosec } \pi x$ **d.** $2x - \cot(\pi - x)$

e. $2 \tan^3 x$ **f.** $3 \cot 2x$ **g.** $\text{cosec}^2 x$ **h.** $\tan x + \cot x$

i. $\sqrt[3]{x} \cos x$ **j.** $\sin 2x \cos x$ **k.** $x\sqrt{1 - 2x}$ **l.** $\sin x \tan x$

m. $3x \sin x$ **n.** $2x(2x + 1)^4$ **o.** $\dfrac{x + 1}{x - 2}$ **p.** $\dfrac{4 + x}{4 - x}$

q. $\dfrac{x^2}{\sin 2x}$ **r.** $\dfrac{x}{\sqrt{1 - x}}$ **s.** $\dfrac{x}{\cos x}$ **t.** $x^3(1 - \cos x)$

u. xe^{2x} **v.** $(x + 3) \ln(x + 3)$

2. **a.** If $y = 2x - \cot(\pi - x)$, find $y'(-1)$. *Hint:* Use answer to **1. d**.

b. If $y = \frac{4 + x}{4 - x}$, find $y'(-4)$. *Hint:* Use answer to **1. p**.

3. Find the derivatives of each of the following:

a. $2x \sec x$ **b.** $\frac{3 \tan x}{x}$ **c.** $(2x + 3)^3(1 - x)^4$ **d.** $\left(\frac{x-1}{2-x}\right)^2$

e. $\frac{1}{(2x-3)^3}$ **f.** $2x^3 \sin\sqrt{x}$ **g.** $\frac{x^2}{\sqrt{\sin 2x}}$

4. If $f(x) = \frac{\sin 2x}{x}$, find $f'(\pi)$.

Unit 12.3 Calculus

Topic 9: Antidifferentiation

Antidifferention is also called integration; it is the opposite of differentiation and is used, for example, to find areas or calculate change in problems relating to physical, biological, economic and social situations. The focus of Topic 9 is to:

- Find indefinite integrals of expressions given in expanded form that have natural number exponents.

Introduction

Antidifferentiation, as the name suggests, is the opposite process to differentiation. Antidifferentiation is often called **integration**. When an expression is *antidifferentiated* or *integrated*, the result is called the **antiderivative** or **integral**.

Example A

Q. Find the antiderivative of $3x^2$.

A. If the expressions x^3 or $x^3 + 1$ or $x^3 - 7$ are differentiated, $3x^2$ results.

Generally the derivative of $x^3 + C$, where C is any fixed number, is $3x^2$. Thus the antiderivative of $3x^2$ is $x^3 + C$, where C is called the **constant of integration**.

Example B

The antiderivative of x^5 is $\frac{x^6}{6} + C$, because if $\frac{x^6}{6} + C$ is differentiated, x^5 results.

The examples above can be generalised to give the following result:

The antiderivative of x^n is $\frac{x^{n+1}}{n+1} + C$.

Note: The result may be checked by differentiating $\frac{x^{n+1}}{n+1} + C$ and confirming that the derivative is x^n.

Example C

The antiderivative of x^{12} is $\frac{x^{13}}{13} + C$.

The integral of x^{-9} is $-\frac{x^{-8}}{8} + C$. [since $\frac{x^{-8}}{-8} = -\frac{x^{-8}}{8}$]

Further rules for antidifferentiation

The *sum* or *difference* of two functions is antidifferentiated by antidifferentiating each function separately.

Example D

The antiderivative of $x^3 - x^2$ is equal to the antiderivative of x^3 minus the antiderivative of x^2.

The antiderivative of x^3 is $\frac{x^4}{4} + C_1$ and the antiderivative of x^2 is $\frac{x^3}{3} + C_2$, where C_1 and C_2 are the constants of integration.

The antiderivative of $x^3 - x^2$ is written $\frac{x^4}{4} - \frac{x^3}{3} + C$ because the difference between the two constants C_1 and C_2 is another constant, C.

Expressions of the type ax^n where a is a constant (such as 5, $\frac{1}{2}$ or –3) can be antidifferentiated by antidifferentiating x^n and then multiplying the result by a.

The antiderivative of ax^n is $\dfrac{ax^{n+1}}{n+1} + C$.

Example E

The antiderivative of $4x^8$ is the antiderivative of x^8 multiplied by 4, which is $\dfrac{4x^9}{9} + C$.

Notation

Integrating a function means finding its antiderivative or integral. The symbol $\int$ represents the general integral, antiderivative, or **indefinite integral** of a function. $\int f(x)\,dx$ is read 'the integral of f with respect to x'.

Example F

1. $\int 5x^5\,dx = \dfrac{5}{6}x^6 + C$ [since $\dfrac{5x^6}{6} = \dfrac{5}{6}x^6$]

2. $\int \dfrac{x^2}{8}\,dx = \int \dfrac{1}{8}x^2\,dx$ [writing in ax^n form]

$= \dfrac{1}{8} \times \dfrac{x^3}{3} + C$ [integrating to get $\dfrac{ax^{n+1}}{n+1} + C$]

$= \dfrac{x^3}{24} + C$

3. $\int 6 - \dfrac{2x^5}{3}\,dx = \int 6 - \dfrac{2}{3} \times x^5\,dx$ [writing in ax^n form]

$= 6x - \dfrac{2}{3} \times \dfrac{x^6}{6} + C$ [integrating]

$= 6x - \dfrac{x^6}{9} + C$ [simplifying]

Note: The expression $\int f(x)dx$ always includes the 'dx', even though it is usually clear that x is the variable in the integration. However, if there is more than one letter involved in the expression to be integrated, such as when some letters are constants, it is vital to say which letter is the variable by using the full and correct expression.

Example G

Q. Integrate yx with respect to x, ie, find $\int yx\,dx$.

A. $\int yx\,dx = \dfrac{yx^2}{2} + C$ [y is treated as a constant]

Note: By comparison, integrating yx *with respect to* y, gives $\dfrac{y^2x}{2} + C$.

$\int yx\,dy = \dfrac{y^2x}{2} + C$ [x is treated as a constant]

Sometimes expressions need to be expanded before they can be integrated (there is no general rule for integrating products or **quotients** of functions).

Example H

$$\int (x^2 + 1)(x + 2)\,dx = \int (x^3 + 2x^2 + x + 2)\,dx \qquad \text{[expanding]}$$
$$= \frac{x^4}{4} + \frac{2x^3}{3} + \frac{x^2}{2} + 2x + C$$

Unit 12.3 Activity 9A: Antidifferentiation

1. Antidifferentiate each of the following functions, with respect to x (treat all other letters as constants):

a. x **b.** x^3 **c.** $2x^4$ **d.** $x^{11} + x^{12}$
e. $x^6 - x^5 + x^4$ **f.** $2x^7 + 3x^6$ **g.** $4x^3 - 2x^5$ **h.** $2x^3 - 3x^4$
i. $x(x - 3)$ **j.** 2 **k.** 3 **l.** $x - 3$
m. $x^2 - 3x + 4$ **n.** $(x + 3)(x + 4)$ **o.** $(2x + 3)^2$ **p.** mx^2
q. $(2x + 3)(3x^2 + 4)$ **r.** $(2x^2 + 5)^2$ **s.** a^2 **t.** $\frac{x}{b}$
u. $cx + \frac{x^2}{d}$ **v.** $ax^2 + bx + c$ **w.** $\frac{x^2}{a} + \frac{x}{b} + \frac{1}{c}$ **x.** $(ax^2 + bx + c)^2$
y. $(ax + x^2)^2$

2. Integrate the following functions with respect to x:

a. $3x$ **b.** $3x^4$ **c.** $3x^7$ **d.** $2x^8$
e. $x^{13} - x^{10}$ **f.** $2x^7 + 8x^5$ **g.** $x^4 - 4x^3 + 3x$ **h.** $(x^2 + 1)(x + 3)$
i. $(2x + 5)^2$ **j.** $Ex + D$

3. Find the following:

a. $\int 4x\,dx$ **b.** $\int \frac{1}{3}x^2\,dx$ **c.** $\int 7x^3\,dx$
d. $\int 2r\,dr$ **e.** $\int 8r^8\,dr$ **f.** $\int 5t^6\,dt$
g. $\int 8\,dx$ **h.** $\int a\,dx$ **i.** $\int t^5\,dx$
j. $\int 2t^5\,dt$ **k.** $\int (2x + 4)\,dx$ **l.** $\int (x^3 - 3x^2 + 5x + 2)\,dx$
m. $\int (4x + 3)^2\,dx$ **n.** $\int (2x^2 + 1)^2\,dx$ **o.** $\int x^2(x + 3)\,dx$

Unit 12.3 Calculus

Topic 10: Finding the constant of integration

Topic 10 builds on the coverage of antidifferentiation introduced in the previous Topic. In Topic 10 we:

- Find the equation from the gradient function.
- Solve rates of change problems (such as kinematics).

Finding the constant of integration

With enough information, the constant of integration can be evaluated.

Example A

Q. Given that $f'(x) = 2x + 3$ and $f(2) = 5$, find $f(x)$.

A.

$f(x) = 2x + 3$

$\therefore\ f(x) = x^2 + 3x + C$ [integrating]

$\therefore\ f(2) = 4 + 6 + C$ [substituting 2 for x in $x^2 + 3x + C$]

$= 10 + C$

but $10 + C = 5$ [since $f(2) = 5$]

$\therefore\ C = -5$

$\therefore\ f(x) = x^2 + 3x - 5$

Example B

Q. $\frac{dy}{dx} = 3x^2 - x + 2$ and $y = 6$ when $x = 2$. Find an expression for y in terms of x.

A.

$\frac{dy}{dx} = 3x^2 - x + 2$

$\therefore\ y = x^3 - \frac{1}{2}x^2 + 2x + C$ [integrating]

$\therefore\ 6 = 8 - 2 + 4 + C$ [substituting $y = 6$ and $x = 2$]

$\therefore\ C = -4$

$\therefore\ y = x^3 - x^2 + 2x - 4$

Example C

Q. The rate at which the volume V of a solid changes with respect to time, t, is given by the expression $10 - 2t$. The volume is initially 50. Find an expression for the volume of the solid at time t.

A. Using the information given:

$\frac{dV}{dt} = 10 - 2t$ [the rate of change is the derivative]

$V = 10t - t^2 + C$ [integrating]

$50 = 0 - 0 + C$ [substituting $V = 50$ when $t = 0$ ('initially' means $t = 0$)]

$\therefore\ 50 = C$

$\therefore\ V = 10t - t^2 + 50$

Unit 12.3 Activity 10A: Finding the constant of integration

1. $f'(x) = x$, and $f(0) = 0$. Find an expression for $f(x)$.
2. $g'(x) = x^2 + x$ and $g(0) = 2$. Find an expression for $g(x)$.
3. $\frac{dy}{dx} = 3x^2$ and $y = 4$ when $x = 1$. Find an expression for y in terms of x.
4. $\frac{dy}{dx} = 4x^3 + 2x$ and $y = 5$ when $x = 2$. Find an expression for y in terms of x.
5. $y' = 2x^3 + 1$ and $y = 7$ when $x = 2$. Find an expression for y in terms of x.
6. $f'(x) = 3x + 2$, $f(3) = 4$. Find an expression for $f(x)$.
7. $g'(x) = 5x^2$, $g(4) = 11$. Find an expression for $g(x)$.
8. $\frac{dy}{dx} = 2x^2 - x + 3$, $y = 16$ when $x = 6$. Find an expression for y in terms of x.
9. $\frac{dy}{dx} = 3x - x^2$, $y = 42$ when $x = 12$. Find an expression for y in terms of x.
10. $y' = 2x^2 - 3x + 2$. The point (1, 5) lies on the graph of this function. Find y in terms of x.
11. $y' = x + 3$. The point (2, 1) lies on the graph of this function. Find y in terms of x.
12. $\frac{dE}{dr} = -2r$ and E is 5 when r is 4. Find the equation relating E to r.
13. **a.** A balloon has volume increasing at the rate of $2t + 1$ cubic centimetres per second. Initially the volume is 12 cm^3. Find the volume at time t, where t is in seconds.
    b. The balloon will burst when the volume is 210 cm^3. Find the time when the balloon will burst.
14. **a.** The water in a tank is escaping at the rate of $60 + 2t$ cubic centimetres per second. Initially the volume was 11 000 cm^3. Find an expression for the volume at time t.
    b. Find the time when the tank becomes empty.
15. **a.** The rate at which fuel is used by a vehicle after the first kilometre of travel is $5 + \frac{6}{x^2}$ litres per kilometre when it has travelled x km. After travelling for 1 km the vehicle has 150 litres of fuel. Find an expression for the amount of fuel it has when it has travelled x km.
    b. When will the volume of fuel be 100 litres?

Distance, speed and acceleration

If the position, s, of an object is expressed as a *function of time*, t, then the velocity, v, of the object is the *instantaneous rate of change of the distance* with respect to time and is written $v = \frac{ds}{dt}$. Similarly, the acceleration, a, is the *instantaneous rate of change of the object's velocity* with respect to time and is written $a = \frac{dv}{dt}$.

Integrating the expressions $v = \frac{ds}{dt}$ and $a = \frac{dv}{dt}$ gives:

$$\boxed{s = \int v\,dt \ \text{ and } \ v = \int a\,dt}$$

Note: Speed is the absolute value of velocity, therefore speed ≥ 0.

Example D

Q. A car moves so that its distance in kilometres from a town is given by $s = 80t(2 - t)$ where t is time in hours.

1. Find the distance of the car from the town after 1.25 hours.
2. Find an expression for the velocity at time t.
3. Find the velocity of the car after 45 minutes.
4. Find when the car is at the greatest distance from the town and the magnitude of that distance.
5. Find the velocity of the car when $t = 1.5$ hours and explain why it has a negative value.
6. Show that the car has a constant acceleration and find its value.

A. **1.** The distance from the town when $t = 1.25$ is:

$s = 80 \times 1.25 \times (2 - 1.25)$ [substituting $t = 1.25$ into $s = 80t(2 - t)$]

$= 75$ km

2. The car's velocity is $v = \frac{ds}{dt}$

$\therefore\ v = \frac{d}{dt}(80t(2 - t))$ [substituting $80t(2 - t)$ for s]

$= \frac{d}{dt}(160t - 80t^2)$ [expanding the bracket]

$= 160 - 160t$ [differentiating]

3. The velocity when $t = 45$ minutes is found by substituting in $v = 160 - 160t$:

$v = 160 - 160 \times 0.75$ [since 45 minutes = 0.75 hours]

$= 40$ km per hour

4. The greatest distance of the car from the town will occur when $\frac{ds}{dt} = 0$.

$\therefore\ 160 - 160t = 0$

$\therefore\ t = 1$ hour [solving the equation]

$\therefore$ the maximum distance $= 80 \times 1 \times (2 - 1)$ [substituting in $80t(2 - t)$]

$= 80$ km

5. The velocity when $t = 1.5$ hours is:

$v = 160 - 160 \times 1.5$ [substituting $t = 1.5$]

$= -80$ km per hour

The *negative* sign shows that the car is moving back towards the town (at a speed of 80 km per hour).

6. The acceleration of the car is given by:

$a = \frac{dv}{dt}$

$= \frac{d}{dt}(160 - 160t)$

$= -160$ km per hour^2, which is constant.

Example E

Q. A toy car moves with velocity $16t - 2t^2$ cm per second for 8 seconds. Initially it is 5 cm away from its base.

1. Find the velocity when 3 seconds have elapsed.
2. Find an expression for the acceleration at time t.
3. Find an expression for the distance of the toy from its base at time t.

A. Let s be the distance, v the velocity, and a the acceleration.

1. After 3 seconds $v = 16 \times 3 - 2 \times 3^2$ [substituting $t = 3$ in $v = 16t - 2t^2$]

$= 30$ cm per second

2. $a = \dfrac{dv}{dt}$

$= \dfrac{d}{dt}(16t - 2t^2)$ [since $v = 16t - 2t^2$]

$= 16 - 4t$ [differentiating]

3. $s = \int v\,dt$

$\therefore\ s = \int (16t - 2t^2)\,dt$ [substituting $v = 16t - 2t^2$]

$\therefore\ s = 8t^2 - \dfrac{2t^3}{3} + C$

Initially (when $t = 0$) $s = 5$ hence:

$5 = 0 - 0 + C$ [substituting $t = 0$ into $s = 8t^2 - \dfrac{2t^3}{3} + C$]

$\therefore\ C = 5$

$\therefore\ s = 8t - \dfrac{2t^3}{3} + 5$

Unit 12.3 Activity 10B: Distance, velocity and acceleration

1. A man moves so that his distance from his house after t minutes is $50t - 5t^2$ metres.
   - **a.** How far is he after 2 minutes?
   - **b.** How far is he after 4 minutes?
   - **c.** How far is he after 8 minutes?
   - **d.** When does he return to his house?
   - **e.** What is his velocity after t minutes?
   - **f.** What is his velocity after 3 minutes?
   - **g.** What is the greatest distance he is away from his house?
   - **h.** What is his acceleration after time t?
   - **i.** What is his acceleration after 3 minutes?

2. A stone is projected vertically upwards with a speed of 30 metres per second. Its height h after t seconds is given by $h = 30t - 5t^2$.
   - **a.** How high is it after 2 seconds?
   - **b.** How high is it after 1 second?
   - **c.** How long does it take to return to the ground?
   - **d.** What is the velocity after t seconds?
   - **e.** What is the velocity after 4 seconds?
   - **f.** What is the velocity when it returns?
   - **g.** What is the greatest height it reaches?
   - **h.** What is the acceleration after time t?

3. A rocket is moving from an observation point with a velocity given by $100 - 4t$ metres per second.
 a. What is the velocity after 5 seconds?
 b. When does the rocket come to rest?
 c. When time is zero the rocket is 100 m from the observation point. Find the distance of the rocket from the observation point when time is t.
 d. Find the distance the rocket is away from the observation point when 3 seconds have gone by.
 e. What is the acceleration of the rocket?

4. A toy moves with a velocity given by $50 - 3t$ cm per second. Initially it is 10 cm from its operator.
 a. Find a formula for the distance from the operator after t seconds.
 b. Find the distance from the operator after 4 seconds.
 c. Find the distance from the operator after 6 seconds.
 d. When does the toy come to rest?
 e. What is the acceleration of the toy?

5. A toy initially has a velocity of 5 cm per second. It is subject to an acceleration of 3 cm per second^2.
 a. What is its velocity after t seconds?
 b. What is its velocity after 10 seconds?
 c. When is its speed 45 cm per second?

6. The distance s of an object from an observer at time t is given by $s = 3t^2 - t^3$.
 a. Find the velocity at time t.
 b. Find the acceleration at time t.
 c. What is the velocity when $t = 2$?
 d. What is the acceleration when $t = 1$?
 e. Solve the equation $s(t) = 0$.

7. $v = 2t - t^2$ describes the velocity in metres per second of an object from an observer who initially held it, where t is the time in seconds.
 a. How far away is it after 1 second?
 b. What is the velocity at time $t = 3$?
 c. When does it come to rest?
 d. What is its maximum distance from the observer in the first 3 seconds?

Unit 12.3 Calculus

Topic 11: Integration of trigonometric functions

Topic 11 extends the coverage of antidifferentiation and builds on the information presented in Topics 9 and 10.

Introduction

The first three basic trigonometric integrals to be considered are $\sin x$, $\cos x$ and $\sec^2 x$, and their integrals are approached through derivatives. In all trigonometric integrations involving the following rules, remember the angle must always be in radians.

Example A

1. $\frac{d}{dx}(\sin 3x) = 3\cos 3x$

It follows that:

$\int \cos 3x \, dx = \frac{1}{3}\sin 3x + c$

2. $\frac{d}{dx}(\sin(2x - \pi)) = 2\cos(2x - \pi)$

It follows that:

$\int \cos(2x - \pi)\, dx = \frac{1}{2}\sin(2x - \pi)$

This leads to the first general formula:

$$\int \cos(ax + b)\, dx = \frac{1}{a}\sin(ax + b) + c, \text{ where } a, b \text{ and } c \text{ are constants}$$

Example B

$\int \cos\left(4x + \frac{\pi}{2}\right) dx = \frac{1}{4}\sin\left(4x + \frac{\pi}{2}\right) + c$

Two more results will follow in a similar way (note the negative sign):

$$\int \sin(ax + b)\, dx = \frac{-1}{a}\cos(ax + b) + c$$

$$\int \sec^2(ax + b)\, dx = \frac{1}{a}\tan(ax + b) + c$$

Example C

1. $\int 3\sin 4x \, dx = -3 \cdot \frac{1}{4}\cos 4x + c$

$= \frac{-3}{4}\cos 4x + c$

2. $\int \sec^2 3x \, dx = \frac{1}{3}\tan 3x + c$

3. $\int 5\sec^2(\pi - 2x)\, dx = 5 \cdot \frac{1}{(-2)}\tan(\pi - 2x) + c$

$= \frac{-5}{2}\tan(\pi - 2x) + c$

Consider now a definite integral:

Example D

Q. Evaluate $I = \int_{\frac{\pi}{2}}^{\pi} (\sin 2x + \cos (3x - \pi))\, dx$.

A. $I = \left[\frac{-1}{2} \cos 2x + \frac{1}{3} \sin (3x - \pi)\right]_{\frac{\pi}{2}}^{\pi}$ [integrating using the previous rules]

$= \left[\frac{-1}{2} \cos 2\pi + \frac{1}{3} \sin (3\pi - \pi)\right] - \left[\frac{-1}{2} \cos \pi + \frac{1}{3} \sin (\frac{3\pi}{2} - \pi)\right]$

[ensure the calculator is in radian mode]

$= \left[\frac{-1}{2} . 1 + \frac{1}{3} . 0\right] - \left[\frac{-1}{2} . (-1) + \frac{1}{3} . 1\right]$

$= \left[\frac{-1}{2}\right] - \left[\frac{1}{2} + \frac{1}{3}\right]$

$= \frac{-4}{3}$

There are three more derivatives to remember in order to obtain antiderivatives:

$$\frac{d}{dx}(\sec x) = \sec x . \tan x$$
$$\frac{d}{dx}(\operatorname{cosec} x) = -\operatorname{cosec} x . \cot x$$
$$\frac{d}{dx}(\cot x) = -\operatorname{cosec}^2 x$$

The formulae for the antiderivatives in these three cases are not given (they would look far too complicated). Instead three examples are given to illustrate the concepts:

Example E

1. $\int \sec 2x \tan 2x \, dx = \frac{1}{2} \sec 2x + c$

2. $\int 3 \operatorname{cosec}^2 (\pi x)\, dx = \frac{-3}{\pi} \cot (\pi x) + c$

3. $\int \operatorname{cosec} 3x \cot 3x \, dx = -\frac{1}{3} \operatorname{cosec} 3x + c$

Note: The angle must be in radians for integration rules to work, but the angle does not have to be expressed in terms of π.

Example F

Q. Evaluate $\int_{-1}^{2} \cos (2x - 3)\, dx$.

A. $\int_{-1}^{2} \cos (2x - 3)\, dx = \left[\frac{1}{2} \sin (2x - 3)\right]_{-1}^{2}$ [by implication the angles are in radians]

$= \frac{1}{2} [\sin (4 - 3)] - \frac{1}{2} [\sin (-2 - 3)]$

$= \frac{1}{2} [\sin 1 - \sin (-5)]$ [using **common factor** $\frac{1}{2}$]

$= -0.059$ (3 dp)

Unit 12.3 Activity 11A: Trigonometric integrals

1. Work out the following integrals:

a. $\int 3 \sec^2 4x \, dx$ **b.** $\int -2 \sin 2x \, dx$ **c.** $\int 4 \sec^2 \frac{x}{3} \, dx$

d. $\int \cos\left(x - \frac{\pi}{3}\right) dx$ **e.** $\int \operatorname{cosec}^2 3x \, dx$ **f.** $\int 3 \operatorname{cosec} 2x \cot 2x \, dx$

g. $\int \left(\cos \frac{x}{2} - \frac{1}{2} \sin 2x\right) dx$ **h.** $\int (3 \cos 4x + 2 \sin 3x) \, dx$

i. $\int \sin (1 - 2x) \, dx$ **j.** $\int 2 \cos (-x) \, dx$ **k.** $\int 2 \sec 4x \tan 4x \, dx$

2. Evaluate the following definite integrals:

a. $\int_0^{\frac{\pi}{2}} \cos 2x \, dx$ **b.** $\int_0^{\pi} (\sin x - \cos x) \, dx$ **c.** $\int_{\pi}^{2\pi} \sin 3x \, dx$

d. $\int_0^2 (3 + \cos 2x) \, dx$ **e.** $\int_1^2 \sin \frac{x}{2} \, dx$ **f.** $\int_0^{\frac{\pi}{2}} 2 \sin\left(3x - \frac{\pi}{2}\right) dx$

g. $\int_0^{\pi} 4 \sec^2 \pi x \, dx$ **h.** $\int_{\frac{\pi}{6}}^{\frac{\pi}{3}} \sec x \tan x \, dx$ **i.** $\int_1^2 (2x + 3 \operatorname{cosec}^2 4x) \, dx$

j. $\int_2^3 \tan 2x \, dx$ **k.** $\int_1^2 \cot x \, dx$ **l.** $\int_{\frac{\pi}{3}}^{\pi} (1 + \sin 2x) \, dx$

3. Arrange the following integrals in order of size (smallest first):

$L = \int_0^{\frac{\pi}{2}} \cos \theta \, d\theta$ $M = \int_0^{\frac{\pi}{2}} 2 \sin 2\theta \, d\theta$ $N = \int_0^{\frac{\pi}{2}} \cos \frac{1}{2}\theta \, d\theta$

4. Find limits a, b which make $\int_b^a \sin 2x \, dx = 0$.

5. The velocity in m s^{-1} of an object after t seconds is given by $v = \sin \pi t$.

a. Explain what the definite integral $\int_{\frac{1}{2}}^{2} \sin \pi t \, dt$ represents.

b. Evaluate this definite integral and express your answer in the correct units.

6. The motion of a pendulum is defined by the formula $a = -4 \sin 2t$ where a is the acceleration in m s^{-2} and t is the time in seconds.

a. Find a formula for the velocity of the pendulum as a function of time, if when $t = 0$, velocity = 2 m s^{-1}.

b. What is the maximum velocity of the pendulum?

7. The mean value of $f(x)$ for values of x from a to b is given by the formula:

mean value of $f(x) = \frac{1}{(b - a)} \int_a^b f(x) \, dx$.

Use this information to solve the following problem:

The range of a projectile fired with an initial velocity of $u = 55$ m s^{-1} at an angle of elevation θ is given by the formula: Range $= \frac{u^2}{g} \sin 2\theta$.

Find the mean range of the projectile as θ varies from 0 to $\frac{\pi}{2}$, and given that $g = 10$.

Unit 12.3: Calculus

Topic 12: Definite integrals and area

This last Topic in Unit 12.3 brings together the information presented in previous Topics and looks at some practical applications of what has been covered in the Unit:

- Demonstrate an understanding of the relationship between integrals and areas.
- Find a simple area under a graph.
- Find areas (including simple compound areas) under polynomial graphs.
- Apply calculus techniques to solve problems.

Definite integrals

The **definite integral** of f with respect to x between the limits of $x = a$ and $x = b$ is written $\int_a^b f(x)\,dx$. If $F(x)$ is an antiderivative of f, then:

$$\int_a^b f(x)\,dx = [F(x)]_a^b = F(b) - F(a)$$

Definite integrals are calculated in the following way:

Example A

$$\begin{aligned}\int_{-1}^{3} 2x\,dx &= [x^2 + C]_{-1}^{3} && \text{[integrating } 2x\text{]}\\ &= (3^2 + C) - ((-1)^2 + C) && \text{[substituting into } x^2 + C\text{]}\\ &= 3^2 - (-1)^2 && \text{[simplifying]}\\ &= 8\end{aligned}$$

Note: The constant of integration always cancels out during the calculation of a definite integral, as the example above shows. For this reason it is unnecessary to include it during the calculation.

Example B

Q. Find $\int_1^2 x^2\,dx$.

A. $$\begin{aligned}\int_1^2 x^2\,dx &= \left[\frac{x^3}{3}\right]_1^2 && \text{[integrating (no constant needed)]}\\ &= \frac{2^3}{3} - \frac{1^3}{3} && \text{[substituting]}\\ &= \frac{7}{3}\end{aligned}$$

Unit 12.3 Activity 12A: Definite integrals

1. Calculate the following definite integrals:

a. $\int_{2}^{4} x\,dx$ **b.** $\int_{1}^{5} x^2\,dx$ **c.** $\int_{1}^{2} 2t^3\,dt$ **d.** $\int_{0}^{4} \frac{u^2}{5}\,du$

e. $\int_{-1}^{1} (x^3 + 2x^2)\,dx$ **f.** $\int_{2}^{3} 2\,dx$ **g.** $\int_{-4}^{2} 3\,du$ **h.** $\int_{-3}^{6} dx$

i. $\int_{-1}^{2} x(2x + 1)\,dx$ **j.** $\int_{-2}^{1} (u + 3)^2\,du$ **k.** $\int_{-3}^{2} (r^3 - 2r^2 + 3r + 1)\,dr$

l. $\int_{-1}^{2\frac{1}{2}} x^2\,dx$ **m.** $\int_{\frac{1}{2}}^{1\frac{1}{2}} \frac{t^3}{4}\,dt$ **n.** $\int_{-3}^{2} (p^2 + 2p - 3)\,dp$ **o.** $\int_{1.2}^{2.8} (r^2 - 2)\,dr$

2. Find: **a.** $\int_{-1}^{1} (x + 4)(x - 7)\,dx$ **b.** $\int_{1.5}^{2.5} x(x + 6)\,dx$

3. a. If $\int_{1}^{3} ax^2\,dx = 13$, find the value of a.

b. If $\int_{-1}^{p} (5x - 2)\,dx = 27.5$, find p, where $p > 0$.

Areas under curves using definite integrals

Let $y = f(x)$ be a function whose graph is a smooth and continuous curve lying above the x-axis.

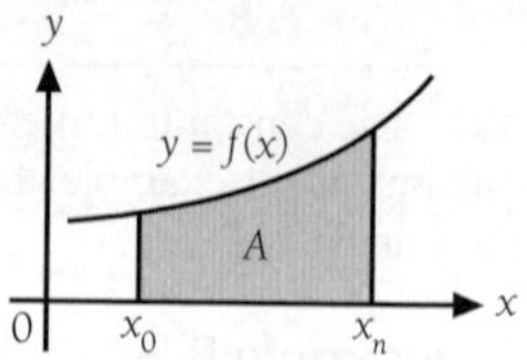

Let A be the area of the shaded region between the curve and the x-axis, from $x = x_0$ to $x = x_n$.

The relationship between the area, A, and the definite integral, $\int_{x_0}^{x_n} f(x)\,dx$, will now be established.

Let P (x, y) be any point on the curve. Suppose P was originally at P_0 (x_0, y_0) and finally takes up the position at P_n (x_n, y_n) as it moves along the curve. As the point moves from P_0 to a general position P, suppose it passes over an area a, which is drawn shaded.

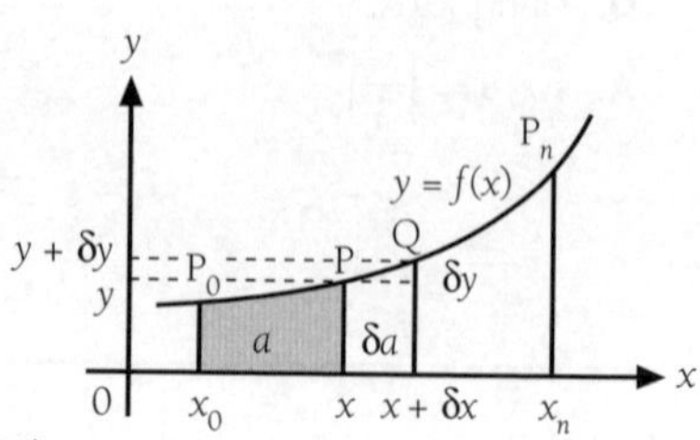

If P moved a small distance further along the curve, to Q $(x + \delta x, y + \delta y)$ say, the small horizontal increment is denoted by δx, and the corresponding small vertical increment in its position is denoted by δy. The increment in area is denoted by δa, and this is the area between the curve and the x-axis from x to $x + \delta x$.

Now concentrate on part of the curve near P and draw the lower and upper rectangles, which are shown separately below, on the right.

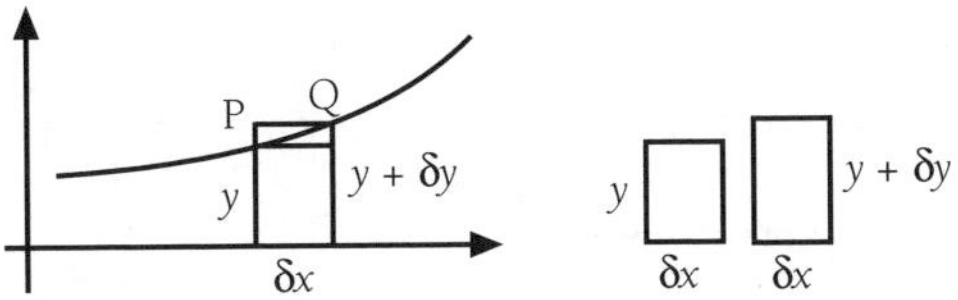

The area δa lies between the areas of the lower and upper rectangles, ie $y\delta x < \delta a < (y + \delta y)\,\delta x$. By making δx small enough, and consequently δy, the area δa can be made as close to $y\delta x$ as desired. In other words, for very small values of δx and δy the areas of the two rectangles are virtually the same and so:

$\delta a = y\delta x$

$\dfrac{\delta a}{\delta x} = y$ [dividing]

In the limit as $\delta x \to 0$,

$$\lim_{\delta x \to 0}\left(\frac{\delta a}{\delta x}\right) = \lim_{\delta x \to 0}(y)$$

$$\frac{\mathrm{d}a}{\mathrm{d}x} = y$$

$$a = \int y\,\mathrm{d}x \text{ or } a = \int f(x)\,\mathrm{d}x$$ [expressing the derivative as an integral]

So, area can be represented by an *indefinite* integral. Now to find A, the *whole* area. Suppose the antiderivative of $f(x)$ is $F(x)$, then
$a = F(x) + c$ where c is the constant of integration.

When $x = x_0$, $a = 0$, so, $0 = F(x_0) + c$ ... (1)

When $x = x_n$, $a = A$, so, $A = F(x_n) + c$ ... (2)

(2) – (1) gives $A - 0 = F(x_n) - F(x_0)$

ie $A = F(x_n) - F(x_0)$

But $F(x_n) - F(x_0)$ is an expression for the definite integral: $\int_{x_0}^{x_n} f(x)\,\mathrm{d}x$.

The logical result which follows is that the area, $A = \int_{x_0}^{x_n} f(x)\,\mathrm{d}x$. This gives rise to a result called the **Fundamental Theorem of Calculus**.

$$A = \int_{x_0}^{x_n} f(x)\,\mathrm{d}x$$
$$= F(x_n) - F(x_0), \text{ where } F \text{ is the antiderivative of } f$$

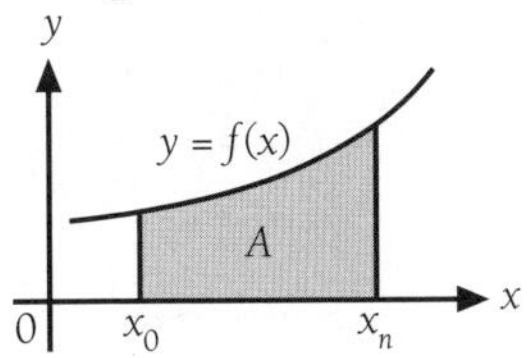

This formula, which relates the definite integral to the antiderivative, provides a method of evaluating definite integrals (and hence areas).

Note: It is best not to give square units for area because the area could represent other quantities, eg work.

Example C

Q. Find the area of the shaded region shown in the graph:

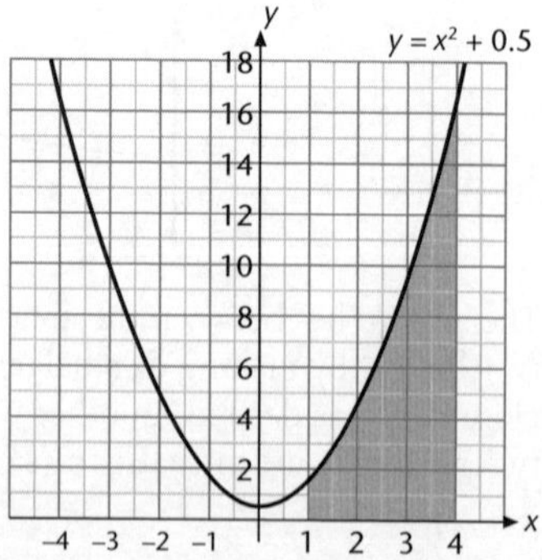

A. Area $= \int_1^4 (x^2 + 0.5)\,dx$

$= \left[\frac{x^3}{3} + 0.5x\right]_1^4$ [integrating]

$= \left(\frac{64}{3} + 2\right) - \left(\frac{1}{3} + \frac{1}{2}\right)$ [substituting into $\frac{x^3}{3} + 0.5x$]

$= 22\frac{1}{2}$

If the graph of $y = f(x)$ has negative y-values between $x = a$ and $x = b$, then the definite integral is the *negative* of the shaded area.

If A is the area below the x-axis then:

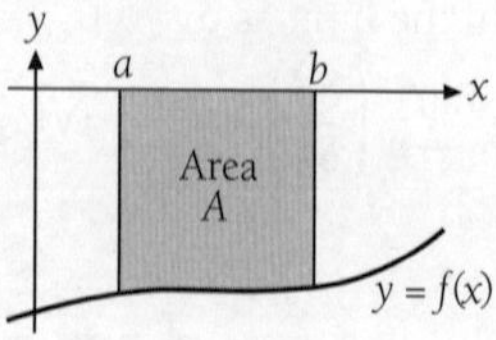

$$A = -\int_a^b f(x)\,dx$$

or, using the absolute value function:

$$A = \left|\int_a^b f(x)\,dx\right|$$

It is advisable to sketch a graph before finding the area required.

Example D

The area of the shaded region shown is:

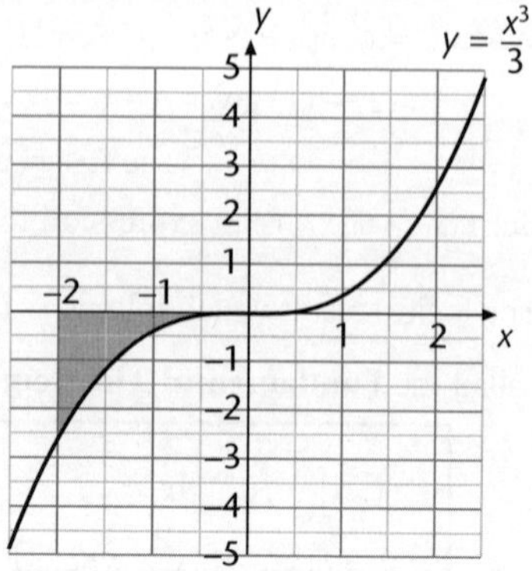

$A = \left|\int_{-2}^{0} \frac{x^3}{3}\,dx\right|$ [all points on the graph between –2 and 0 are below the x-axis]

$= \left|\left[\frac{x^4}{12}\right]_{-2}^{0}\right|$ [integrating]

$= \left|0 - \frac{4}{3}\right|$ [substituting]

$= \frac{4}{3}$ [since $\left|-\frac{4}{3}\right| = \frac{4}{3}$]

When the function has both negative and positive values as shown, the shaded area between the graph and the x-axis is made up of an area above the axis, labelled A_1, and an area below the axis, labelled A_2.

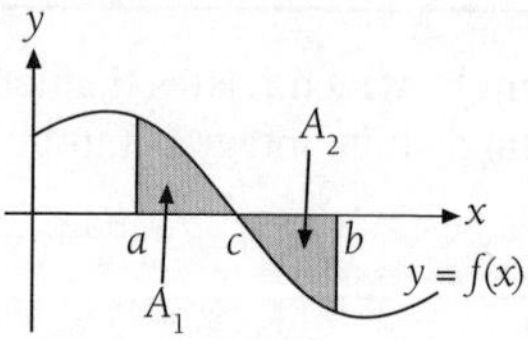

The shaded area is $A_1 + A_2 = \int_a^c f(x)\,dx + \left|\int_c^b f(x)\,dx\right|$

Example E

The area between $y = 4 - x^2$, $x = 1$ and $x = 3$ is

$$\int_1^2 (4 - x^2)\,dx + \left|\int_2^3 (4 - x^2)\,dx\right| = \left[4x - \frac{x^3}{3}\right]_1^2 + \left|\left[4x - \frac{x^3}{3}\right]_2^3\right|$$

$$= \frac{5}{3} + \frac{7}{3}$$

$$= 4$$

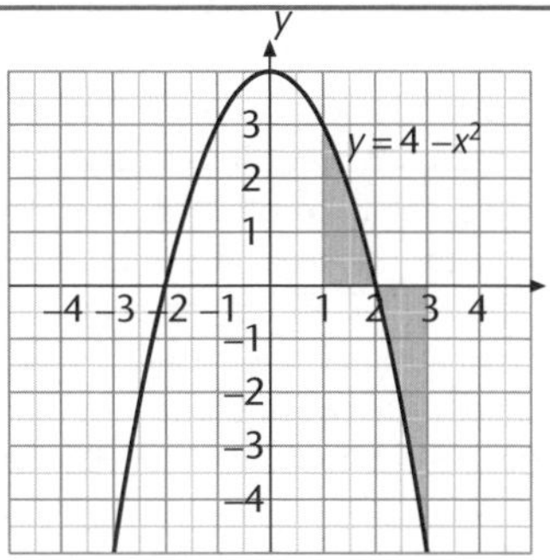

Unit 12.3 Activity 12B: Areas under graphs

1. The graph shown is a sketch of $y = x^2 - 2x + 3$. Three areas A_1, A_2 and A_3 are shaded.
   Find the values of A_1, A_2, A_3.

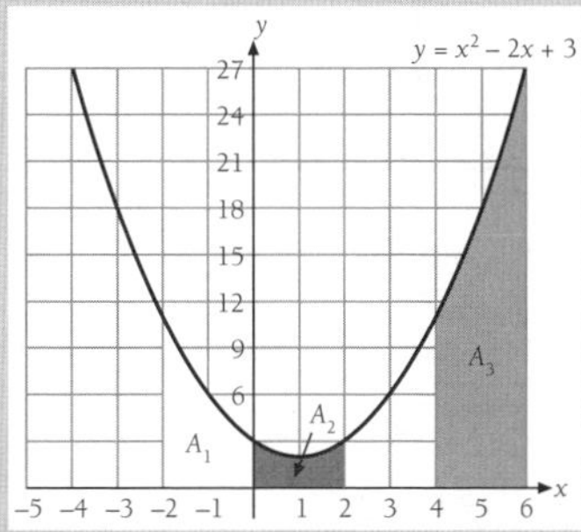

2. Find the area between:
   a. $y = x + 1$, $x = 1$, $x = 3$ and the x-axis.
   b. $y = 2x - 1$, $x = 2$, $x = 4$ and the x-axis.
   c. $y = x^2$, $x = 2$, $x = 5$ and the x-axis.
   d. $y = 4 - x^2$, $x = 1$, $x = 2$ and the x-axis.
   e. $y = 4x - x^2$ and the x-axis.
   f. $y = x^2 - 3x - 4$ and the x-axis.
   g. $y = x^3$, $x = -1$, $x = 2$ and the x-axis.
   h. $x = 1$, $y = (x - 2)(x - 4)$, $x = 3$ and the x-axis.
   i. $x = 0$, $y = 4 - x^2$, $x = 3$ and the x-axis.

Unit 12.3 Activity 12C: Problem solving using definite integrals

1. For the diagram to the right, find what fraction of the rectangle is shaded.

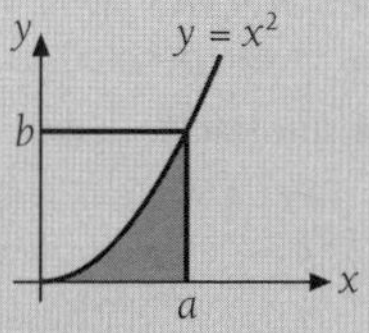

2. Find the value of k so that the shaded area is half the area of the largest rectangle in the diagram alongside.

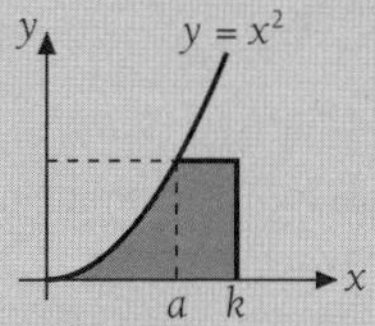

3. Find the value of k so that the area between the curve $y = k - x^2$ and the x-axis is 2.

4. Find the value of k so that the area contained between $y = x^k$, $x = 0$, $x = 1$ and the x-axis is 3.

Areas between curves

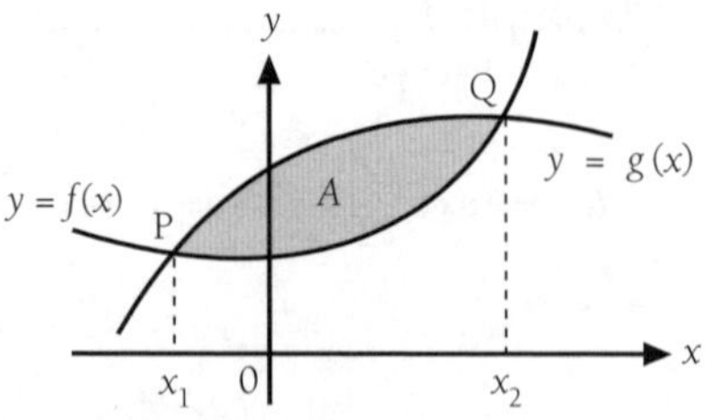

Suppose two curves $y = f(x)$ and $y = g(x)$ intersect at P and Q and the area enclosed between them is required. The required area is shaded and called A. It is necessary to find the x-coordinates of the points of intersection P and Q, because these will be used as the limits. They are found by solving the equations $y = g(x)$ and $y = f(x)$ simultaneously. Suppose the **roots** of the simultaneous equations are x_1 and x_2. The sketches show the separate graphs. The areas between each individual graph and the x-axis are called A_1 and A_2.

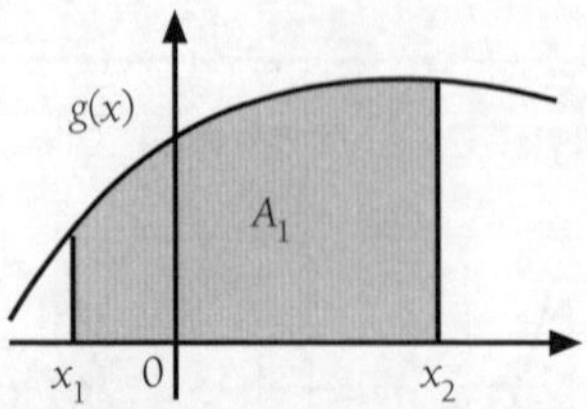

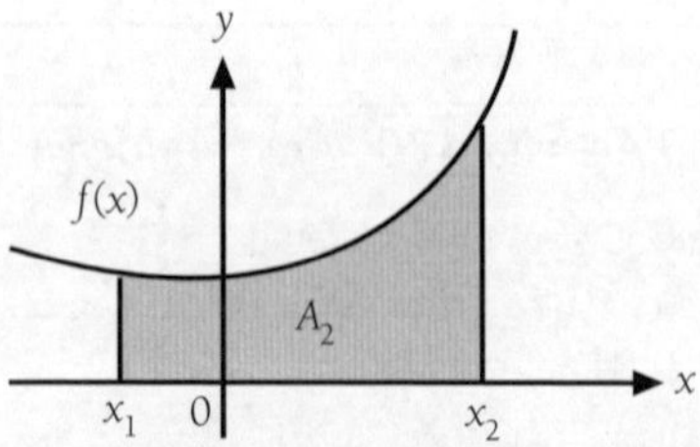

The area between the two curves is given by

$A = A_1 - A_2$

$A = \int_{x_1}^{x_2} g(x)\,dx - \int_{x_1}^{x_2} f(x)\,dx$

$= \int_{x_1}^{x_2} (g(x) - f(x))\,dx$

Subtraction of the functions is possible since the limits are the same for both integrals.

Example F

Q. The equation of the curve shown alongside is $f(x) = -x^2 + 6$. Find the shaded area.

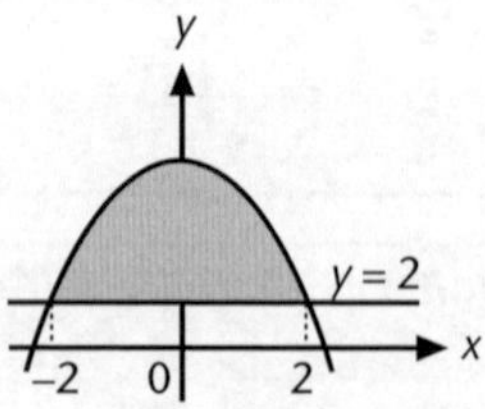

A. The shaded area is between the curve $y = -x^2 + 6$ and the line $y = 2$.

Solving these equations simultaneously to find the points of intersection of the straight line and the curve is the first part of the problem.

Substituting $y = 2$ in $y = -x^2 + 6$ gives:

$2 = -x^2 + 6$

$x^2 = 4$ [rearranging]

$x = \pm 2$ [these are the limits for the definite integral]

The formula for the area between two curves is

$A = \int_{x_1}^{x_2} (g(x) - f(x))\,dx$

The shaded area is

$$A = \int_{-2}^{2}(-x^2 + 6 - 2)\,dx \quad \text{[substituting } g(x) = -x^2 + 6 \text{ and } f(x) = 2 \text{ and limits]}$$

$$= \int_{-2}^{2}(-x^2 + 4)\,dx \quad \text{[simplifying]}$$

$$= \left[-\frac{x^3}{3} + 4x\right]_{-2}^{2} \quad \text{[antidifferentiating]}$$

$$= \left[-\frac{2^3}{3} + 8\right] - \left[-\frac{(-2)^3}{3} - 8\right] \quad \text{[substituting the limits } x = 2 \text{ and } x = -2]$$

$$= 10\frac{2}{3}$$

The shaded area is $10\frac{2}{3}$

Example G

Q. Calculate the shaded area in the figure.

A. The two functions $y = 2x$ and $y = \tan x$ are subtracted in order to find the shaded area. ($2x$ is written first in the subtraction because $y = 2x$ lies above $y = \tan x$ for the shaded part of the graph.) Thus,

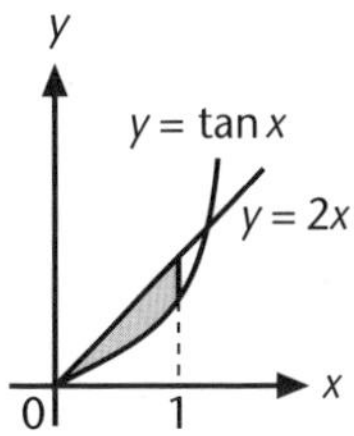

$$A = \int_{0}^{1}(2x - \tan x)\,dx$$

$$= \int_{0}^{1}\left(2x - \frac{\sin x}{\cos x}\right)dx \quad \left[\text{using } \tan x = \frac{\sin x}{\cos x}\right]$$

$$= \left[x^2 - \ln|\cos x|\right]_0^1 \quad \text{[antidifferentiating]}$$

$$= \left[1^2 - \ln|\cos 1|\right] - \left[0^2 - \ln|\cos 0|\right] \quad \text{[substituting the limits]}$$

$$= 1.62 \text{ (2 dp)} \quad \text{[using calculator]}$$

The shaded area is 1.62.

Investigation

The diagram shows the cross-section of a lean-to shed against a garage, and PQ is the straight roofing iron. OP is the wall of the garage and QR is a vertical pillar. John has decided to replace the straight roofing iron with a new roof in the shape of a parabola. The mathematical model for the shed uses OP as the y-axis, and OR as the x-axis. The distance OR is 4 metres, and the equation of the curve of the new roof is $y = \sqrt{x}$. PQ is a tangent to the parabola at the point Q.

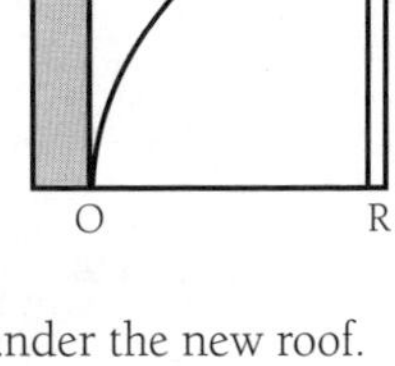

John's son, Bill, claims that there will be a 10% reduction in the space under the new roof. Check out Bill's claim.

Unit 12.3 Activity 12D: Areas

1. Using the definite integral, calculate the shaded areas between the curves and the x-axis:

a.

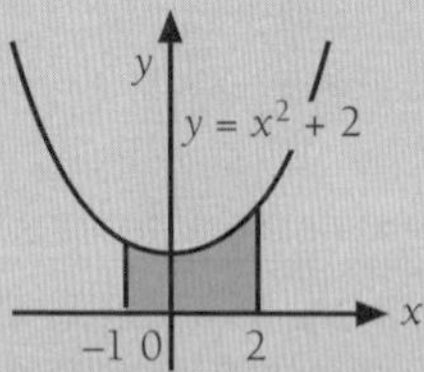

b.

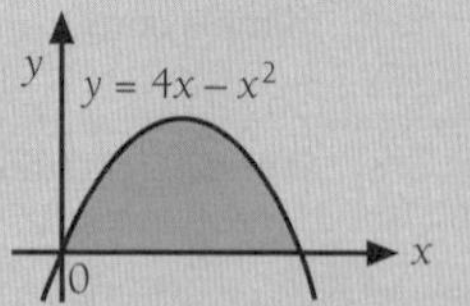

c.

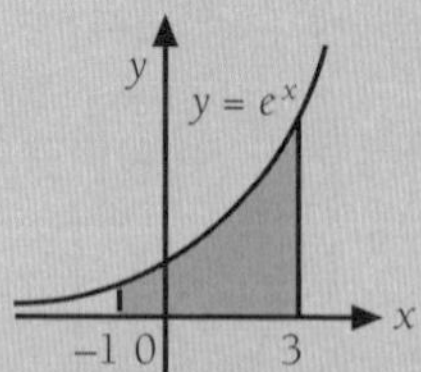

d.

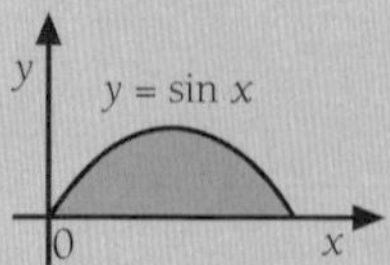

e.

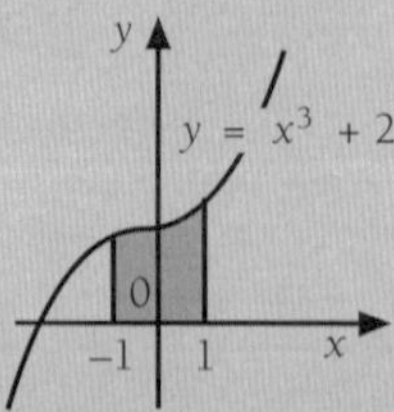

f.

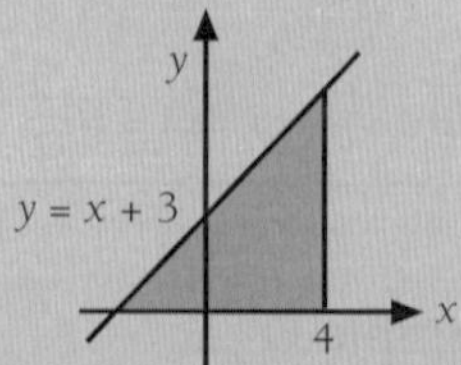

g.

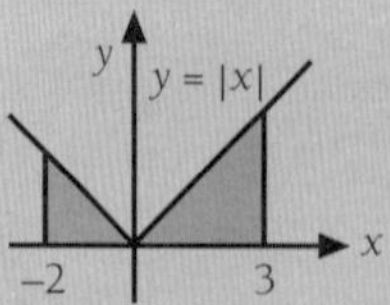

h.

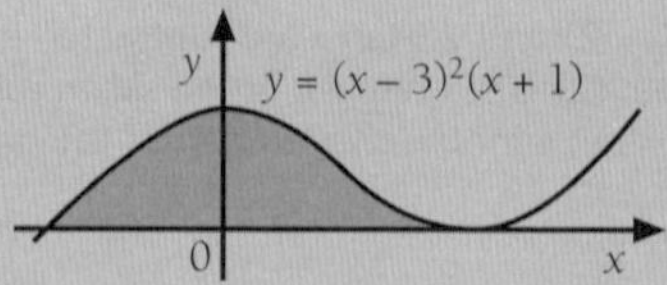

i.

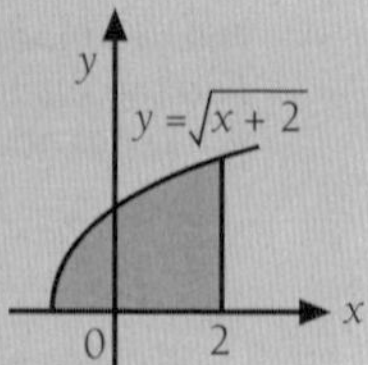

2. Find the areas shaded between the curve and the y-axis:

a.

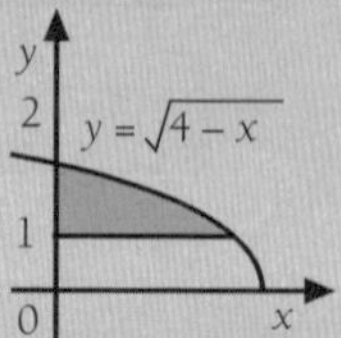

b.

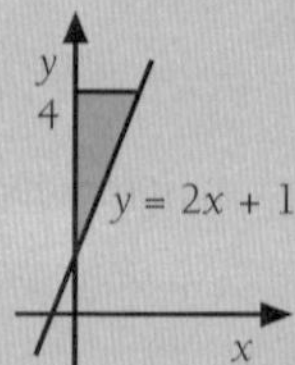

c.

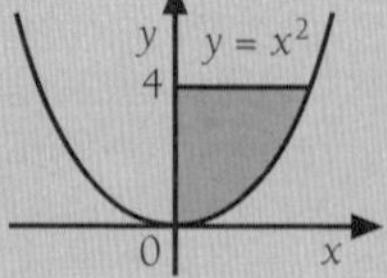

d. Verify your answer to **2. b.** by using the formula for the area of a triangle.

3. Find the shaded areas between the curves and the straight lines:

a.

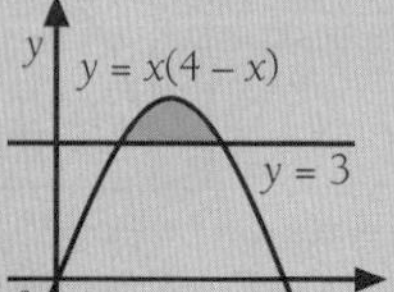

b.

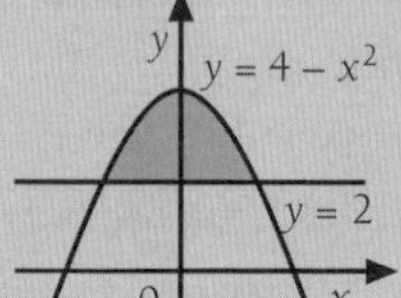

c.

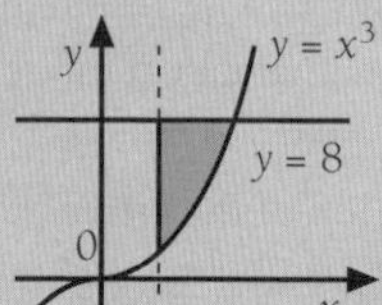

4. In the diagrams alongside, the equation of both curves is $y = \cos x$. If the two shaded areas are equal in area, find a.

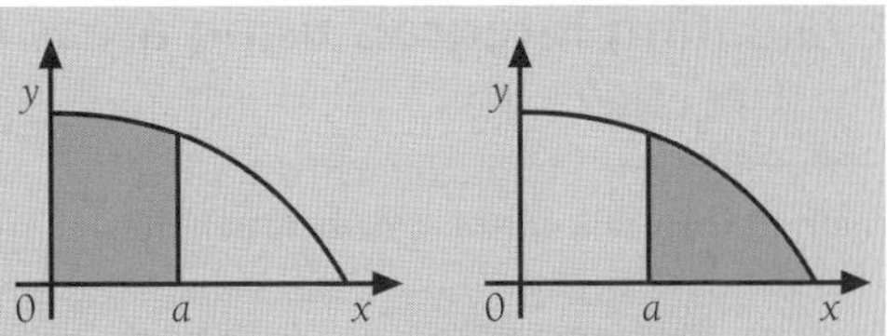

5. The diagram shows the entrance to a tunnel. The brick part is to be painted. Axes are drawn on the figure and the dimensions are as shown. The roof of the tunnel is in the shape of a parabola whose equation is $y = 4 - x^2$.

Calculate the area of the bricks to be painted.

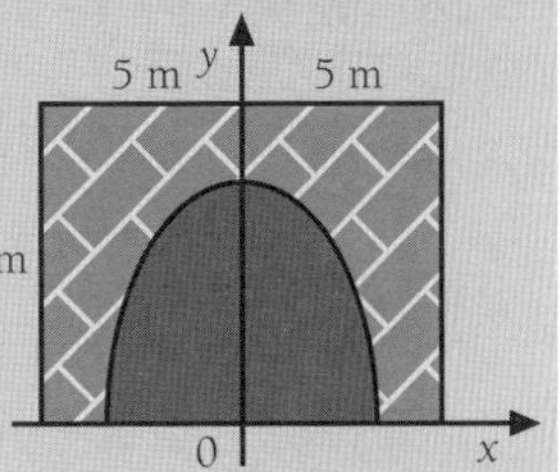

6. The Greek mathematician Archimedes, who lived from about 287 BC to 212 BC, knew that the ratio of the areas A to B as shown in the graph is 2:1.

Using calculus (which was not known to Archimedes), prove this result.

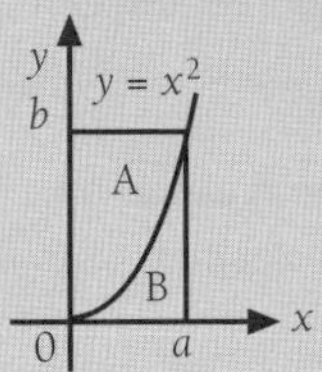

7. Find the shaded areas between curves:

a.

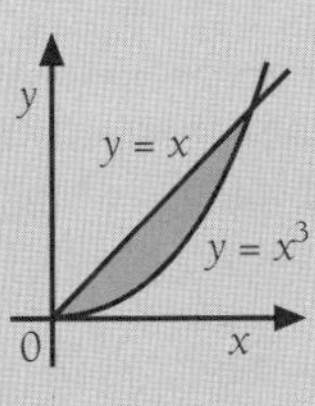

b.

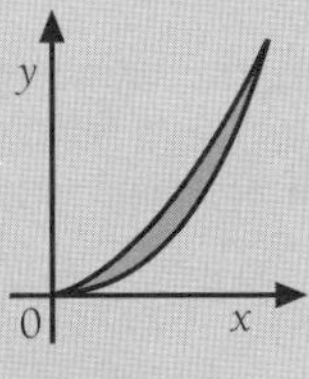

One of the curves is $y = x^3$ and the other curve is $y = x^4$. Distinguish between them and find the shaded area.

c.

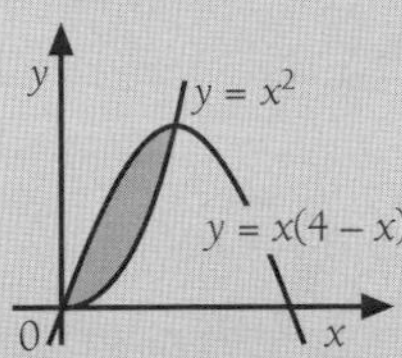

d.

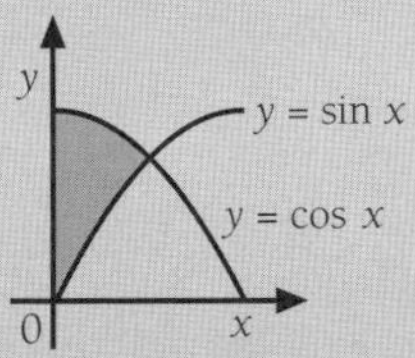

e.

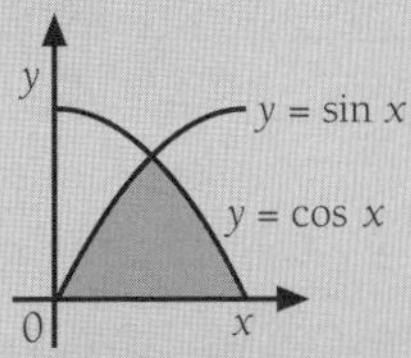

8. Sketch the graphs of $y = 2x^2$ and $y = 3 - x^2$ and find the area enclosed between them.

9. Sketch the graphs of $y = x^2$ and $y = \sqrt{x}$ in the first quadrant. Find the area between the two curves.

10. Find the ratio of the areas in which the line $y = x$ divides the area enclosed by curve $y = 4x - x^2$ and the x-axis.

11. Find the areas between the curves and the x-axis for the domains given:

a. $f(x) = x^3 + \sin x$, for $\frac{\pi}{2} \le x \le \pi$.

b. $f(x) = 3x + \tan x$, for $0 \le x \le \frac{\pi}{4}$.

Evaluating integrals using areas

The graph of $f(x) = \cos x$ is shown.

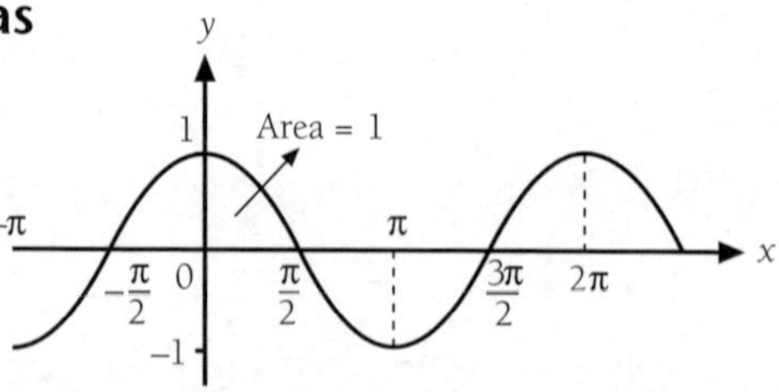

In the following discussion when 'area' is referred to, it will be the area between the curve and the x-axis.

- The area between $x = 0$ and $x = \frac{\pi}{2}$ is:

$$\int_0^{\frac{\pi}{2}} \cos x \, dx = [\sin x]_0^{\frac{\pi}{2}} \qquad \text{[integrating]}$$
$$= \sin \frac{\pi}{2} - \sin 0 = 1$$

- Now, by symmetry, the *area* between $x = \frac{\pi}{2}$ and $x = \pi$ will also equal 1, so $\int_{\frac{\pi}{2}}^{\pi} \cos x \, dx = -1$, and the integral does not need to be worked out to know this.

- The next definite integral can also be evaluated without integrating, using the 'algebraic sum' idea.

  $\int_0^{\pi} \cos x \, dx = 1 + -1 = 0$

  But, the *area* between $x = 0$ and $x = \pi$ is, of course $1 + 1 = 2$.

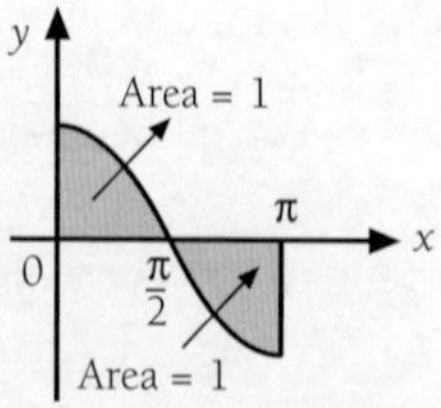

Unit 12.3 Activity 12E: Integrals using areas

1. a. i. Find the shaded area.

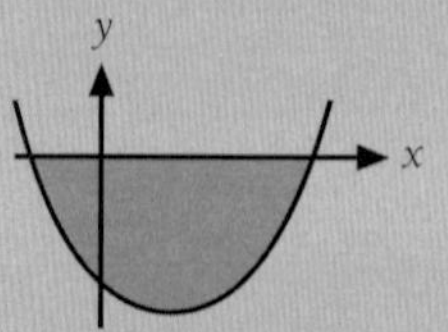

$y = (x - 2)(x + 1)$

ii. Explain why the definite integral used in your answer to part **i** gives a negative answer.

b. i. Find the shaded area.

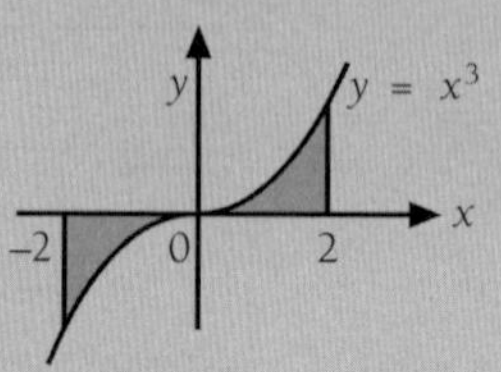

ii. Evaluate $\int_{-2}^{2} x^3 \, dx$, without integrating.

2. Evaluate the definite integral $\int_0^{\frac{\pi}{2}} \cos x \, dx$.

Using your result, evaluate the following definite integrals (a sketch of $y = \cos x$ will be helpful).

a. $\int_{-\frac{\pi}{2}}^{0} \cos x \, dx$ **b.** $\int_{\frac{3\pi}{2}}^{2\pi} \cos x \, dx$ **c.** $\int_0^{\frac{3\pi}{2}} \cos x \, dx$

d. $\int_0^{\pi} \cos x \, dx$ **e.** $\int_0^{2\pi} \cos x \, dx$

3. Suppose $f(x)$ is a periodic, odd function of period 4, whose graph is as shown.

Given that $\int_{-2}^{0} f(x)\,dx = -5$, evaluate the following definite integrals:

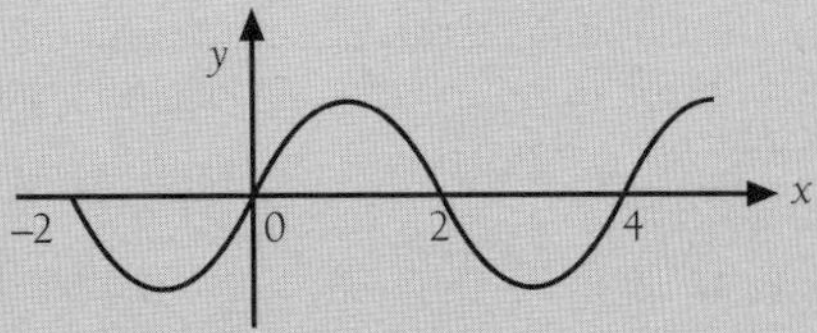

a. $\int_{2}^{4} f(x)\,dx$ **b.** $\int_{0}^{4} f(x)\,dx$ **c.** $\int_{-2}^{4} f(x)\,dx$

4. Each of the graphs below has equation $y = \sin x$. Find the shaded areas:

a.

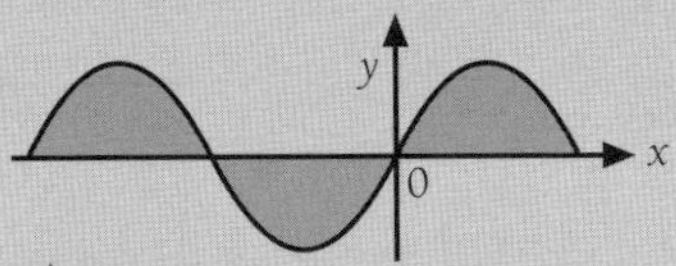

b.

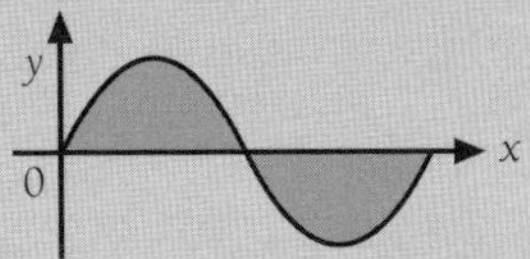

c.

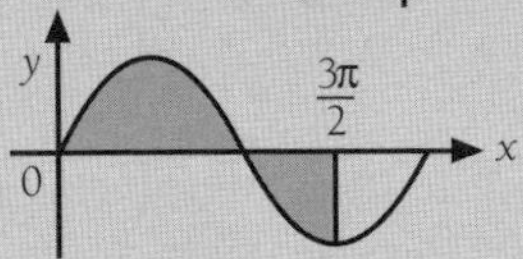

5. $y = f(x)$ is defined by the curve drawn alongside. The areas of the regions are A_1, A_2, and A_3. Express each of these definite integrals in terms of the areas.

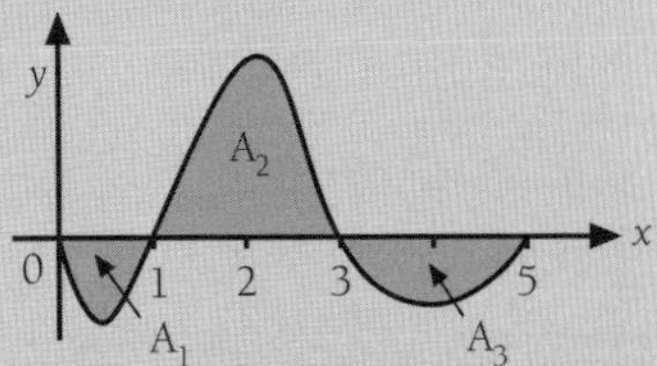

a. $\int_{0}^{1} f(x)\,dx$ **b.** $\int_{0}^{3} f(x)\,dx$

c. $\int_{1}^{5} f(x)\,dx$

6. Give an expression for the total area of the shaded region as the sum of two definite integrals.

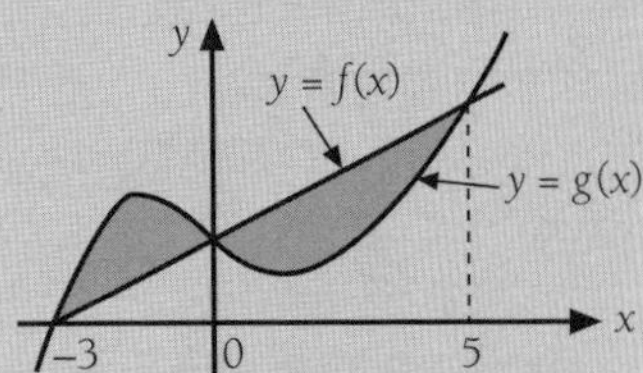

7. $f(x)$ is a periodic, odd function such that $f(x) = f(x + 4)$ and is defined by its graph. If $\int_{0}^{2} f(x)\,dx = k$, evaluate:

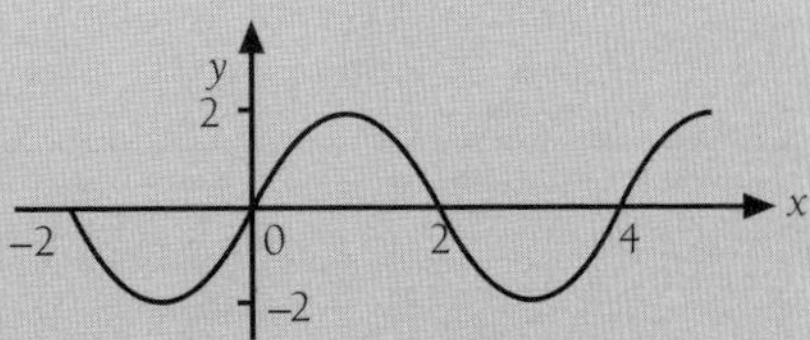

a. $\int_{0}^{4} f(x)\,dx$ **b.** $\int_{-2}^{4} f(x)\,dx$

c. $\int_{2}^{0} f(x)\,dx$

8. In this question, use the function $f(x)$ defined in question **7**.

a. Sketch the graph of $y = f(-x)$ and evaluate the following definite integrals:

i. $\int_0^2 f(-x)\,dx$ **ii.** $\int_2^4 f(-x)\,dx$ **iii.** $\int_{-2}^4 f(-x)\,dx$

b. Sketch the graph of $y = -f(x)$ and evaluate these definite integrals:

i. $\int_0^2 -f(x)\,dx$ **ii.** $\int_{-2}^2 -f(x)\,dx$

c. Sketch the graphs of $y = f(x) + 2$ and evaluate these definite integrals:

i. $\int_0^2 (f(x) + 2)\,dx$ **ii.** $\int_{-2}^4 (f(x) + 2)\,dx$

d. Sketch the graph of $y = f(x + 2)$ and evaluate these definite integrals:

i. $\int_0^2 f(x + 2)\,dx$ **ii.** $\int_{-2}^2 f(x + 2)\,dx$

9. The graph of the function f is shown on the right for $0 \le x \le 6$.

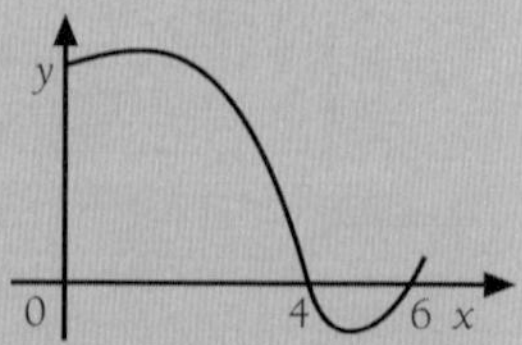

$\int_0^k f(x)\,dx$ attains its greatest value when k is equal to which of the following values?

A. 0 **B.** 1 **C.** 2

D. 4 **E.** 6

10. The figure shows the graph of $y = f(x)$. A_1 and A_2 are the areas of the regions shown, with $0 < A_2 < A_1$.

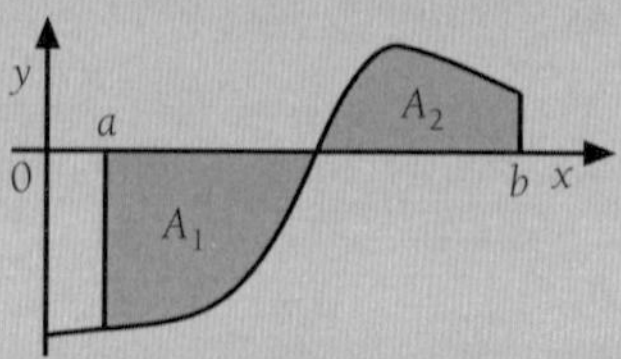

The value of $\int_a^b f(x)\,dx$ is equal to which of the following?

A. $A_2 - A_1$ **B.** $A_2 + A_1$ **C.** $A_1 - A_2$

D. $|A_1 - A_2|$ **E.** $\frac{A_1 + A_2}{2}$

Appendix

Mathematical symbols

Sets

$b \in A$	b is an element of A
$c \notin A$	c is not an element of A
U or ξ	the universal set
$\varnothing$ or { }	the empty or null set
A´	the complement of A
$A \subset B$	A is a subset of B
$A \not\subset B$	A is not a subset of B
$A \cup B$	the union of A and B, ie, the set containing elements in A or B (or in both)
$A \cap B$	the intersection of A and B, ie, the set containing elements in both A and B
$n(A)$	the number of elements in set A

Number sets

N	Natural numbers
W	Whole numbers
J	Integer numbers
Q	Rational numbers
Q´	Irrational numbers
R	Real numbers

Mathematical formulae

Area

$\frac{1}{2}bh$	triangle	a^2	square (side a)
bh	rectangle or parallelogram	$\frac{1}{2}(a+b)h$	trapezium
$\frac{1}{2}d_1d_2$	rhombus or kite	πr^2	circle
$\frac{\theta}{360} \times \pi r^2$	sector (θ in degrees)	$4\pi r^2$	sphere (surface area)
$\frac{1}{2}r^2\theta$	sector (θ in radians)	$2\pi r(r+h)$	closed cylinder (surface area)

Volume

l^3	cube (side l)	Ah	prism (A = base area)
lbh	cuboid	$\frac{1}{3}Ah$	pyramid (A = base area)
$\frac{1}{2}bhl$	triangular prism	$\frac{4}{3}\pi r^3$	sphere
$\pi r^2 h$	cylinder	$\frac{1}{3}\pi r^2 h$	cone

Sequences and series

Arithmetic series

$a + (a + d) + (a + 2d) + (a + 3d) + \ldots$

$t_n = a + (n - 1)d$

$S_n = \frac{n}{2}(2a + (n - 1)d)$

Geometric series

$a + ar + ar^2 + ar^3 + \ldots$

$t_n = ar^{n-1}$

$S_n = \frac{a(1 - r^n)}{(1 - r)}, r \neq 1$

$S_\infty = \frac{a}{1 - r}$ for $|r| < 1$

Coordinate geometry

Distance

$d = \sqrt{(x_1 - x_2)^2 + (y_1 - y_2)^2}$

Midpoint of a line

midpoint $\left(\frac{x_1 + x_2}{2}, \frac{y_1 + y_2}{2}\right)$

Gradient

$m = \frac{y_2 - y_1}{x_2 - x_1}$

Lines

$y = mx + c$

$y - y_1 = m(x - x_1)$

Perpendicular and parallel lines

$y = m_1x + c_1$ and $y = m_2x + c_2$ are parallel if $m_1 = m_2$ and perpendicular if $m_1 \times m_2 = -1$

Circles

$x^2 + y^2 = r^2$	circle with centre (0, 0) and radius r
$(x - a)^2 + (y - b)^2 = r^2$	circle with centre (a, b) and radius r

Perimeter

$C = \pi d$ or $2\pi r$	circumference of a circle
$\frac{\theta}{360} \times 2\pi r$	arc length of a sector (θ in degrees)
$r\theta$	arc length of a sector (θ in radians)

Trigonometry

$\sin\theta = \frac{o}{h}$	o = opposite	
$\cos\theta = \frac{a}{h}$	h = hypotenuse	
$\tan\theta = \frac{o}{a}$	a = adjacent side	
$h^2 = a^2 + o^2$	Pythagoras' theorem	

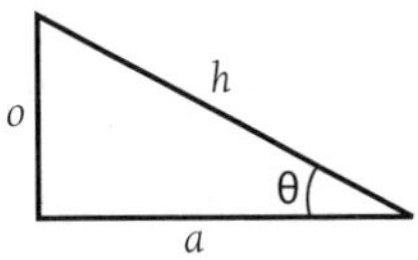

Special angles

θ	0°	30°	45°	60°	90°
sin θ	0	$\frac{1}{2}$	$\frac{1}{\sqrt{2}}$	$\frac{\sqrt{3}}{2}$	1
cos θ	1	$\frac{\sqrt{3}}{2}$	$\frac{1}{\sqrt{2}}$	$\frac{1}{2}$	0
tan θ	0	$\frac{1}{\sqrt{3}}$	1	$\sqrt{3}$	

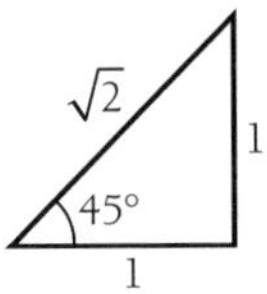

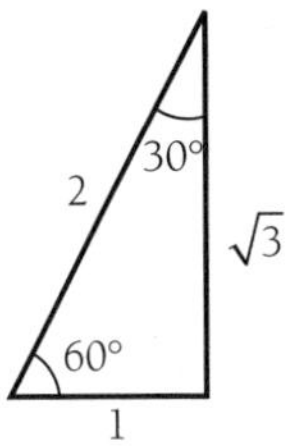

$$\text{cosec}\,\theta = \frac{1}{\sin\theta}$$

$$\sec\theta = \frac{1}{\cos\theta}$$

$$\cot\theta = \frac{1}{\tan\theta}$$

$$\cot\theta = \frac{\cos\theta}{\sin\theta}$$

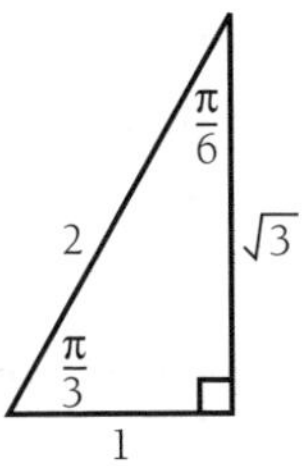

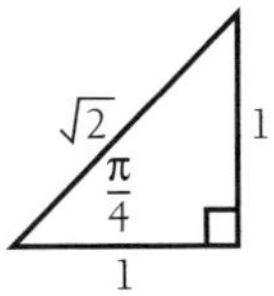

Sine rule

$$\frac{a}{\sin A} = \frac{b}{\sin B} = \frac{c}{\sin C}$$

Cosine rule

$$c^2 = a^2 + b^2 - 2ab\cos C$$

Identities

$\cos^2\theta + \sin^2\theta = 1$

$\tan^2\theta + 1 = \sec^2\theta$

$\cot^2\theta + 1 = \text{cosec}^2\,\theta$

Compound angles

$\sin(A \pm B) = \sin A\cos B \pm \cos A\sin B$

$\cos(A \pm B) = \cos A\cos B \quad \sin A\sin B$

$\tan(A \pm B) = \frac{\tan A \pm \tan B}{1 \quad \tan A\tan B}$

Products

$2\sin A\cos B = \sin(A+B) + \sin(A-B)$

$2\cos A\sin B = \sin(A+B) - \sin(A-B)$

$2\cos A\cos B = \cos(A+B) + \cos(A-B)$

$2\sin A\sin B = \cos(A-B) - \cos(A+B)$

Sums

$\sin C + \sin D = 2\sin\frac{C+D}{2}\cos\frac{C-D}{2}$

$\sin C - \sin D = 2\cos\frac{C+D}{2}\sin\frac{C-D}{2}$

$\cos C + \cos D = 2\cos\frac{C+D}{2}\cos\frac{C-D}{2}$

$\cos C - \cos D = -2\sin\frac{C+D}{2}\sin\frac{C-D}{2}$

Double angles

$\sin 2A = 2\sin A\cos A$

$\cos 2A = \cos^2 A - \sin^2 A$

$= 2\cos^2 A - 1$

$= 1 - 2\sin^2 A$

$\tan 2A = \frac{2\tan A}{1-\tan^2 A}$

General solutions

If $\sin\theta = \sin\alpha$, then $\theta = n\pi + (-1)^n\alpha$

If $\cos\theta = \cos\alpha$, then $\theta = 2n\pi \pm \alpha$

If $\tan\theta = \tan\alpha$, then $\theta = n\pi + \alpha$

where n is any integer

Trigonometric identities

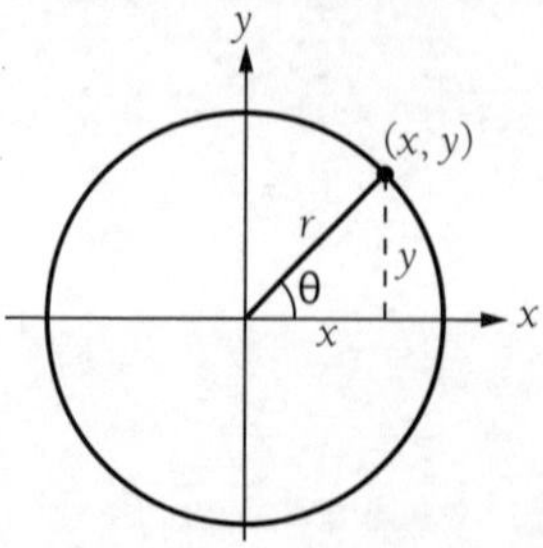

$\sin(90^\circ \pm \theta) = \cos\theta$

$\cos(90^\circ \pm \theta) = \mp\sin\theta$

$\sin(180^\circ \pm \theta) = \mp\sin\theta$

$\cos(180^\circ \pm \theta) = -\cos\theta$

$\sin(360^\circ - \theta) = -\sin\theta$

$\cos(360^\circ - \theta) = \cos\theta$

Triangles

Sine rule: $\frac{a}{\sin A} = \frac{b}{\sin B} = \frac{c}{\sin C}$ or $\frac{\sin A}{a} = \frac{\sin B}{b} = \frac{\sin C}{c}$

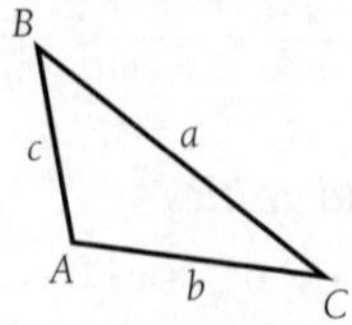

Cosine rule: $a^2 = b^2 + c^2 - 2bc\cos A$ or $\cos A = \frac{b^2+c^2-a^2}{2bc}$

Area of triangle: Area $= \frac{1}{2}ab\sin C$

Calculus

$$\frac{d}{dx}(ax^n) = nax^{n-1}$$

Differentiation of a sum

$$(f + g)' = f' + g'$$

Integration

Function $y = f(x)$	Indefinite integral $\int y\,dx = \int f(x)\,dx$	Definite integral $\int_a^b y\,dx = \int_a^b f(x)\,dx$
0	k	0
c	$cx + k$	$c(b - a)$
$x^n\ (n \neq -1)$	$\frac{x^{n+1}}{n+1} + k$	$\frac{b^{n+1} - a^{n+1}}{n+1}$

Differentiation

$y = f(x)$	$\frac{dy}{dx} = f'(x)$
$\ln x$	$\frac{1}{x}$
e^{ax}	ae^{ax}
$\sin x$	$\cos x$
$\cos x$	$-\sin x$
$\tan x$	$\sec^2 x$
$\sec x$	$\sec x \tan x$
$\operatorname{cosec} x$	$-\operatorname{cosec} x \cot x$
$\cot x$	$-\operatorname{cosec}^2 x$

Integration

$y = f(x)$	$\equiv f(x)\,dx$
x^n	$\frac{1}{n+1}x^{n+1} + c$
$\frac{1}{x}$	$\ln \lvert x \rvert + c$
$\frac{f'(x)}{f(x)}$	$\ln \lvert f(x) \rvert + c$

Product rule

$$(f \cdot g)' = f \cdot g' + g \cdot f'$$

or if $y = uv$ then $\frac{dy}{dx} = u\frac{dv}{dx} + v\frac{du}{dx}$

Quotient rule

$$\left(\frac{f}{g}\right)' = \frac{g.f' - f.g'}{g^2}$$

or if $y = \frac{u}{v}$ then $\frac{dy}{dx} = \frac{v\frac{du}{dx} - u\frac{dv}{dx}}{v^2}$

Composite function or chain rule

$(f(g))' = f'(g) \cdot g'$

or if $y = f(u)$ and $u = g(x)$ then $\frac{dy}{dx} = \frac{dy}{du} \cdot \frac{du}{dx}$

Volume of revolution

$y = f(x)$ between $x = a$ and $x = b$ rotated about the x-axis.

Volume $= \int_a^b \pi y^2 \, dx$

Answers

Answers for many questions include a 'Marking Guide':

- **A** ('Achievement', meaning 'satisfactory achievement').
- **M** ('Merit', meaning 'high achievement').
- **E** ('Excellence', meaning 'very high achievement').

The Marking Guide has been made by the authors and the publishers and is not an official guide, but we hope it will be a help to students who are striving for the best possible results.

Unit 12.1 Patterns and Algebra

Unit 12.1 Activity 1A: Definition and notation of sets (page 2)

1. **a.** **i.** {2, 3, 5, 7, 11, 13, 17, 19}

ii. Answers will vary. Discuss with your teacher.

iii. {100, 121, 144, 169, 196}

b. They are all finite.

c. **i.** n(P) = 8

ii. Answers will vary. Discuss with your teacher.

iii. n(R) = 5

2. **a.** F **b.** T **c.** T **d.** F **e.** T **f.** F

g. F **h.** T **i.** F **j.** F **k.** T **l.** T

m. T **n.** T

3. ∅, {a}, {b}, {c}, {a,b}, {a, c}, {b, c}, {a, b, c,}

4. **a.** 32 **b.** {p, q, r, s}, {p, q, r, t}, {p, q, s, t}, {p, r, s, t}, {q, r, s, t}

Unit 12.1 Activity 2A: Universal sets, complements and Venn diagrams (page 5)

1. **a.** A = {2, 4, 6, 8, 10}, B = {2, 3, 5, 7}, C = {3, 6, 9}

b. **i.** {2} **ii.** {6} **iii.** {3}

iv. {2, 3, 5, 7} **v.** { } **vi.** { }

c. **i.** {2, 3, 4, 5, 6, 7, 8, 10} **ii.** {2, 3, 4, 6, 8, 9, 10} **iii.** {2, 3, 5, 6, 7, 9}

iv. {1, 2, 3, 4, 5, 6, 7, 8, 9, 10} **v.** {2, 3, 5, 7}

d.

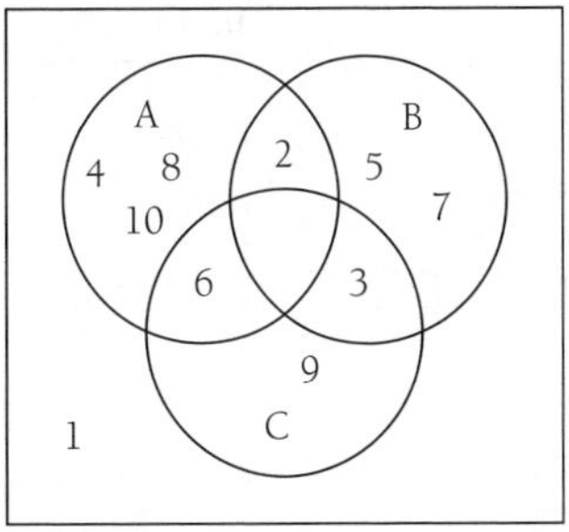

e. A' = {1, 3, 5, 7, 9} B' = {1, 4, 6, 8, 9, 10} C' = {1, 2, 4, 5, 7, 8, 10}

f. **i.** {1, 2, 3, 5, 7, 9} **ii.** {1, 2, 3, 4, 5, 6, 7, 8, 10} **iii.** {1, 4, 8, 10}

iv. {3, 5, 7} **v.** {1, 4, 6, 8, 9, 10}

2. **a.** V = { a, e, i, o, u} ; W = {c, o, n, s, a, t}

b. **i.** {a, o,} **ii.** {e, i, u} **iii.** {n, t}

iv. {c, o, n, s, a, t} **v.** { }

c. **i.** {a, e, i, o, u, c, n, s, t} **ii.** {a, e, i, o, u, n, t} **iii.** {c, o, n, s, a, t, e, i, u}

iv. ξ **v.** {c, o, n, s, a, t}

d. **i.** {a, e, i, o, u, c, n, s, t} **ii.** { } **iii.** { }

iv. {u, n, i, t, e, a, o} **v.** {u, n, i, t, e} **vi.** {a, e, i, o, u, n, t}

vii. {u, n, i, t, e} **viii.** {i, e, u} **ix.** { }

x. {a, e, i, o, u, c, n, s, t} **xi.** {a, e, i, o, u, c, n, s, t}

xii. {a, e, i, o, u, n, t}

3. **a.** **i.** {12} **ii.** {6, 12, 18}

iii. {12} **iv.** {12}

b. $R \subset P$

c.

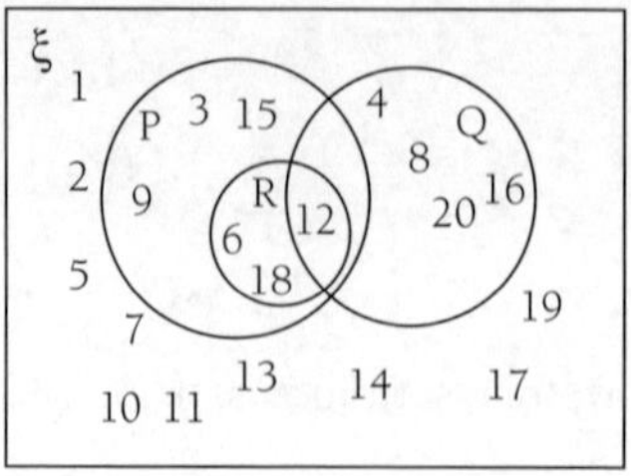

Unit 12.1 Activity 3A: Field laws for sets (page 9)

1. **a.**

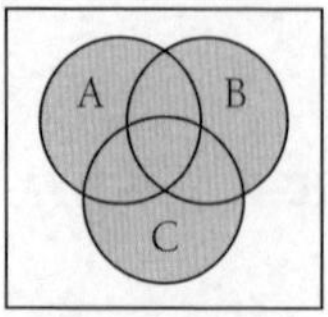

b.

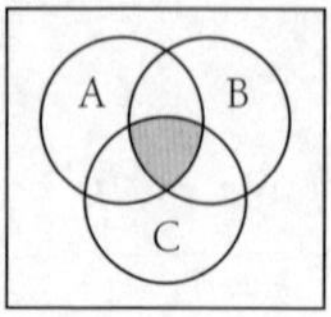

c.

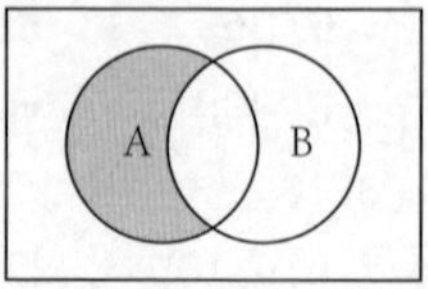

d.

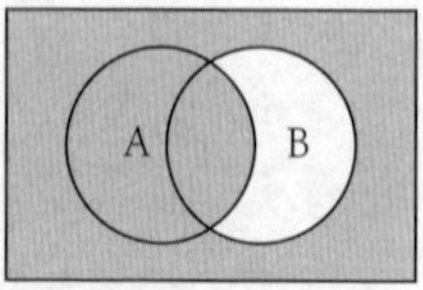

e.

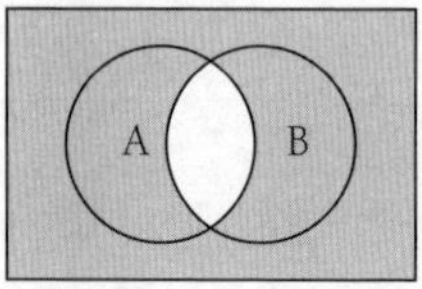

f.

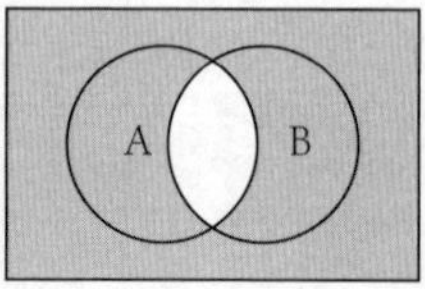

g.

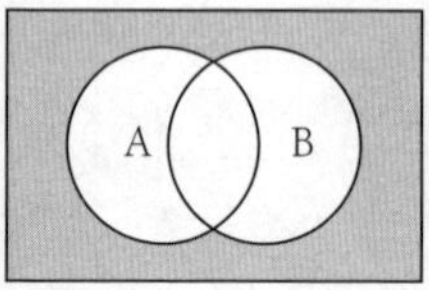

h.

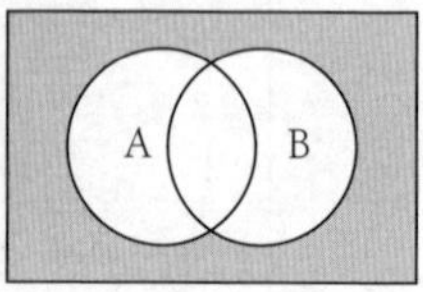

i. $(A \cup B)' = A' \cap B'$

$(A \cap B)' = A' \cup B'$

2. **a.**

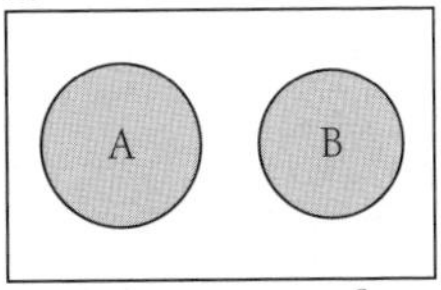

b.

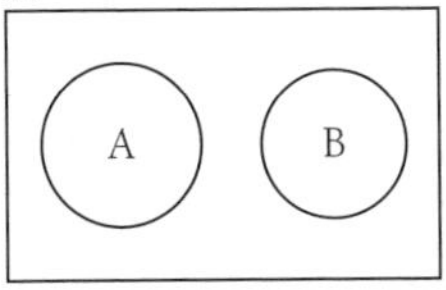

c.

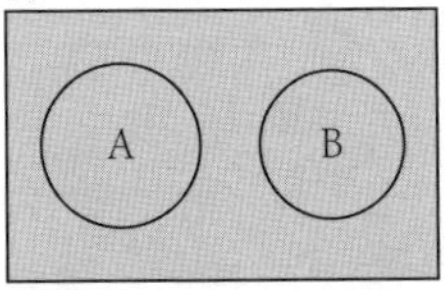

3. **a.**

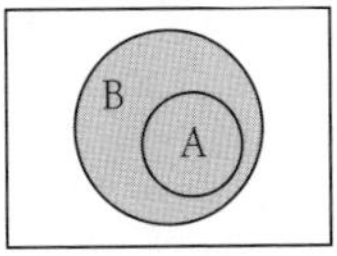

b.

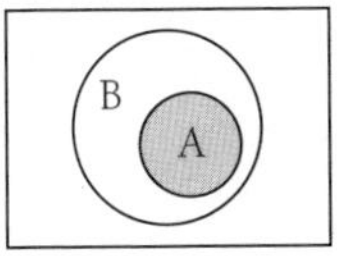

c.

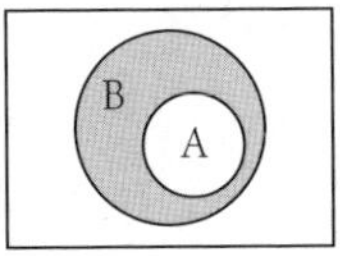

d.

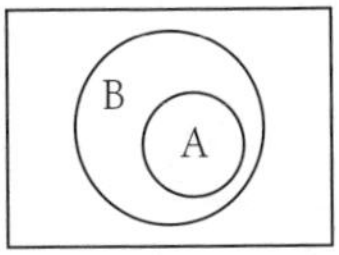

4. **a.**

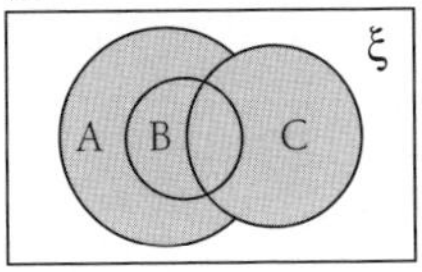

b.

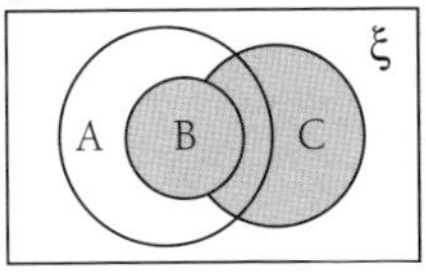

c.

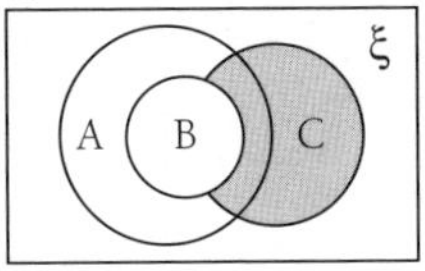

d.

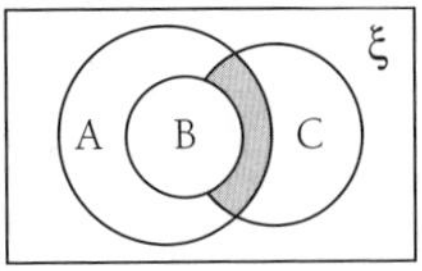

5.

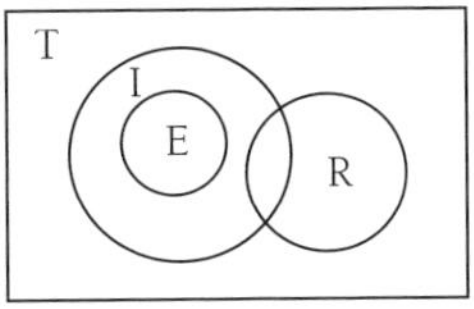

6.

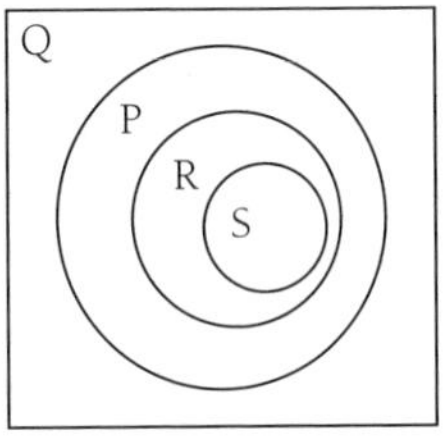

7. **a.** $(A \cup B)'$ **b.** $B' \cap A$ **c.** $(A \cup B) \cap C$

d. $(A \cup C)'$ **e.** $(A \cap B) \cup (B \cap C) \cup (A \cap C)$ **f.** $A' \cap (B \cap C)$

Unit 12.1 Activity 4A: Using sets to solve problems (page 13)

1. **a.** 13 **b.** 11 **c.** 4 **d.** 20

e. 3 **f.** 13 **g.** 12 **h.** 14

i. 22 **j.** 0 **k.** 5 **l.** 7 **m.** 10

2. **a.** 5 **b.** 10 **c.** 1

d. 25 **e.** 36 **f.** 13 **g.** 44

3.

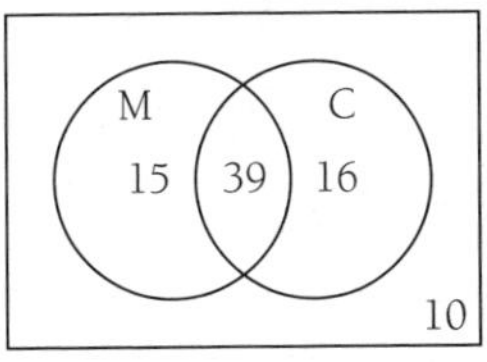

a. 39 **b.** 16 **c.** 15 **d.** 10

4.

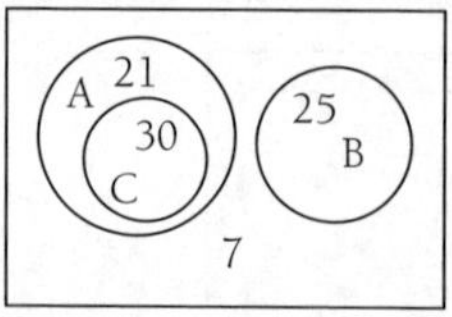

25 students have purchased textbook B.

5.

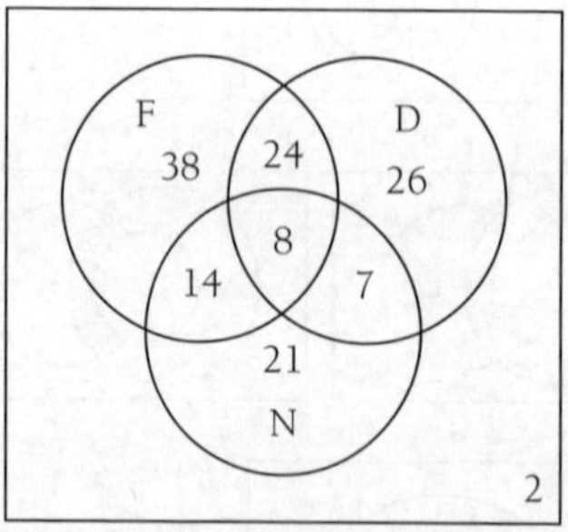

a. 8 **b.** 84 **c.** 65 **d.** 50

6.

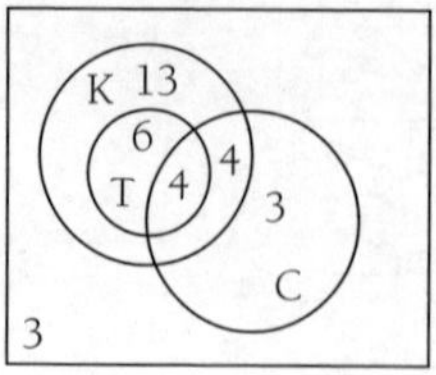

a. 4 **b.** 30 **c.** 11 **d.** 17

Unit 12.1 Activity 5A: Finding terms of a sequence (page 16)

1. a. 6, 7, $n + 1$ **b.** 4, 5, $n - 1$ **c.** 15, 18, $3n$ **d.** 13, 15, $2n + 1$

e. 5, 7, $2n - 5$ **f.** 16, 32, 2^{n-1} **g.** 81, 243, 3^{n-1} **h.** 162, 486, $2(3^{n-1})$

i. $\frac{5}{6}, \frac{6}{7}, \frac{n}{n+1}$ **j.** $\frac{4}{5}, \frac{5}{6}, \frac{n-1}{n}$ **k.** 4, 2, $12 - 2n$ **l.** 1, $\frac{1}{2}$, $16(2^{1-n})$

m. 0.20, 0.17, $\frac{1}{n}$ (to 2 dp) **n.** 25, 36, n^2 **o.** $\frac{1}{25}, \frac{1}{36}, \frac{1}{n^2}$

p. 11, 13, $2n - 1$ **q.** 26, 37, $n^2 + 1$ **r.** 2.24, 2.45, $\sqrt{n}$ (to 2 dp)

2. a. 1, 2, 3, 4, n **b.** 3, 4, 5, 6, $n + 2$ **c.** 3, 5, 7, 9, $2n + 1$

3. a. 6, 8, 10, 12, $2n + 4$ **b.** 5, 6, 7, 8, $n + 4$ **c.** 9, 12, 15, 18, $3n + 6$

Unit 12.1 Activity 5B: Listing sequences from formulae (page 17)

1. **a.** 2, 5, 10, 17 **b.** 5, 7, 9, 11 **c.** 1, 8, 27, 64
   d. 3, 7, 13, 21 **e.** $2, \frac{5}{2}, \frac{10}{3}, \frac{17}{4}$ **f.** 2, 2.25, 2.37, 2.44
   g. $-\frac{1}{2}, \frac{1}{2}, -\frac{3}{8}, \frac{1}{4}$ **h.** 0, 5, 8, 17
2. **a.** 30, 42, 58.8, 82.32 **b.** 12, 13.2, 14.52, 15.972 **c.** 41, 48, 55, 62
   d. $\sqrt{2}$, 2, 4, 16 **e.** 6, 12, 24, 48

Unit 12.1 Activity 5C: Limits of sequences (page 18)

1. **a.** No limit **b.** 0 **c.** 1 **d.** 0 **e.** 0
   f. No limit **g.** 1 **h.** No limit **i.** $\frac{1}{2}$ **j.** $\frac{1}{2}$
2. **a.** $91, 85, 81, 78\frac{1}{3}, 76\frac{5}{9}, 75\frac{10}{27}$ **b.** 73
   c. As a consequence of her diet she loses 27 kg.
3. **a.** $n + 2$ **b.** $8n + 50$ **c.** $\frac{8n + 50}{n + 2}$
   d. as n gets very large the 50 and 2 in the formula become negligibly small, so
   $\frac{8n + 50}{n + 2} \rightarrow \frac{8n}{n}$ = K8 per hour (**M/E**)

Unit 12.1 Activity 5D: Sigma notation (page 19)

1. **a.** 13 **b.** 5.4321 **c.** $\frac{47}{60}$ **d.** 0 **e.** 34 **f.** 45
   g. 49 **h.** 2 500 (**A**)
2. **a.** $\sum_{n=1}^{3} 3n$ **b.** $\sum_{n=2}^{4} 2n$ or $\sum_{n=1}^{3} 2n + 2$ **c.** $\sum_{n=5}^{8} n$ **d.** $\sum_{n=1}^{4} n^2$ **e.** $\sum_{n=1}^{4} 2^{n+1}$
3. **a.** –2 **b.** 36 **c.** 30 **d.** 64 **e.** 55
4. 2 (**M**) **5.** 1 (**M**)

Unit 12.1 Activity 6A: Terms of arithmetic sequences (page 22)

1. **a.** 17, 59, $3n - 1$ (**A**) **b.** 35, 51, $4n + 7$ (**A**) **c.** –6, –34, $6 - 2n$ (**A**)
   d. $3\frac{1}{2}, 6\frac{1}{2}, \frac{n + 1}{2}$ (**A**) **e.** 2.5, 3.1, $0.1n + 1.9$ (**A**)
2. **a.** –4, 0, 4, 8, 12, 16, $4n - 8$ (**M**) **b.** –14, –9, –4, 1, 6, 11, 16, 21, $5n - 19$ (**M**)
   c. –1, 1, 3, 5, 7, 9, 11, 13, $2n - 3$ (**M**) **d.** 12, 9, 6, 3, 0, –3, –6, –9, $15 - 3n$ (**M**)
   e. –201 (**M**) **f.** 41st term (**M**) **g.** 14th term (**M**) **h.** –51 (**M**)
   i. 276 (**M**) **j.** 13 (**M**)

3. a. 135 months (***A***) **b. i.** After 20 months (***A***) **ii.** October 2013 (***A***)

c. After 102 months (same number at 101 months). (***M***)

4. a.

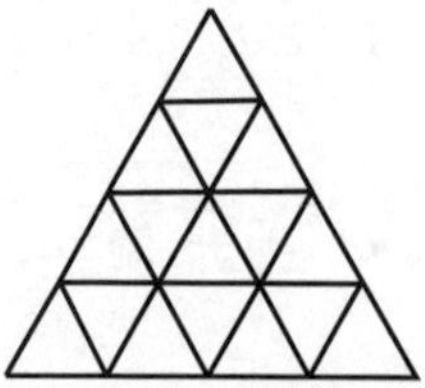

b. L_n: 1, 3, 5, 7, 9, $2n - 1$

J_n: 1, 4, 9, 16, n^2

S_n: 1, 2, 3, 4, n

P_n: 2, 3, 4, 5, $n + 1$

W_n: 3, 6, 10, 15

c. L_n, S_n, P_n (***E***)

Unit 12.1 Activity 6B: Partial sums of arithmetic sequences (page 24)

1. 610 (***A***) **2.** 60 300 (***A***) **3.** 6 475 (***A***) **4.** 120 (***A***)

5. 320 (***A***) **6.** 710 (***A***)

7. a. 7 **b.** 3 **c.** 100 **d.** 15 550 (***A***)

8. a. 50 **b.** 7 325 **c.** 5 **d.** 24 (***A***)

9. –967.5 (***A***) **10.** 477.5 (***A***) **11.** 999 (***A***) **12.** 1 524 (***A***)

13. 10 822.5 (***A***) **14.** 3 925 (***A***) **15.** –12 570 (***A***) **16.** 5 406 (***M***)

17. 1 113 (***M***) **18.** 11 (***M***) **19.** 5 (***M***) **20.** 2 (***M***)

21. 1 944 (***M***) **22.** 4 242 (***M***) **23.** 1 092 km (***E***)

Unit 12.1 Activity 7A: Terms of a geometric sequence (page 28)

1. a. $r = 2$ **b.** not GP **c.** $r = 0.5$ **d.** $r = 1.2$ **e.** $r = -1$ **f.** not GP

g. $r = -2$ **h.** not GP

2. a. $r = 3, t_5 = 405$ **b.** $r = -3, t_5 = -405$ **c.** $r = \frac{1}{3}, t_5 = \frac{5}{3}$

d. $r = 0.3, t_6 = 0.729$ **e.** $r = -\frac{10}{3}, t_6 = -1\,000$

3. a. $t_4 = 256, t_5 = 1\,024, t_n = 4^n$ (***A***)

b. $t_4 = 375, t_5 = -1\,875, t_n = -3(-5)^{n-1}$ (***A***)

c. $t_4 = 4.1472, t_5 = 4.976\,64, t_n = 2(1.2)^{n-1}$ (***A***)

d. $t_4 = 1.5625, t_5 = 0.390\,625, t_n = 100(0.25)^{n-1}$ (***A***)

e. $t_5 = -6.25, t_6 = 3.125, t_n = -100(-0.5)^{n-1}$ (***A***)

4. a. $\frac{1}{8}, \frac{1}{128}, 4\left(\frac{1}{2}\right)^{n-1}$ (***A***) **b.** $1\,944, 52\,488, 8(3)^{n-1}$ (***A***)

c. $512, -4\,096, -(-2)^{n-1}$ (***A***) **d.** $\frac{-8}{243}, \frac{-128}{1\,594\,323}, 3\left(-\frac{2}{9}\right)^{n-1}$ (***A***)

e. $93\frac{3}{4}, 2\,343\frac{3}{4}, \frac{3}{4}(5)^{n-1}$ (***A***)

5. a. $\frac{5}{9}, \frac{5}{3}, 5, 15, 45, 135, \frac{5}{9}(3)^{n-1}$ (**A**) b. 3, 6, 12, 24, 48, 96, $3(2)^{n-1}$ (**A**)

c. $1\frac{7}{9}, 20\frac{1}{4}, \frac{32}{27}(1.5)^{n-1}$ (**A**) d. $\frac{54}{125}, \frac{54}{125}\left(\frac{5}{3}\right)^{n-1}$ (**A**)

e. $\frac{4}{729}, 4\left(\frac{1}{3}\right)^{n-1}$ (**A**) f. $\frac{11}{8}, \frac{11}{8}(2)^{n-1}$ (**A**)

g. $-12, 24\left(-\frac{1}{2}\right)^{n-1}$ (**A**) h. 6 (**M**) i. Eighth term (**M**)

j. Ninth term (**M**) k. Tenth term (**M**) l. 3.6 (**M**)

6. a. K5 000, K5 550, K6 160.50, K6 838.16, K7 590.35 (**A**) b. 51 years (**M**)

7. 3 170 000 (3 sf) (**M**)

8. a. K732 (**A**) b. The business would sell the computer during the sixth year (between the end of the fifth and the end of the sixth). (**M**)

Unit 12.1 Activity 7B: Partial sums of geometric sequences (page 30)

1. a. 1 062 880 b. 9.999 36 c. 6 825

2. a. 12 b. 1.5 c. 8 d. 591.093 75

3. 3 069 (**A**) 4. 113.33 (**A**) 5. 4 (**M**) 6. 5 (**M**)

7. 453.32 (**M**) 8. 44 (**M**) 9. K608 171 (**M**)

10. a. 147.75 kg, 145.53 kg, 143.35 kg, 141.20 kg, 139.08 kg b. 10.92 kg (**A**)

11. 522 kg (**A**) 12. 23 875 (**A**) 13. 3 840.90 (**A**) 14. 11 (**M**)

15. 10 (**M**)

Unit 12.1 Activity 7C: Sum to infinity of a geometric sequence (page 32)

1. b, c, e (**A**)

2. a. 3 (**A**) b. 6 (**A**) c. $4\frac{2}{7}$ (**A**) d. $-2\frac{10}{11}$ (**A**) e. $\frac{2}{3}$ (**A**)

3. a. $12\frac{1}{2}$ (**A**) b. $10\frac{10}{13}$ (**A**) c. $40\frac{1}{2}$ (**M**) d. 172.23 (**M**) e. 32 (**M**)

4. $\frac{3}{8}$ (**M**) 5. 75 000 (**M**) 6. K4 000 (**M**)

7. a. K406 500, K203 250, K101 625 (**A**) b. K813 000 (**M**)

8. a. 19 683 tonnes, 6 561 tonnes, 2 187 tonnes, 729 tonnes, 243 tonnes (**A**)

b. 29 524.5 tonnes (**M**)

9. 1 300 000 (**M**)

10. a. 1 g b. 5 g, 1 g, 0.2 g, 0.04 g, 0.008 g

c. Each term when divided by its predecessor gives a constant ratio of 0.2 (**A**)

d. $5(0.2)^{n-1}$ g (**A**) e. 6.25 g (**M**)

Unit 12.1 Activity 8A: Applications of geometric sequences (page 39)

1. a. 6 480, 6 998.40, 7 558.27

   b. 6 998.4/6 480 = 1.08; 7 558.27/6 998.4 = 1.08; $V_n = 6\ 480 \times 1.08^{n-1}$

   c. K11 105.58 d. After 10 years.

2. K24 885.74

3. a. 16 675/18 125 = 0.92 = 15 341/16 675

   b. 10 990 c. 1 998 d. 6 179

4. a. $B_n = 9 \times 2^{n-1}$ b. 288 c. 6 days

5. a. K30, K33, K36.30; $A_n = 30 \times 1.1^{n-1}$ b. K295.49 c. K8 973.80

6. a. K91 800, K78 030, K66 325.50; $91\ 800 \times 0.85^{n-1}$

   b. K47 920.17 c. In the 10th year (or after 9 years)

7. a. 6 000, 7 200, 8 640; $6\ 000 \times 1.2^{n-1}$ b. 4 years c. 155 752

8. a. 12.96 cm b. 2.3056 m c. 2.5 m

9. a. 83 cm b. 11.72 m c. 26.6 m

10. K5 004.89, K9.78

Unit 12.1 Activity 8B: Contrasting arithmetic and geometric sequences (page 43)

1. a.

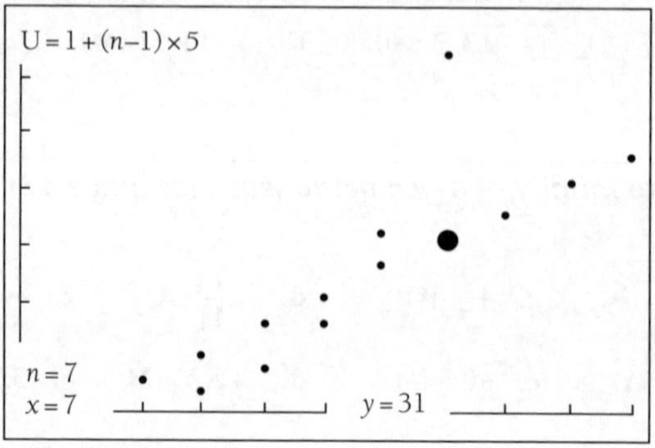

   b. $n = 6$

2. a. B b. E c. D

   d. C e. F f. A

3. Compound; K96.60

4. **a.** $A_n = 5 + 2n$; $B_n = 5 \times 1.2^n$

b.

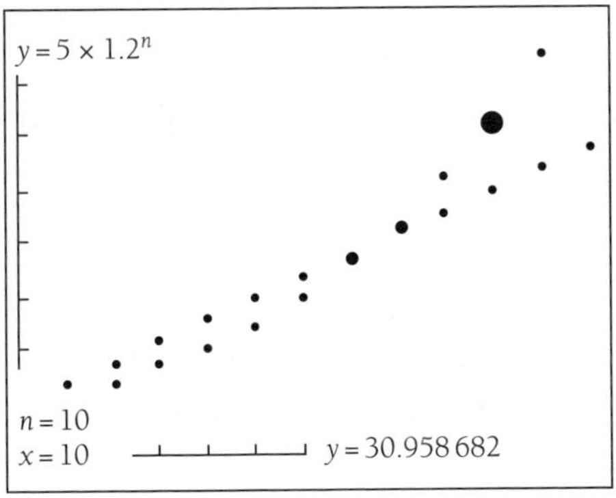

c. After his 18th birthday. **d.** K10.15 **e.** A: K8 580 ; B: K8 359.11

5. **a.** A: 80 toea; B: 81 toea **b.** A: K2.85; B: K2.85

c. $A_n = 1.05 - 0.05n$; $B_n = 0.95^{n-1}$ **d.** Never

e. After 20 minutes **f.** A: K10.50: B: K20

6. A: K3 500, K2 750, K2 000, K1 250, K500

B: K4 521.80, K2 607.50, K1 503.60, K867.07, K500

Unit 12.1 Activity 9A: Expanding (page 47)

1. **a.** $1 + 5x + 10x^2 + 10x^3 + 5x^4 + x^5$ (**E**)

b. $1 - 8a + 28a^2 - 56a^3 + 70a^4 - 56a^5 + 28a^6 - 8a^7 + a^8$ (**E**)

c. $x^5 + 5x^4y + 10x^3y^2 + 10x^2y^3 + 5xy^4 + y^5$ (**E**)

d. $x^6 - 6x^5y + 15x^4y^2 - 20x^3y^3 + 15x^2y^4 - 6xy^5 + y^6$ (**E**)

e. $1 - 18y + 135y^2 - 540y^3 + 1\,215y^4 - 1\,458y^5 + 729y^6$ (**E**)

f. $32 + 240x + 720x^2 + 1\,080x^3 + 810x^4 + 243x^5$ (**E**)

g. $256 - 256x + 96x^2 - 16x^3 + x^4$ (**E**) **h.** $1 - \frac{5x}{2} + \frac{10x^2}{4} - \frac{10x^3}{8} + \frac{5x^4}{16} - \frac{x^5}{32}$ (**E**)

2. **a.** $1 - 18x + 144x^2 - 672x^3$ (**E**) **b.** $x^7 + 14x^6y + 84x^5y^2 + 280x^4y^3$ (**E**)

c. $x^{10} - 20x^9y + 180x^8y^2 - 960x^7y^3$ (**E**)

3. **a.** $a^4 + 4a^3b + 6a^2b^2 + 4ab^3 + b^4$ (**E**) **b.** Proofs (**E**).

4. **a.** $1 - 8x + 24x^2 - 32x^3 + 16x^4$ (**E**) **b.** $1 - 7x + 16x^2 - 8x^3 - 16x^4 + 16x^5$ (**E**)

5. $1 - x - 4x^2$ (**E**) **6.** $1 - x + x^2 - x^3$ and $1 - 5x + 11x^2$ (**E**)

7. **a.** $(1 + x)^5$ (**E**) **b.** $(1 - 2x)^6$ (**E**) **c.** $(2 + 3x)^4$ (**E**)

Unit 12.1 Activity 9B: Binomial series (page 50)

1. **a.** $1\,512\,x^3$ (**E**) **b.** $3\,360x^4$ (**E**) **c.** $672x^2y^5$ (**E**) **d.** $1\,120x^4y^4$ (**E**)
e. $1\,120x^4$ (**E**) **f.** $15x^4$ (**E**)

2. **a.** 126 (**E**) **b.** 1 120 (**E**) **c.** $\frac{35}{16}$ (**E**) **d.** 4 860 (**E**)
e. $\binom{n}{4}a^{n-4}$ (**E**) **f.** –560 (**E**)

3. **a.** $\binom{8}{r}2^{8-r}x^r$ (**E**) **b.** $\binom{8}{r}(-2x)^r$ (**E**) **c.** $\binom{9}{r}x^{9-2r}$ (**E**)

4. **a.** 2 160 (**E**) **b.** –20 (**E**) **c.** 7 920 (**E**) **5.** **a.** 45 (**E**) **b.** $20\sqrt{x^9}$ (**E**)

6. 14 (**E**) **7.** 1.09 (**E**) **8.** –3 (**E**) **9.** Proof. (**E**)

10. $41 + 29\sqrt{2}$ (**E**) **11.** 152 (**E**) **12.** 128 (**E**)

Unit 12.1 Activity 9C: Revision questions (page 51)

1. $(x + 1.5)^2 - 3.25$ (**A**) **2.** $-x^5 + 2x^4 + 9x^3 - 17x^2 - 8x + 15$

3. $a = \sqrt{2}$, $b = -\sqrt{2}$ or vice versa (**A**) **4.** $\frac{a^2 + a + 1}{a(1 + a)}$ **5.** $x = 3$ (**A**)

6. $x = -5, 2$ (**A**) **7.** $a = -5; b = 3$ (**A**) **8.** $x = \frac{zy}{y - z}$

9. $x = \sqrt{3} - 1, -\sqrt{3} - 1$ (**A**)

10. **a.** $x, x \neq -1$ **b.** No simpler form. **c.** $\frac{1}{2^{n+1}}$ **d.** $\frac{x - 3}{x + 1}$

11. $x = -3, -9$ (**A**) **12.** $x = -3, 1$ (**A**) **13.** $x = -2, 1, 3$ (**A**)

14. **a.** $11 + 6\sqrt{3}$ (**A**) **b.** $28 - 14\sqrt{3}$ (**A**)

15. **a.** $R = 2$ (**A**) **b.** $x = 3, 1 \pm \sqrt{3}$ (**A**)

16. **a.** **i.** $x = -1, 5$ (**A**) **ii.** $x = 2, 4$ (**A**) **iii.** $x = 4$ (**M**)
b. $\log 3x^2$ (**A**) **c.** $-\sqrt{3}$ (**A**)

17. **a.** $x = 8$ (**A**) **b.** $x = 1.5$ (**A**) **c.** $x = 1.609$ (3 dp) (**A**)

18. 4.879 (**A**) **19.** $s = 1, \frac{t - 1}{2}$ (**E**) **20.**

x	0	1	1	–1	–1
y	0	1	–1	1	–1
z	0	1	–1	–1	1

(**M**)

21. $x = \frac{b^2 + 2b - 2}{b - 1}$

22. $A = B = C = D = 1$ (**M**)

23. $a = 1; b = 3; c = 3; d = -11$ (**M**)

24. $a = \frac{1 - x}{1 + x}; y = \frac{2x}{1 + x^2}$ **25.** $x = 4$ (**M**) **26.** Proof. (**M**) **27.** $p = 14$ (**M**)

28. Proof. (**M**) **29.** 5 376 (**E**) **30.** **a.** $a = n; b = \frac{n(n - 1)}{2}$ (**E**) **b.** 1.771 (**E**)

Unit 12.1 Activity 10A: Linear equations (page 56)

1. **a.** 2.5 **b.** 4 **c.** $11\frac{1}{2}$ **d.** 3 **e.** 5 **f.** –1 **g.** –5
h. $-15\frac{1}{2}$ **i.** –4 **j.** $1\frac{19}{21}$

2. 58.8 kg **3.** K8.80 **4.** Wamsie spent K49.00, Hosea K24.50, David K13.00.

5. 124 km **6.** $583\frac{1}{3}$ L

Unit 12.1 Activity 10B: Simultaneous linear equations in two unknowns (page 61)

1. a. $x = -1, y = 2$ **b.** $x = 4, y = -1$ **c.** No solutions. **d.** $x = 1\frac{1}{3}, y = 0$

e. $x = 2, y = 1$ **f.** Infinite set of solutions: any (x, y) where $3x - 4y = 30$.

2. a. $x = -23, y = -6$
b. $x = -3, y = -2$
c. $x = \frac{10}{7}, y = \frac{5}{3}$
d. $x = \frac{5}{2}, y = -2$

3. James gets K11.50, Gandhi K10.

4. 54 km

5. 14 tricycles, 11 bicycles.

6. 11 500 **7.** 12 **8. a.**

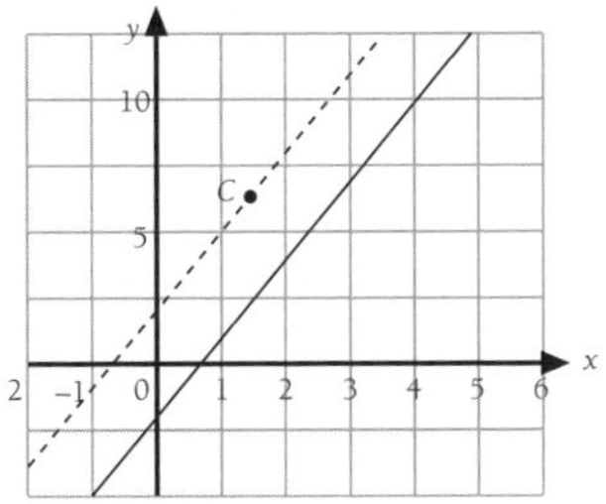

b. $a = -3, b = 6$

9. If $k = 0$ the lines $y = 2$ and $y = 1$ are parallel horizontal lines (which have no point of intersection).

10. a. $x = 5.5, y = 1$ **b.** $c = 10$
c. If $c \neq 10$ then the lines are not parallel and thus they have a unique point of intersection.

11. $a = 4.5, b = -1.5$

Unit 12.1 Activity 11A: Simultaneous linear equations in three unknowns (page 68)

1. a. $x = 2, y = -1, z = 3$ (**A**) **b.** $x = 1\frac{1}{4}, y = 1\frac{1}{2}, z = 2$ (**A**)

c. Infinitely many solutions: any (x, y, z) where $y = \frac{10 - 3x}{7}$ and $z = \frac{5x + 2}{7}$ (**A**)

2. a. $x = 4, y = \frac{1}{2}, z = -2$ (**A**) **b.** $x = \frac{1}{3}, y = \frac{1}{2}, z = -1$ (**A**)

c. No solutions. (**A**) **d.** $x = 0.1, y = 0.2, z = 0.5$ (**A**)

3. $x = 3, y = 4, z = 5$. Thus the number of children can be 3 or 4. (**A**)

4. Infinitely many solutions: any (x, y, z) where $y = 7 - 5x$ and $z = 12 - 9x$. Hence Jason cannot find out how many jobs he will have. (**A**)

5. $d = 9, e = 2, f = 1$. Hence the meal contains 9 units of D, 2 units of E and 1 unit of F. (**A**)

6. a. $A = 4, B \neq 19$ (**E**) **b.** $A = 4, B = 19$ (**E**)

7. **a.** $3x - z = 0$ (**E**) **b.** $2z - 6x = -2$ or $3x - z = 1$ (**E**)

c. Attempting to solve the equations in **a.** will result in $0 = -2$ (or $0 = 1$) which is impossible. Hence there are no solutions. (**E**)

d. Each plane is parallel to the line of intersection of the other two planes, as shown. (**E**)

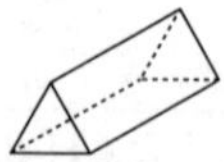

8. **a.** $x = 1, y = 0, z = -3$ (**E**) **b.** $A = 1, B = -3$ and $C = 11$ is one possible set. (**E**)

c. $A = -6, B = 4, C \neq -10$ (**E**)

9. The system of equations is consistent, with at least one solution ((0, 0, 0) is a solution). (**E**)

10. **a.** $p = -8, q = 6$ (**E**) **b.** $p = -8, q \neq 6$ (**E**) **c.** $p \neq -8$, q any value (**E**)

Unit 12.1 Activity 11B: Simultaneous linear equations and word problems (page 71)

1. 5, 7, 12.5 (**M**) **2.** K15.50, K14.00, K13.50 (**M**) **3.** 60, 38, 22 (**M**)

4. 15, 10, 6 (**M**) **5.** K250, K302, K338 (**M**) **6.** 16, 8, 6 (**M**)

7. 22, 4, 10 (**M**) **8.** K0.35, K0.90, K0.40 (**M**) **9.** 5, 3, 6 (**M**)

10. 3, 5, 13 (**M**) **11.** 75, 65, 80 (**M**) **12.** $A = 3, B = -1, C = 2$ (**M**)

13. **a.** $20a + 5b + 80c = 140$, $2a + b + 3c = 11$, $4a + 2b + 8c = 24$ (**M**)

b. $a = 2, b = 4, c = 1$ (**M**)

14. **a.** **i.** $0.01p + 0.1q + r = 2.025$, $0.04p + 0.2q + r = 2.9$, $0.09p + 0.3q + r = 3.425$ (**M**)

ii. $p = -17.5, q = 14, r = 0.8$ (**M**)

b. $s = -10, t = 8, u = 2$ (**M**)

15. $a + b + c = 200\,000$ $a - b - c = 3200$ $b = 2c$

A sells K101 600, B sells K65 600, C sells K32 800 (**M**)

16. $a + b + c = 460$ $3a + 4.5b + 1.9c = 1\,547.5$ $180a + 200b + 200c = 88\,500$

175 of burger A, 185 of burger B and 100 of burger C are sold. (**M**)

17. $x + y + z = 20$ $20x + 15y + 10z = 285$ $x + y = z$

The number of K20 books was 7, the number of K15 books was 3 and the number of K10 books was 10. (**M**)

18. $x + y + z = 849$ $x - y - z = 105$ $y - z = 60$

Tracey's board is K477, she invests K216 and spends K156. (**M**)

19. **a.** $30a + 36b + 35c = 1\,035$

b. $a = \dfrac{-639 - 5c}{30}$ $b = \dfrac{279 - 5c}{6}$

c. a can never be positive and $c < 55.8$. So there would be wastage of nitrogen, phosphorus or potassium to achieve compost production, or more could be ordered to make up the shortfall. (**E**)

Unit 12.2 Trigonometry and Vectors

Unit 12.2 Activity 1A: Finding sines, cosines and tangents (page 79)

1. **a.** $-\cos 57°$ **b.** $\sin 44°$ **c.** $-\tan 35°$ **d.** $-\tan 7°$ **e.** $\cos 50°$ **f.** $-\sin 76°$
g. $\tan 34°$ **h.** $-\sin 18°$ **i.** $-\cos 18°$ **j.** $\sin 40°$ **k.** $\cos 32°$ **l.** $-\tan 20°$
m. $\cos 46°$ **n.** $\sin 47°$ **o.** $\cos 53°$

2. **a.** $\frac{1}{\sqrt{2}}$ **b.** $\frac{\sqrt{3}}{2}$ **c.** $\frac{1}{\sqrt{3}}$ **d.** $-\frac{\sqrt{3}}{2}$ **e.** $-\frac{1}{2}$ **f.** $\sqrt{3}$
g. $-\frac{1}{\sqrt{3}}$ **h.** -1 **i.** $\frac{\sqrt{3}}{2}$ **j.** $\frac{1}{\sqrt{2}}$ **k.** $-\frac{1}{\sqrt{2}}$ **l.** $-\frac{1}{\sqrt{2}}$
m. $-\frac{1}{\sqrt{2}}$ **n.** $\frac{\sqrt{3}}{2}$ **o.** $-\frac{\sqrt{3}}{2}$

Unit 12.2 Activity 1B: Radians (page 81)

1. **a.** 0.314 **b.** 0.663 **c.** 2.721 **d.** 4.309 **e.** 7.292

2. **a.** 149° **b.** 195° **c.** 67.6° **d.** 32.7° **e.** 1.72°

3. **a.** 120° **b.** 90° **c.** 100° **d.** 67.5° **e.** 144°
f. 105° **g.** 75° **h.** 22.5° **i.** 540° **j.** 150°

4. **a.** $\frac{2\pi}{5}$ **b.** $\frac{3\pi}{5}$ **c.** $\frac{7\pi}{15}$ **d.** $\frac{9\pi}{4}$ **e.** $\frac{\pi}{8}$
f. $\frac{2\pi}{5}$ **g.** $\frac{3\pi}{5}$ **h.** $\frac{7\pi}{15}$ **i.** $\frac{11\pi}{15}$ **j.** $\frac{3\pi}{20}$

5. 11π or 34.56

Unit 12.2 Activity 1C: Exact trigonometric values for angles in radians (page 82)

1. 1 **2.** $-\frac{1}{\sqrt{2}}$ **3.** $\frac{1}{\sqrt{2}}$ **4.** -1 **5.** $-\sqrt{3}$

6. $-\frac{1}{2}$ **7.** $-\frac{1}{2}$ **8.** 1 **9.** $-\frac{\sqrt{3}}{2}$ **10.** $\frac{1}{\sqrt{3}}$

Unit 12.2 Activity 1D: Basic trigonometric graphs (page 83)

1.

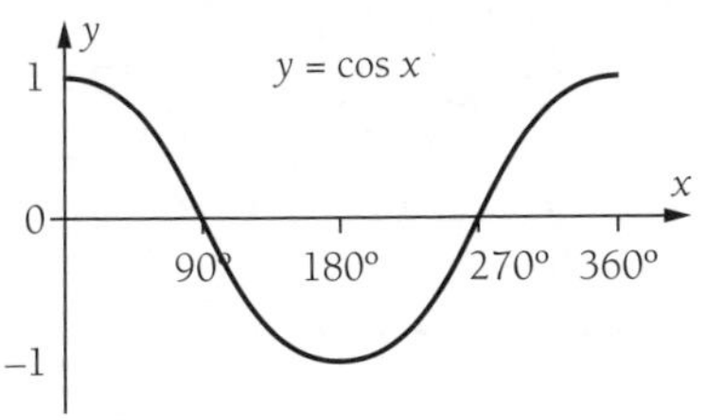

2.

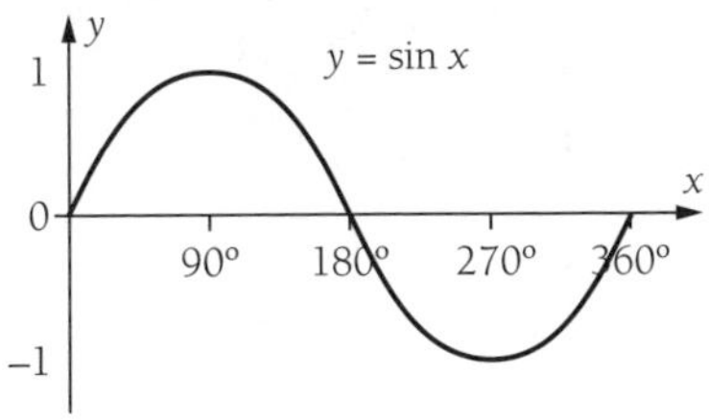

3. **a.** $y = \sin x$ **b.** $0° \le x \le 360°$ or $[0°, 360°]$ **c.** $-1 \le y \le 1$ or $[-1, 1]$
d. point symmetry about (180°, 0) **e.** no

4. a. 0 **b.** 180° **c.** $0° \le x \le 360°$ or $[0°, 360°]$

d. $-1 \le y \le 1$ or $[-1, 1]$ **e.** $y = \cos x$

5. a. $y = \tan x$ **b.** 90°, 270° **c.** $x = 90°, x = 270°$

d. point symmetry about (180°, 0) **e.** about 60°

f. twice **g.** 3 times

Unit 12.2 Activity 1E: Graphs of transformed trigonometric functions (page 87)

1.

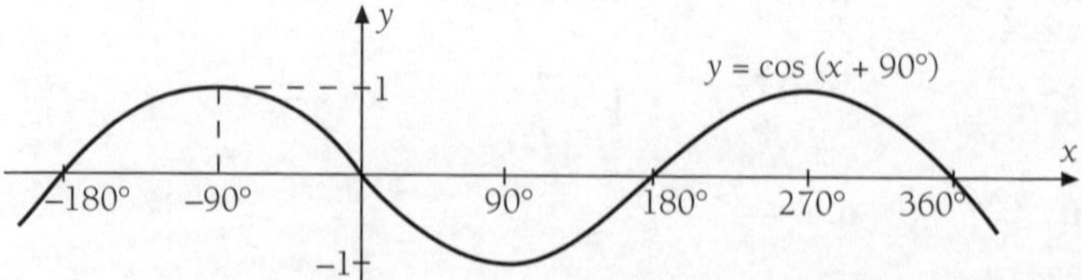

2.

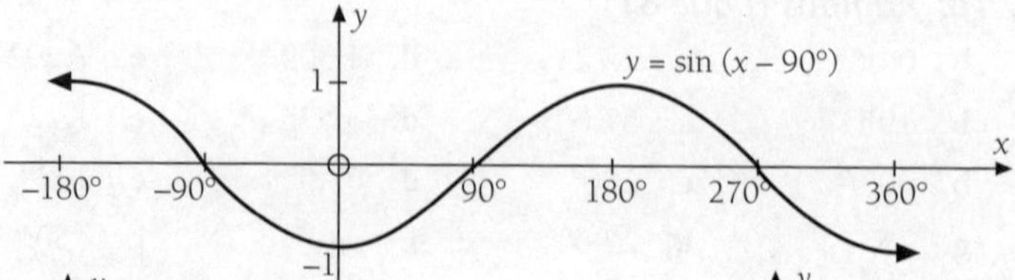

3.

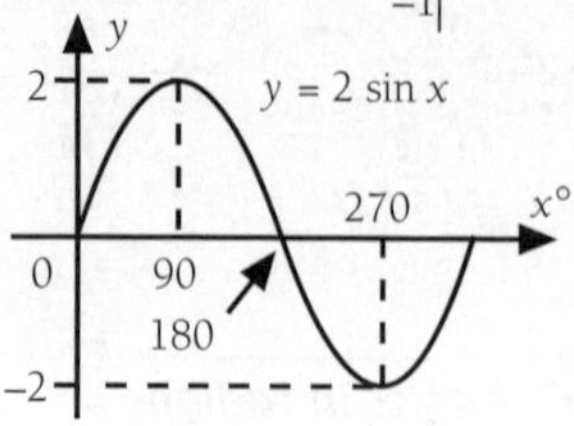

4.

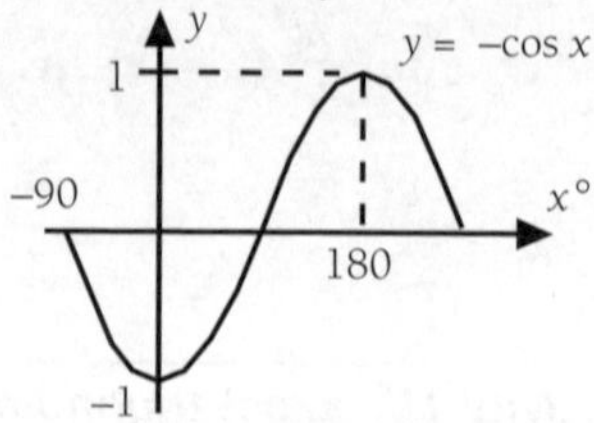

5.

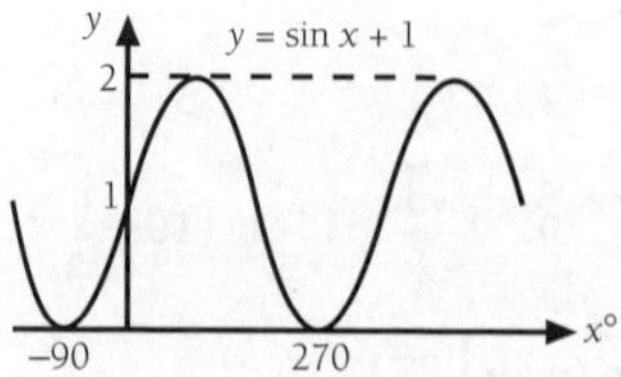

6.

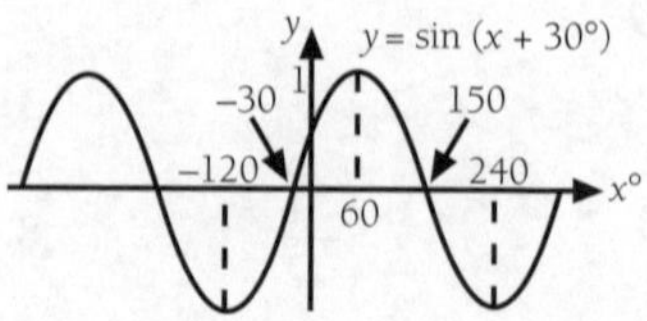

7.

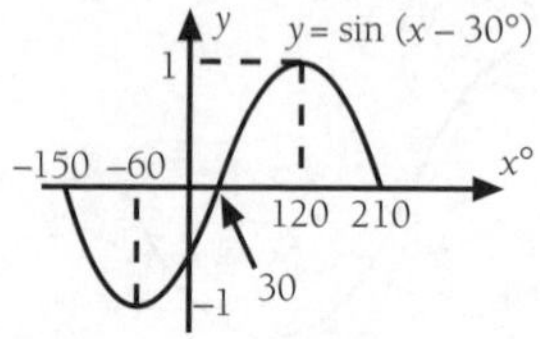

8.

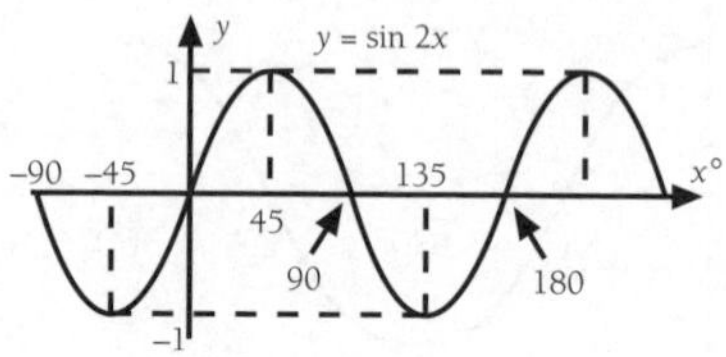

9. $y = 2 \cos x$

10. $y = \sin x + 2$

11. $y = \sin (x - 180°)$ or $y = -\sin x$

12. $y = \cos 2x$

Unit 12.2 Activity 1F: Trigonometric graphs with two transformations (page 88)

1.

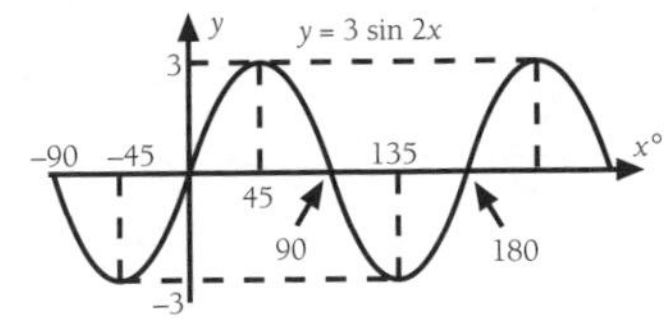

2.

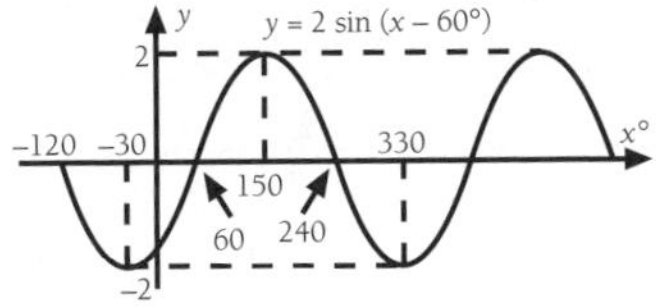

3.

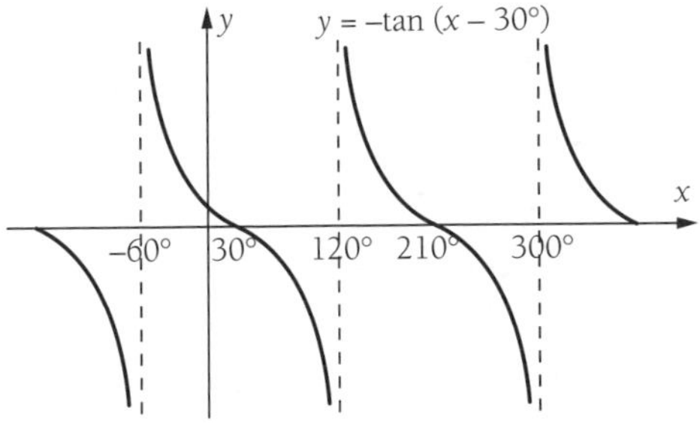

4.

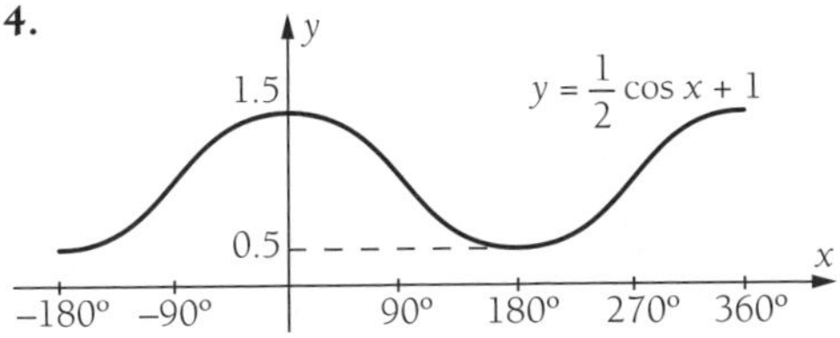

5.

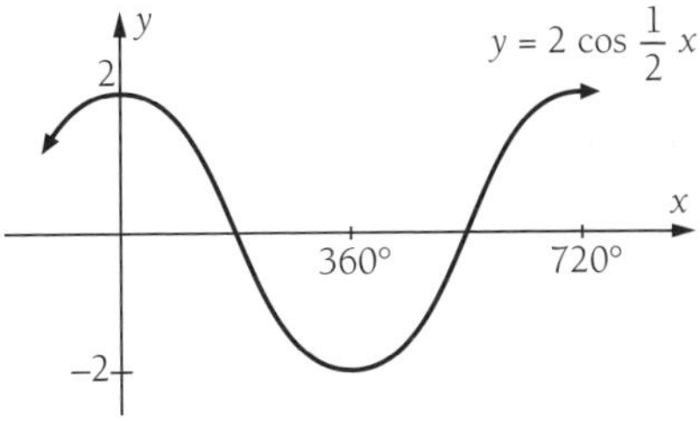

6.

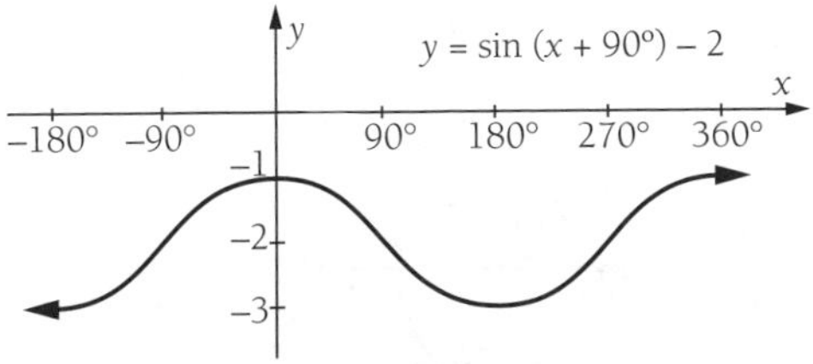

7. $y = -3 \sin x$ (**M**)

8. $y = 4 \cos (x - 20°)$ (**M**)

9. e

10. a.

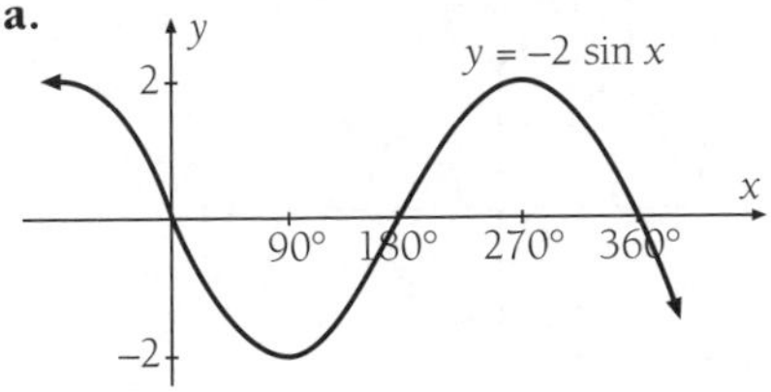

b.

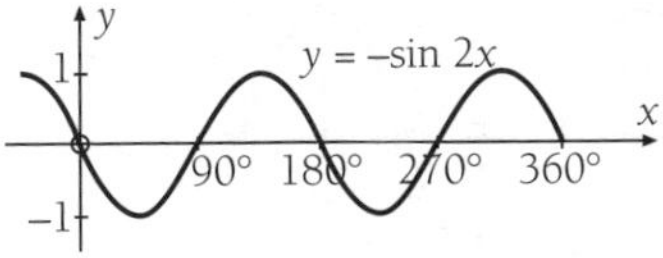

c.

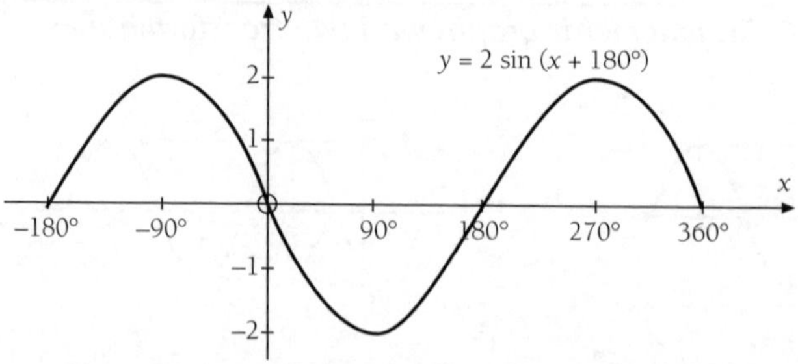

d.

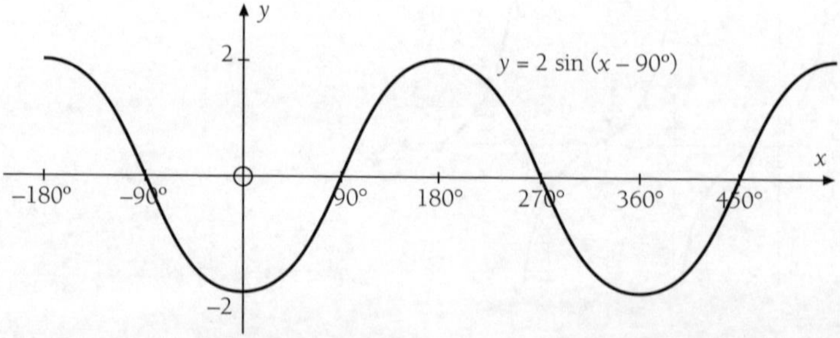

Unit 12.2 Activity 1G: Trigonometric graphs using radians (page 90)

1.

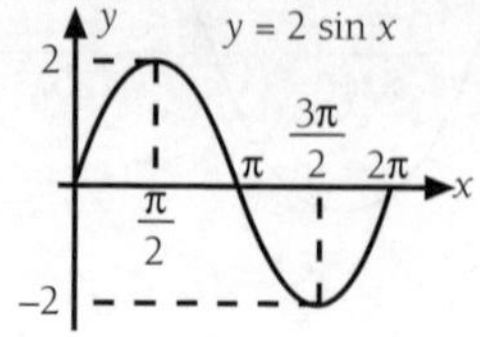

2.

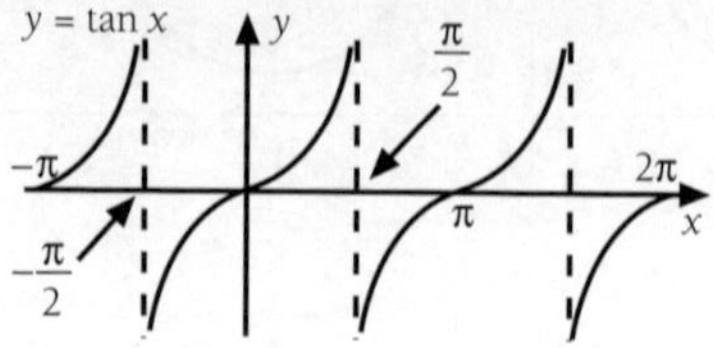

3.

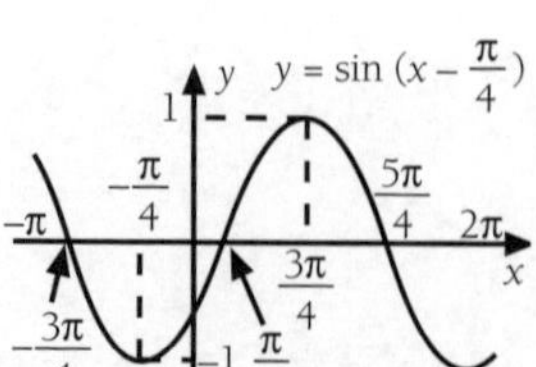

4.

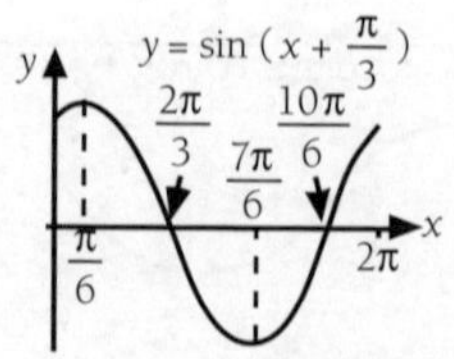

5.

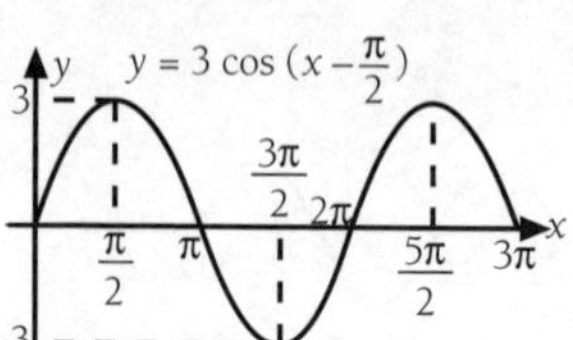

6.

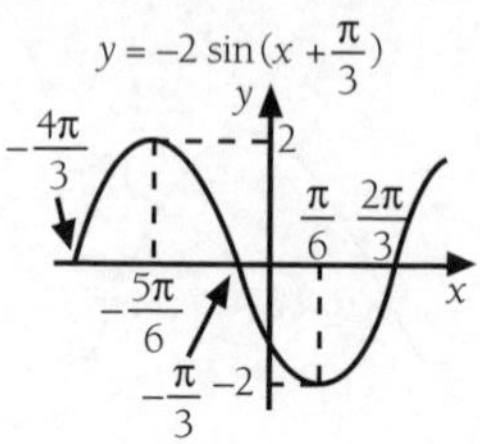

7. 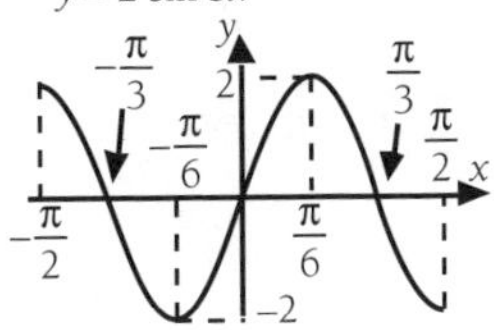

8. 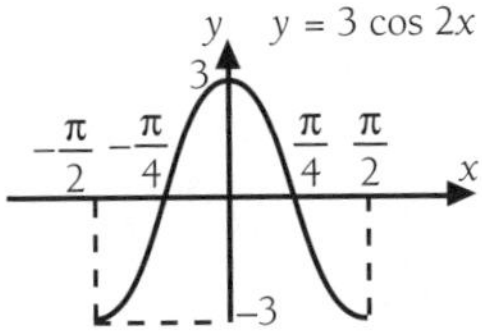

9. a. ii. or $y = \sin 2x$

b. 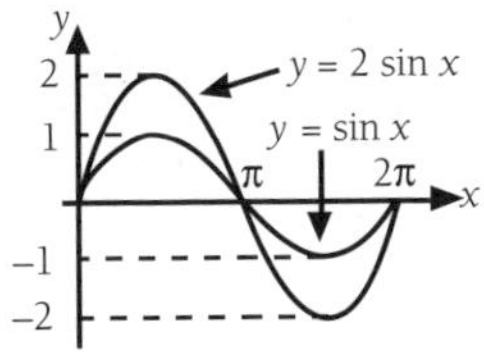

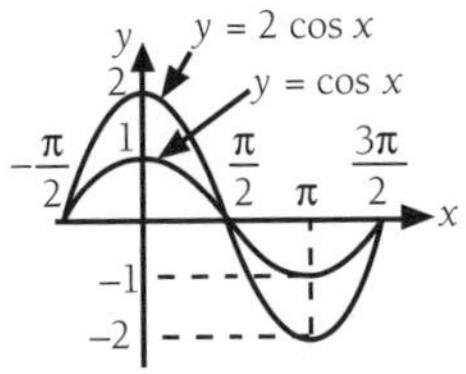

10. a. $g(x) = \cos 2x$, $h(x) = -\sin x$ **b.** 2π **c.** $\left\{\frac{\pi}{2}, \frac{7\pi}{6}, \frac{11\pi}{6}\right\}$

Unit 12.2 Activity 2A: Basic trigonometry (page 95)

1. a. $\frac{-1}{\sqrt{2}}$ **b.** $\frac{-1}{2}$ **c.** $-\sqrt{3}$ **d.** $\sqrt{2}$ **e.** $\frac{2}{\sqrt{3}}$ **f.** $\frac{1}{\sqrt{3}}$

g. $\frac{-\sqrt{3}}{2}$ **h.** $\frac{-1}{\sqrt{2}}$ **i.** $\sqrt{3}$ **j.** $\frac{-2}{\sqrt{3}}$ **k.** $\sqrt{2}$ **l.** $\sqrt{3}$

m. $\frac{-2}{\sqrt{3}}$ **n.** $\sqrt{2}$ **o.** 1 **p.** 1 **q.** 2

2. $\frac{-\left(2+\sqrt{2}\right)}{2}$ **3. a.** −1 **b.** 1 **4.** $\frac{2+\sqrt{2}}{2}$ **5.** $\frac{\pi+2}{2}$

6. a. $\frac{7\pi}{6}$ **b.** $\frac{-2}{\sqrt{3}}$ **c.** $\frac{1}{\sqrt{3}}$ **7.** $(-\sqrt{2}, \sqrt{2})$

8. a. $\alpha + \theta = \pi$ (**A**) **b.** $\theta = 2\pi - \alpha$ (**A**) **c.** $\theta = 3\pi - \alpha$ (**A**) **d.** $\theta = 3\pi - \alpha$

Unit 12.2 Investigations: Practical trigonometry (page 101)

1. Answers will vary. (**M**) **2.** Tidal transformation $h = H \cos K\pi(t - 3)$ (**M**)

3. 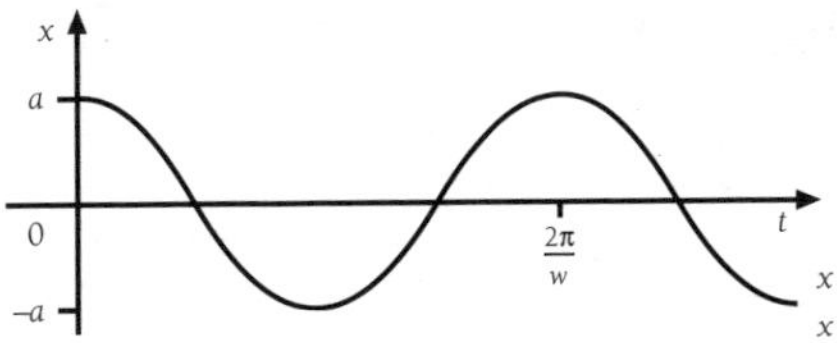

$x = a \cos(wt)$
$x = a \sin\left(\frac{\pi}{2} - wt\right)$ (**M**)

4. a. $y = \sin(x + \frac{\pi}{2})$ ie A = 1, B = –1, C = $\frac{\pi}{2}$, D = 0 (***M***)

b. $y = -\sin(x - \frac{\pi}{2})$ ie A = 1, B = 1, C = $\frac{-\pi}{2}$, D = 0 (***M***)

c. $y = \sin 2(x + \frac{\pi}{4})$ ie A = 1, B = 2, C = $\frac{\pi}{4}$, D = 0 (***M***)

d. $y = 2\sin(x + \frac{\pi}{2})$ ie A = 2, B = 1, C = $\frac{-\pi}{2}$, D = 0 (***M***)

e. $y = -2\sin 2(x + \frac{\pi}{4})$ ie A = –2, B = 2, C = $\frac{-\pi}{4}$, D = 0 (***M***)

f. $y = \frac{1}{2}\sin 3x$ ie A = $\frac{1}{2}$, B = 3, C = 0, D = 0 (***M***)

g. $y = 3\sin(-2(x - \frac{\pi}{4})) - 3$ ie A = 3, B = –2, C = $\frac{-\pi}{4}$, D = –3 (***M***)

h. $y = 0.5\sin(x - \frac{\pi}{2}) - 2$ ie A = 0.5, B = 1, C = $\frac{-\pi}{2}$, D = –2 (***M***)

Note: Other possible values for A, B, C, D may exist.

Unit 12.2 Activity 2B: Transformations of trigonometric graphs (page 102)

1. a.

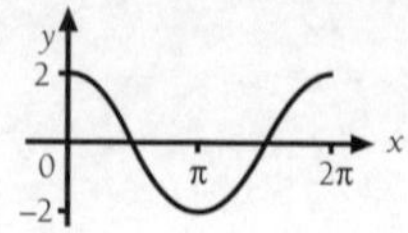

Period = 2π
Amplitude = 2 (***A***)

b.

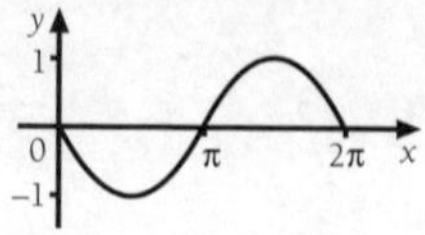

Period = 2π
Amplitude = 1 (***A***)

c.

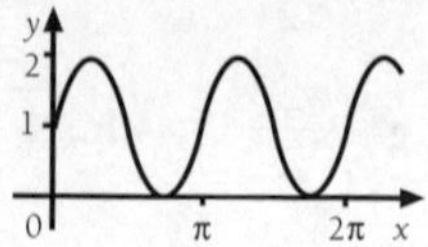

Period = π
Amplitude = 1 (***A***)

d.

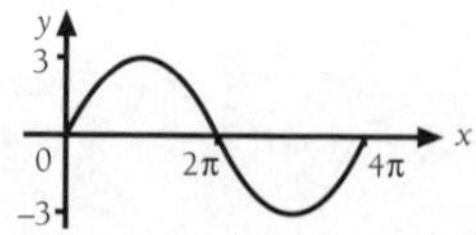

Period = 4π
Amplitude = 3 (***A***)

e.

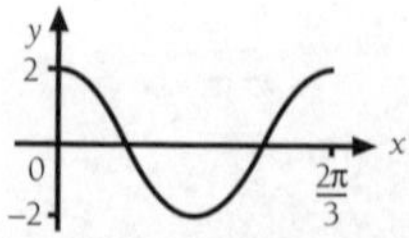

Period = $\frac{2\pi}{3}$
Amplitude = 2 (***A***)

f.

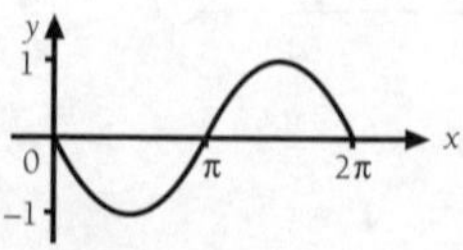

Period = 2π
Amplitude = 1 (***A***)

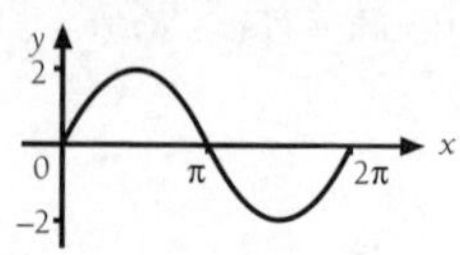

g.

Period = 2π
Amplitude = 2 (***A***)

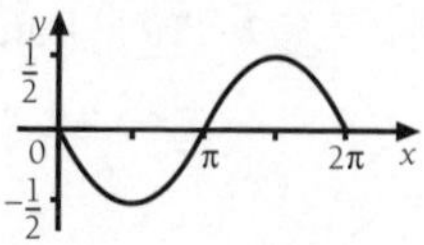

h.

Period = 2π
Amplitude = $\frac{1}{2}$ (***A***)

2. $B = \frac{\pi}{2}$

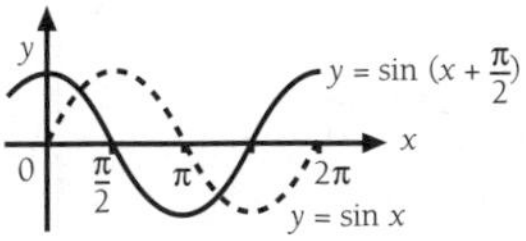

(**A**)

3. **a.**

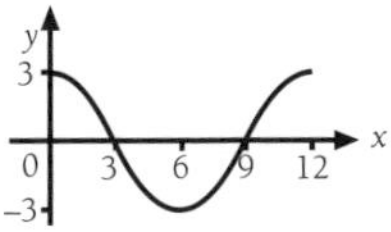

(**A**)

b. From graph, minimum at (6, –3). (**A**)

c. 52 minutes and 53 seconds after 8 am;
and 7 minutes and seven seconds after 11 am;
and 52 minutes and 53 seconds past 8 pm;
and 7 minutes and 7 seconds after 11 pm. (**A**)

4. **a.** f: 3, –3; g: 2, –2 (**A**) **b.** f: 3, g: π (**A**) **c.** f: $\frac{1}{3}$, g: $\frac{1}{\pi}$ cycles per hour (**A**)

5. **a.**

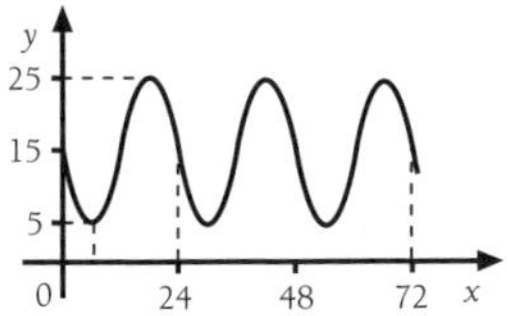

(**A**)

b. Maximum temperature = 25 °C at 6 pm
Minimum temperature = 5 °C at 6 am daily (**A**)

c. Amplitude = 10; Frequency = $\frac{1}{24}$ cycles per hour (**A**)

6. **a.**

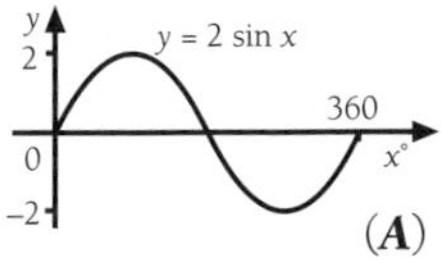

(**A**)

b.

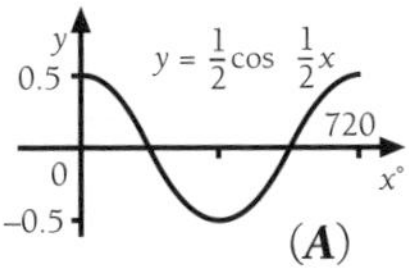

(**A**)

c.

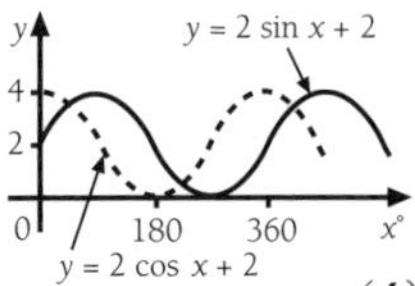

(**A**)

d.

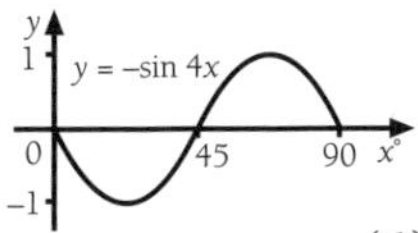

(**A**)

e.

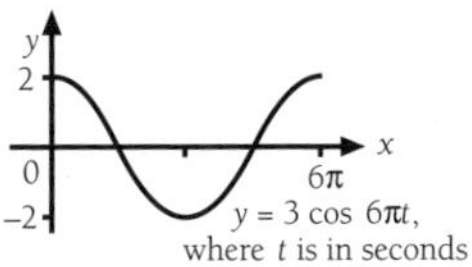

(**A**)

7. **a. i.** Amplitude = 2; Period = 3π (**A**) **ii.** Amplitude = $\frac{1}{2}$; Period = 4 (**A**)

b. i. Frequency = $\frac{1}{3\pi} = \frac{1}{30}$ cycles per second (**A**)

ii. Frequency = $\frac{1}{4} = \frac{1}{12.73}$ cycles per second (**A**)

c. i.

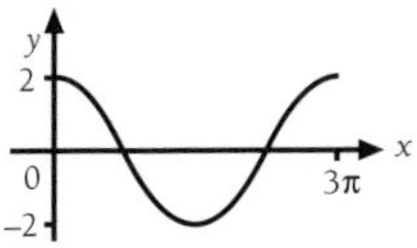

(**A**)

ii.

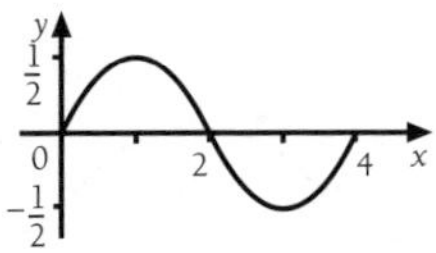

(**A**)

8. a.

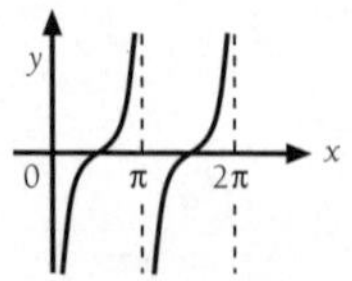

Period = π (**M**)

b.

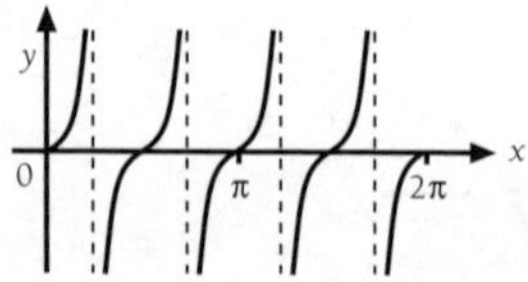

Period = $\frac{\pi}{2}$ (**M**)

c.

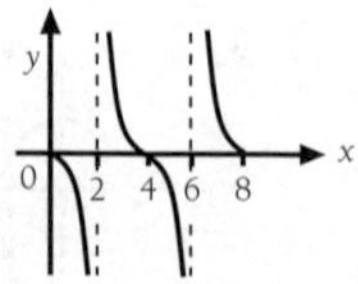

Period = 4 (**M**)

d.

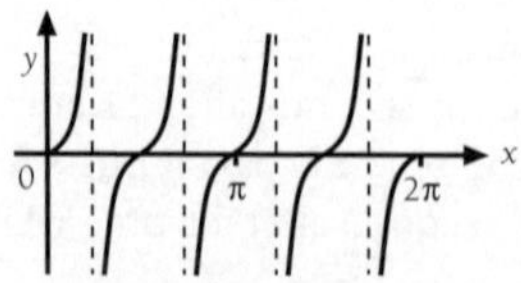

Period = $\frac{\pi}{2}$ (**M**)

9. a. 10 m (***A***) **b.** 1 m (***A***) **c.** 30 seconds (***A***) **d.** 21 m (***A***)

e. 16 m (***A***) **f.** t = 8.5 (2 sf) or t = 21.5 (2 sf) seconds (***A***)

10. a. $y = 30 + 30\cos\left(\frac{\pi x}{80}\right)$ (**M**) **b.** $\frac{dy}{dx}$ is maximum when $x = 40$, $y = 30$ (**M**)

11. a. $y = 2\cos\left(\frac{x}{2}\right)$ (**M**) **b.** $y = 0.5\sin\left(\frac{\pi x}{4}\right)$ (**M**) **c.** $y = -6\sin(4x)$ (**M**)

d. $y = -2\cos(\pi x)$ (**M**) **e.** $y = 2\sin(0.6x^\circ)$ (**M**) **f.** $y = -\cos(9x^\circ)$ (**M**)

12. a. Amplitude = 3; Period = 10 (**M**) **b.** $a = 3, b = \frac{\pi}{5}$ (**M**)

13. a. $y = A\sin B(x + C) + D$, where $A = 4$, $B = \frac{\pi}{6}$, $C = 0$, $D = 6$ (**M**) b. 4 hours (**M**)

14. a.

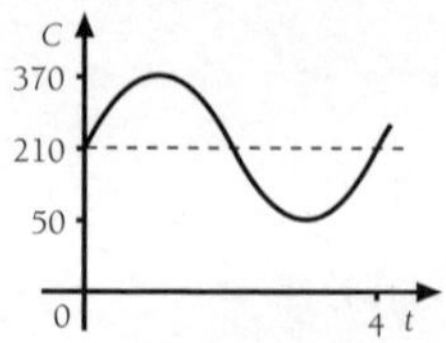

$C = 160\sin\left(\frac{\pi t}{2}\right) + 210$ (**M**)

b. 50 amps (**M**) **c.** $C = 320\sin(\pi t) + 420$ (**M**)

15. a.

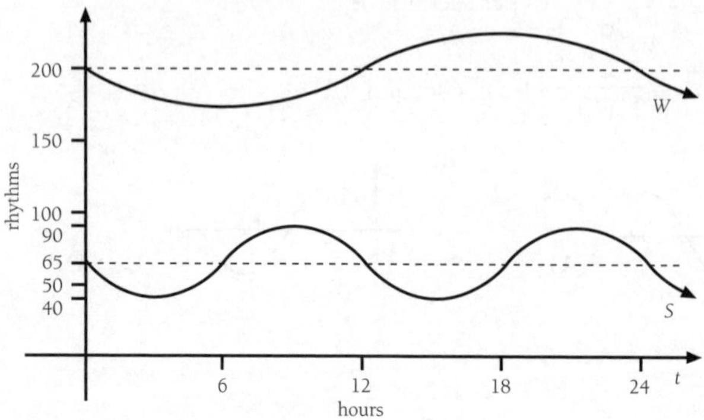

(***E***)

b. Between 6 pm and midnight (both graphs are above their mean level). (***E***)

c. Translate W by +3 hours, ie $W = -30 \sin \frac{\pi}{12}(t - 3) + 200$, so maximum values for S and W coincide at 2100 hours (preferably halve the period of W, or double the period of S). (***E***)

d. Sleeping, eating, etc interfere with the 24-hour cycle. (***E***)

Unit 12.2 Activity 3A: Solving straightforward trigonometric equations (page 109)

1. a. 53.1°, 126.9° (***A***) **b.** 53.1°, 126.9° (***A***) **c.** –306.9°, –233.1° (***A***)
d. –306.9°, –233.1° (***A***) **e.** no solutions (***A***)

2. a. 36.9° (***A***)
b. i. 36.9°, 143.1°, –216.9°, –323.1° (***A***) **ii.** 36.9°, 143.1° (***A***)
iii. 36.9°, 143.1° (***A***) **iv.** no solutions (***A***)

3. a. 113.6° (***A***)
b. i. –246.4°, –113.6°, 113.6°, 246.4° (***A***) **ii.** 113.6°, 246.4° (***A***)
iii. –113.6°, –246.4° (***A***) **iv.** –113.6°, 113.6° (***A***)

4. a., b.

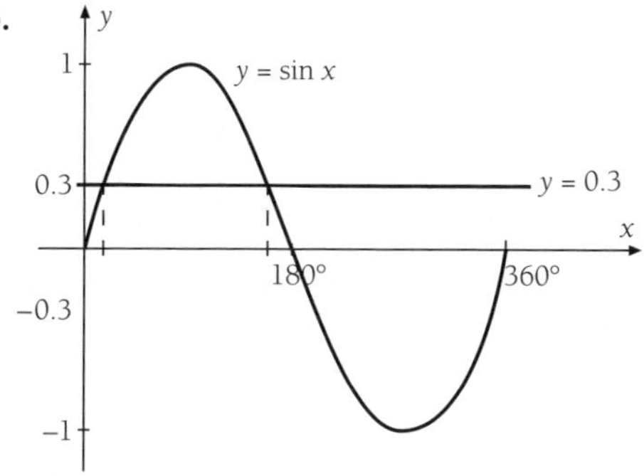

(***A***)

c. 17.5° (***A***)
d. 17.5°, 162.5° (***A***)

5. a., b.

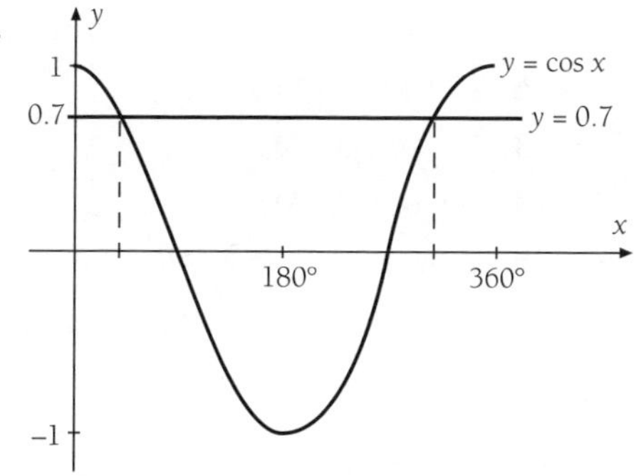

(***A***)

c. 45.6° (***A***)
d. 45.6°, 314.4° (***A***)

6. 107.5°, 252.5° (***A***)

7. a. –11.5°, –168.5°, 191.5°, 348.5° (***A***) **b.** 551.5°, 708.5° (***A***)

8. a., b.

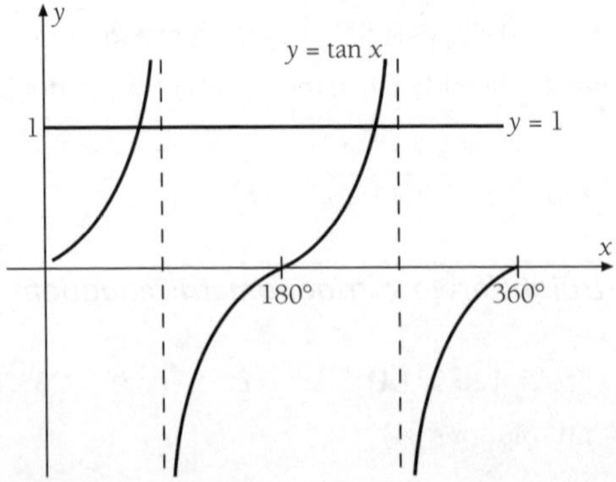

(**A**)

c. 45°, 225° (**A**)

9. 2 $(x = 90°, x = 270°)$ (**A**)

10. 2 $(x = 0°, x = 360°)$ (**A**)

Unit 12.2 Activity 3B: Solving trigonometric equations using radians (page 112)

1. 0.3047, 2.8369 (**A**)
2. 0.7754, 2.3662 (**A**)
3. 1.1593, 5.1239 (**A**)
4. 3.2417, 6.1830 (**A**)
5. 0.4115, 2.7301 (**A**)
6. 1.8755, 4.4077 (**A**)
7. 0.6747, 3.8163 (**A**)
8. 2.5536, 5.6952 (**A**)
9. 0.5404, 3.6820 (**A**)
10. 2.6779, 5.8195 (**A**)
11. d (**A**)
12. c (**A**)
13. b (**A**)
14. d (**A**)
15. b (**A**)
16. $\left\{\frac{7\pi}{6}, \frac{11\pi}{6}\right\}$ (**A**)
17. $\left\{\frac{\pi}{3}, \frac{5\pi}{3}\right\}$ (**A**)
18. $\left\{\frac{\pi}{4}, \frac{3\pi}{4}\right\}$ (**A**)
19. $\left\{-\frac{2\pi}{3}, \frac{\pi}{3}\right\}$ (**A**)
20. $\left\{\frac{3\pi}{4}, \frac{7\pi}{4}\right\}$ (**A**)
21. $\left\{-\frac{\pi}{3}, -\frac{2\pi}{3}\right\}$ (**A**)
22. $\left\{-\frac{\pi}{4}, \frac{\pi}{4}\right\}$ (**A**)

Unit 12.2 Activity 3C: Practical problems using trigonometry (page 113)

1. a. 1 m **b.** 0.99 m **c.** after 1.05 minutes and 5.24 minutes (**A**)

2. a.

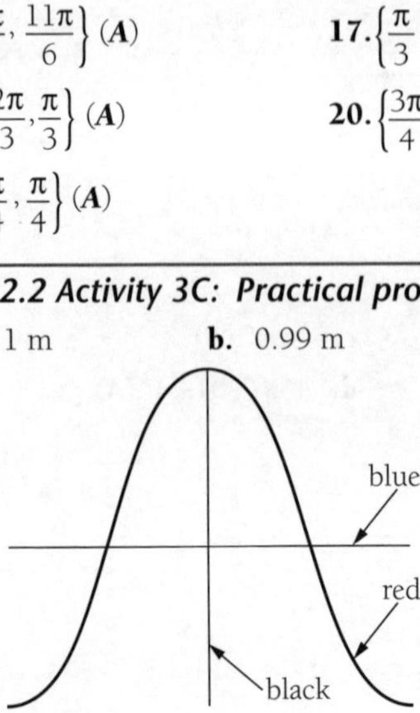

b. –72.5°, 72.5° (**A**)

3. 30 seconds (**M**)

4. 30 seconds (**M**)

5. After 5.36 seconds (**M**)

Unit 12.2 Activity 4A: Simple trigonometric equations (page 117)

1. 18.5°, 161.5° (**A**)
2. 64.4°, 295.6° (**A**)
3. 129.6°, 309.6° (**A**)
4. 117.5°, 242.5° (**A**)
5. 300.4°, 239.6° (**A**)
6. 34.3°, 214.3° (**A**)
7. 71.6°, 288.4° (**A**)
8. 28.9°, 151.1° (**A**)
9. 234.4°, 305.6° (**A**)
10. 115.1°, 295.1° (**A**)
11. 82.2°, 277.8° (**A**)
12. 66.5°, 246.5° (**A**)
13. 225°, 315° (**A**)
14. 60°, 120° (**A**)
15. 150°, 330° (**A**)
16. 270° (**A**)
17. 45°, 315° (**A**)
18. 60°, 240° (**A**)
19. 210°, 330° (**A**)
20. 45°, 135°, 225°, 315° (**M**)
21. 25.7°, 86.3° (**M**)
22. 188.6°, 277.4° (**M**)
23. 113.6°, 246.4° (**M**)
24. 16.6°, 196.6° (**M**)
25. 90°, 180° (**M**)
26. 0°, 60° (**M**)
27. 150°, 270° (**M**)
28. 62.5°, 191.5° (**M**)
29. 24.0°, 204.0° (**M**)
30. 48.0°, 252.0° (**M**)

Unit 12.2 Activity 4B: Trigonometric equations where coefficient of $x \neq 1$ (page 118)

1. **a.** 11.4, 48.6, 131.4, 168.6, 251.4, 288.6 (**M**)
b. 38.8, 141.2, 218.8, 321.2 (**M**)
c. 17.7, 77.7, 137.7, 197.7, 257.7, 317.7 (**M**)
d. 73.8, 163.8, 253.8, 343.8 (**M**)
e. 31.1, 121.1, 211.1, 301.1, 58.9, 148.9, 238.9, 328.9 (**M**)
f. 104.2, 284.2, 165.8, 345.8 (**M**)
g. 18.7, 108.7, 198.7, 288.7 (**M**)
h. 14.9, 104.9, 194.9, 284.9, 30.1, 120.1, 210.1, 300.1 (**M**)
i. 15.4, 344.6 (**M**)
j. 196.4 (**M**)
k. 15, 165, 195, 345
l. 30, 150, 270 (**M**)

2. **a.** 28.2, 61.8 (**M**)
b. 22.9, 97.1, 142.9 (**M**)
c. 160, 520 (**M**)
d. –176.9, –123.1, –56.9, –3.1, 63.1, 116.9 (**M**)
e. 72.2, 107.8, 252.2, 287.8, 432.2, 467.8 (**M**)
f. 24, 69, 114, 159 (**M**)
g. 15, 75 (**M**)
h. 386.7, 693.3 (**M**)
i. 9.2, 99.2, 189.2 (**M**)
j. 3.1, 56.9, 123.1, 176.9 (**M**)
k. ±132.9, ±107.1, ±12.9 (**M**)
l. 30, 120 (**M**)

Unit 12.2 Activity 4C: Trigonometric equations using radians (page 119)

1. **a.** $\left\{\frac{\pi}{12}, \frac{5\pi}{12}\right\}$(**M**)
b. $\left\{\frac{2\pi}{3}, 0\right\}$ (**M**)
c. $\left\{\frac{\pi}{6}, \frac{\pi}{3}, \frac{7\pi}{6}, \frac{4\pi}{3}\right\}$(**M**)
d. $\left\{\frac{\pi}{4}, \frac{5\pi}{12}, \frac{11\pi}{12}, \frac{13\pi}{12}, \frac{19\pi}{12}, \frac{7\pi}{4}\right\}$(**M**)
e. $\left\{\frac{3\pi}{2}\right\}$
f. $\left\{\frac{\pi}{8}, \frac{5\pi}{8}, \frac{9\pi}{8}, \frac{13\pi}{8}\right\}$(**M**)
g. $\left\{\frac{\pi}{3}, \frac{5\pi}{3}\right\}$(**M**)
h. No solution (**M**)

2. **a.** $\left\{\frac{5\pi}{12}, \frac{13\pi}{12}\right\}$(**M**)
b. {1.9127, 3.3233}(**M**)
c. {0.34, 2.8}(**M**)
d. {2.42, 3.86}(**M**)
e. {1.07, 2.08}(**M**)
f. {2.16, 5.3}(**M**)
g. {2.29, 5.43}(**M**)
h. {1.72, 4.56} (**M**)

Unit 12.2 Activity 4D: Applications of trigonometry (page 121)

1. a. 7.5 m **b.** 15 m **c.** 5 m **d.** 6 sec

e. $t = 1$ sec or $t = 5$ sec (**e.** only: ***E***)

2. a. 2 m **b.** 5 m **c.** when $t = 0$ (**c.** only: ***E***)

3. a. 16 cm **b.** 8 cm **c.** 7.5 sec **d.** 1.25 sec and 6.25 sec (**c.** & **d.** only: ***E***)

4. a. 11.5 m **b.** After 6 seconds, height is 15 m. (**b.** only: ***E***)

Unit 12.2 Activity 4E: Piecewise trigonometric functions (page 122)

1. a. $A = 0.5321$ (***E***) **b.** 2.125 seconds (***E***)

2. a. 0.2618, 1.3090, 4.9742, 6.0214 (***E***) **b.** 2.094, 4.1888 (***E***)

Unit 12.2 Activity 5A: Trigonometric identities (page 125)

1. Proofs. (***M***) **2.** Proofs. (***M***)

3. a. $x^2 + y^2 = 4$ (***M***) **b.** $a^2x^2 + a^2b^2 = b^2y^2$ (***M***) **c.** $a^2x^2 = x^2y^2 + y^2$ (***M***)

d. $x^2 + y^2 = 2$ (***M***)

4. a. Proof. (***E***) **b.** $\sin\theta = \dfrac{u^2 - 1}{u^2 + 1}$ (***E***)

Investigation: Compound angle identities (page 126)

BC = $\sin x$, AB = $\cos x$, CE = $\sin x \cos y$, EF = $\cos x \sin y$, CF = $\sin (x + y)$

But CF = CE + EF

$\therefore \sin (x + y) = \sin x \cos y + \cos x \sin y$

Investigation (page 127)

$$\sin \angle AOB = \sin (\angle BOR - \angle AOR)$$
$$= \sin \angle BOR \cos \angle AOR - \cos \angle BOR \sin \angle AOR$$
$$= \frac{a}{R} \times \frac{a}{R} - \frac{b}{R} \times \frac{b}{R} = \frac{a^2 - b^2}{R^2}$$

Investigations (page 129)

1. a. i. $\frac{1}{2}R^2x$ **ii.** $\frac{1}{2}R^2 \sin x$ **iii.** $\frac{1}{2}R^2 \tan x$

b. Proof. **c.** 1

2. $x = 1.9$

Unit 12.2 Activity 5B: Problem solving in trigonometry (page 129)

1. 125.9 m (***E***) **2.** Proof. (***E***) **3.** $\sin\theta = \dfrac{12x}{13\sqrt{x^2 + 10x + 169}}$ (***E***)

4. Proof. (***E***) **5.** Proof. (***E***) **6.** 13.8 cm or 3.2 cm (***E***)

7. 12.5 km; 271.8° (1 dp) (**E**) **8.** Proof. (**E**) **9.** Proof. (**E**)

10. $r(1 - \cos \alpha)$ **11.** Proof. **12.** Proof. (**E**) **13.** Proof. (**E**)

14. Box 14.3923 × 26 = 374.2 cm², saving area 9.8 cm². (**E**)

15. a.

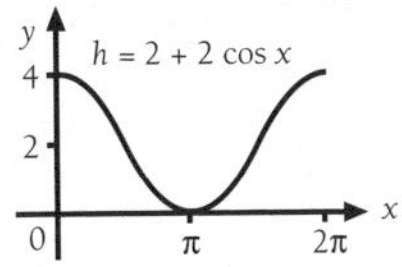

(**E**)

b.

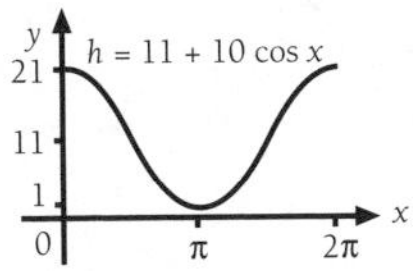

(**E**)

Unit 12.2 Activity 5C: Revision questions (page 131)

1.

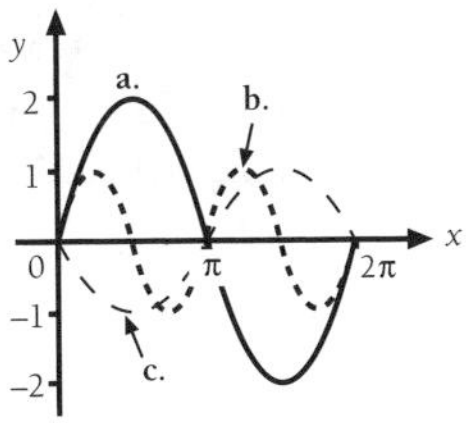

(**A**)

2. a. Proof. (**M**) **b.** Proof. (**M**)

3. a. i. 664.9 m (**A**) **ii.** Proof. (**E**)

iii. Proof. (**E**)

b. $x = \dfrac{d \cos \alpha \sin (\theta + \beta)}{\sin (\alpha - \beta)}$ or $x = \dfrac{d (\sin \theta + \cos \theta \tan \beta)}{\tan \alpha - \tan \beta}$ (**E**)

4. $a + b = 3\pi$ (**A**)

5. a. $(x - 1)^2 + 6$ **b.** $x = 1$ **c.** $t = 2n\pi + \dfrac{\pi}{2}, n \in W$ (**M**) **d.** 6 volts (**M**)

6. Proof. (**M**)

7. a. i. 18 cm (**M**) **ii.** 10 cm (**M**) **iii.** 2 s (**M**)

iv. The period would decrease and the range of the graph would stay the same. (**M**)

b. 0.737 (**A**)

8. a. $\dfrac{-4}{3}; \dfrac{5}{4}$ (**A**) **b.** 2.214 rad (**A**)

9. 2 cos A + 1 (**E**) **10.** Proof. (**M**)

Unit 12.2 Activity 6A: Properties of vectors (page 138)

1. a. $3\underset{\sim}{a}$ will have the same direction and sense of direction as $\underset{\sim}{a}$ but will be three times the length.

b. $-2\underset{\sim}{c}$ will have the same direction but opposite sense of direction as $\underset{\sim}{c}$ and will be two times the length of $\underset{\sim}{c}$.

c.

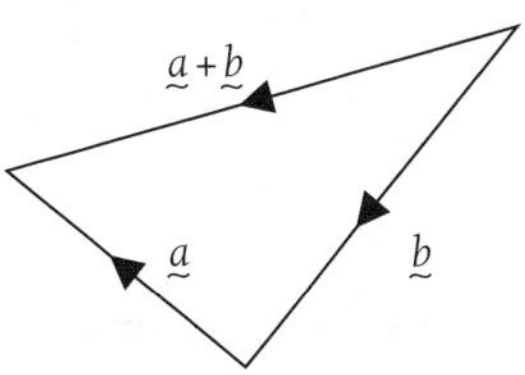

d.

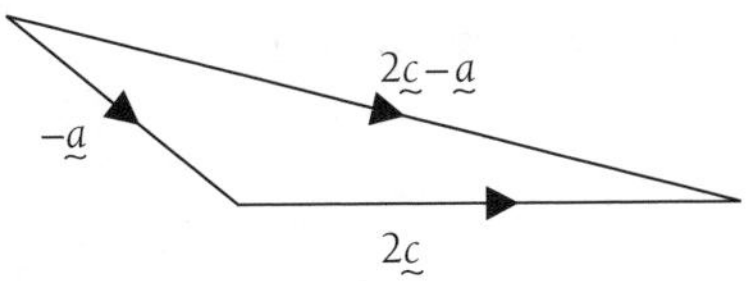

2. a.

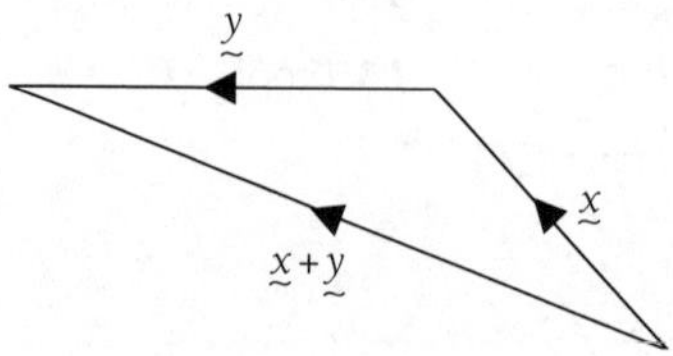

b.

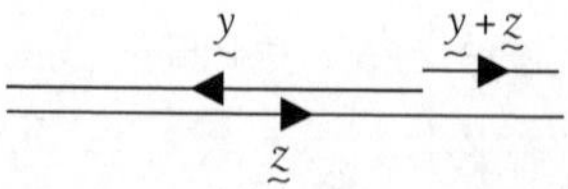

c.

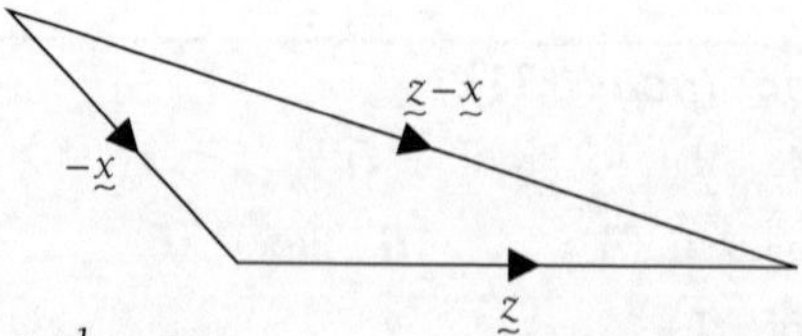

d.

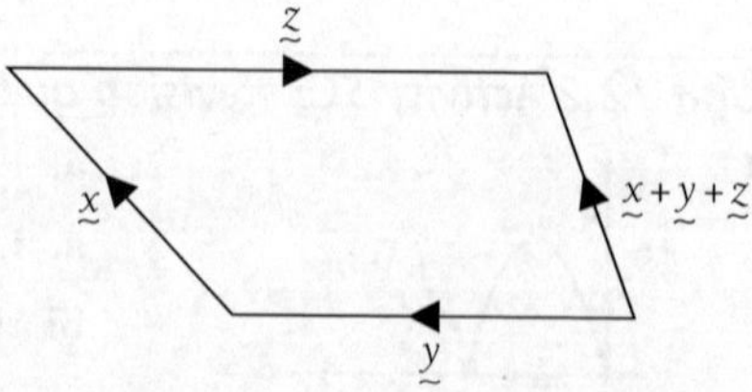

3. a. $\underset{\sim}{a}+\underset{\sim}{b}=-\underset{\sim}{c}$

b. $\underset{\sim}{s}=\underset{\sim}{p}+\underset{\sim}{q}+\underset{\sim}{r}$

c. $\underset{\sim}{p}+\underset{\sim}{q}=\underset{\sim}{r}+\underset{\sim}{s}$

d. $\underset{\sim}{e}+\underset{\sim}{d}=\underset{\sim}{a}-\underset{\sim}{b}+\underset{\sim}{c}$

4. a. i., ii., iii.

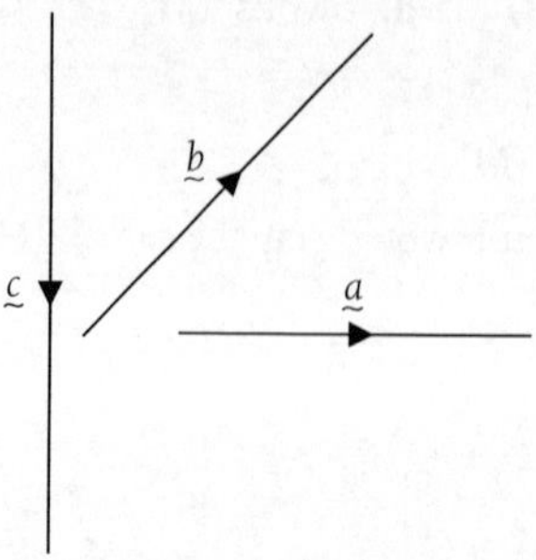

iv.

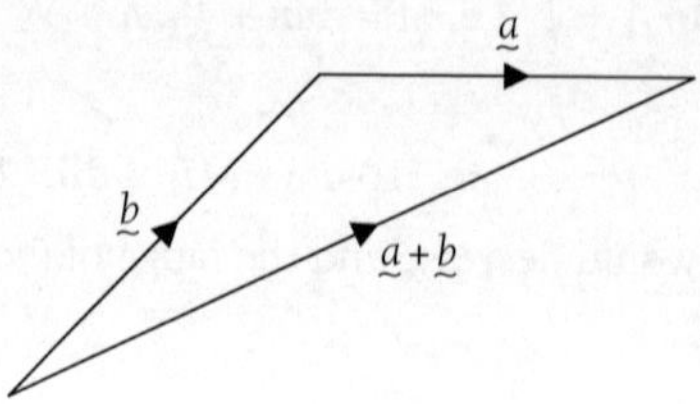

v.

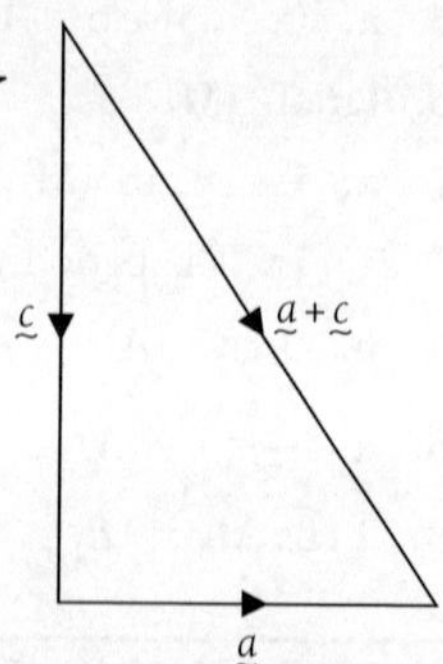

b. Discuss your answers with your teacher.

c. i. 69.32 km/hr, N65.92°E **ii.** 61.03 km/hr, S34.99°E

5. a. Discuss your answers with your teacher.

b. 18.03 kg wt, N56.3°W

c. 18.03 kg wt, S56.3°W

6. a.

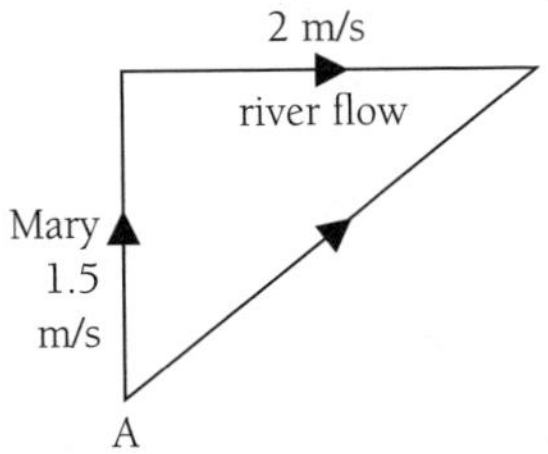

b. 2.5 m/s

c. N53.13°E

7. a.

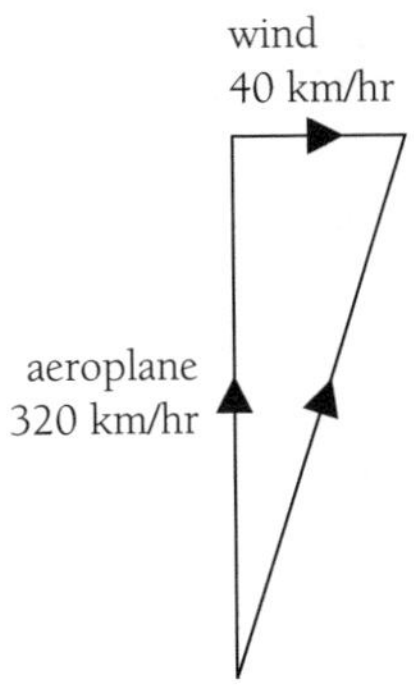

b. 322.49 km/hr

c. N7.13°E

Unit 12.2 Activity 7A: Geometric proofs using vectors (page 142)

1. a. $2\underset{\sim}{b}$ **b.** $\underset{\sim}{a} + 2\underset{\sim}{b}$ **c.** $\underset{\sim}{a} + \underset{\sim}{b}$

2. Discuss your proof with your teacher.

3. $2\overrightarrow{CB}$

4. Discuss your proof with your teacher.

5. a. i. $\underset{\sim}{b} - \underset{\sim}{a}$ **ii.** $\frac{1}{3}(\underset{\sim}{b} - \underset{\sim}{a})$ **iii.** $\frac{1}{3}(2\underset{\sim}{a} + \underset{\sim}{b})$ **iv.** $\frac{1}{2}\underset{\sim}{b} - \underset{\sim}{a}$

v. $\frac{1}{4}\underset{\sim}{b} - \frac{1}{2}\underset{\sim}{a}$ **vi.** $\frac{1}{4}(2\underset{\sim}{a} + \underset{\sim}{b})$

b. P, O and N are collinear points.

6. a. i. $\underset{\sim}{b} - \underset{\sim}{a}$ **ii.** $\frac{1}{2}\underset{\sim}{b} - \underset{\sim}{a}$ **iii.** $\frac{1}{3}(\underset{\sim}{b} - 2\underset{\sim}{a})$ **iv.** $\frac{1}{2}(\underset{\sim}{a} + \underset{\sim}{b})$

v. $\frac{1}{2}(\underset{\sim}{a} + \underset{\sim}{b})$

b. A, G and N are collinear points.

7. Discuss your proof with your teacher.

8. a. i. $\frac{1}{3}(2\underset{\sim}{a} + \underset{\sim}{b})$ **ii.** $\frac{1}{3}(2\underset{\sim}{a} + \underset{\sim}{b})$

b. Points G and H are the same points.

Unit 12.2 Activity 8A: Further properties of vectors (page 147)

1. 3.32 kg wt, 39.81°

2. 6.17 kg wt

3. 135.6°, $2\sqrt{6}$ kg wt

4. a. 6.32 km/hr **b.** 33.33 m

5. a. N3.90°W **b.** 554.15 km/hr

6. 7 034.13 N, 14.62°

7. a. 059° **b.** 15.77 knots

Unit 12.2 Activity 9A: Rectangular components of vectors (page 151)

1. i. a. 10 **b.** $\frac{3}{5}\underset{\sim}{i}+\frac{4}{5}\underset{\sim}{j}$ **ii. a.** $\sqrt{34}$ **b.** $-\frac{3}{\sqrt{34}}\underset{\sim}{i}+\frac{5}{\sqrt{34}}\underset{\sim}{j}$

iii. a. 13 **b.** $\frac{5}{13}\underset{\sim}{i}-\frac{12}{13}\underset{\sim}{j}$ **iv. a.** $\sqrt{10}$ **b.** $\frac{1}{\sqrt{10}}\underset{\sim}{i}-\frac{3}{\sqrt{10}}\underset{\sim}{j}$

2. a. $\overrightarrow{OP}=-2\underset{\sim}{i}+\underset{\sim}{j}$; $\overrightarrow{OQ}=3\underset{\sim}{i}+5\underset{\sim}{j}$ **b.** $5\underset{\sim}{i}+4\underset{\sim}{j}$ **c.** $\sqrt{41}$

d. $\frac{5}{\sqrt{41}}\underset{\sim}{i}+\frac{4}{\sqrt{41}}\underset{\sim}{j}$ **e.** $\underset{\sim}{i}+6\underset{\sim}{j}$ **f.** $11\underset{\sim}{i}+14\underset{\sim}{j}$

3. a. $\underset{\sim}{i}+\underset{\sim}{j}$ **b.** $\sqrt{13}$ **c.** $\frac{1}{\sqrt{10}}\underset{\sim}{i}+\frac{3}{\sqrt{10}}\underset{\sim}{j}$

d. $\underset{\sim}{a}+\underset{\sim}{c}=-\underset{\sim}{i}$ **e.** $\underset{\sim}{b}+\underset{\sim}{c}=-8\underset{\sim}{j}$

4. $\left|\overrightarrow{OP}\right|=\left|\overrightarrow{PQ}\right|=\sqrt{37}$

5. $2\left|\overrightarrow{PQ}\right|=\left|\overrightarrow{RS}\right|$

6. $2\left|\overrightarrow{PQ}\right|=\left|\overrightarrow{QR}\right|$

7. m = −12

8. $\overrightarrow{OM}=\underset{\sim}{i}+2\underset{\sim}{j}$

9. $\overrightarrow{OS}=-2\underset{\sim}{i}+\frac{7}{2}\underset{\sim}{j}$

Unit 12.2 Activity 10A: Component vectors in three dimensions (page 153)

1. i. a. 5 **b.** $\frac{1}{5}\underset{\sim}{i}+\frac{4}{5}\underset{\sim}{j}+\frac{3}{5}\underset{\sim}{k}$ **ii. a.** 3 **b.** $\frac{-2}{3}\underset{\sim}{i}+\frac{2}{3}\underset{\sim}{j}-\frac{1}{3}\underset{\sim}{k}$

iii. a. $5\sqrt{2}$ **b.** $\frac{3}{5\sqrt{2}}\underset{\sim}{i}-\frac{3}{5\sqrt{2}}\underset{\sim}{j}-\frac{1}{\sqrt{2}}\underset{\sim}{k}$ **iv. a.** $\sqrt{74}$ **b.** $\frac{7}{\sqrt{74}}\underset{\sim}{i}-\frac{3}{\sqrt{74}}\underset{\sim}{j}+\frac{4}{\sqrt{74}}\underset{\sim}{k}$

2. a. $\overrightarrow{OP}=-2\underset{\sim}{i}+\underset{\sim}{j}+3\underset{\sim}{k}$; $\overrightarrow{OQ}=-3\underset{\sim}{i}+5\underset{\sim}{j}-\underset{\sim}{k}$ **b.** $5\underset{\sim}{i}+4\underset{\sim}{j}-4\underset{\sim}{k}$

c. $\sqrt{57}$ **d.** $\frac{5}{\sqrt{57}}\underset{\sim}{i}+\frac{4}{\sqrt{57}}\underset{\sim}{j}-\frac{4}{\sqrt{57}}$

e. $\underset{\sim}{i}+6\underset{\sim}{j}+2\underset{\sim}{k}$ **f.** $7\underset{\sim}{i}+3\underset{\sim}{j}-7\underset{\sim}{k}$

3. a. $5\underset{\sim}{i}+\underset{\sim}{j}+5\underset{\sim}{k}$ **b.** $\sqrt{22}$

c. $\underset{\sim}{a}+\underset{\sim}{c}=-\underset{\sim}{i}$ **d.** $\frac{1}{\sqrt{14}}\underset{\sim}{i}+\frac{3}{\sqrt{14}}\underset{\sim}{j}+\frac{2}{\sqrt{14}}\underset{\sim}{k}$

4. $\left|\overrightarrow{OP}\right|=\left|\overrightarrow{PQ}\right|=\sqrt{41}$

5. $\overrightarrow{OM} = \underset{\sim}{i} + 2\underset{\sim}{j} + \frac{1}{2}\underset{\sim}{k}$

6. $\underset{\sim}{b} = -\frac{1}{3}\underset{\sim}{a}$

8. $\overrightarrow{QR} = -2\overrightarrow{PQ}$

9. $p = \frac{1}{6}; = q - \frac{11}{12}$

Unit 12.2 Activity 11A: Scalar product of vectors (page 157)

1. **a.** 3 **b.** –9.23 **c.** 600 **d.** –69.28 **e.** 0 **f.** 0
2. **a.** 226.52 **b.** 0 **c.** –259.81
   d. –320 **e.** 181.04 **f.** –85.41
   g. –181.22 **h.** 165 **i.** 181.04
3. **a.** 5 **b.** 70.35°
4. 88.15°
5. **a.** –2 **b.** 99.27°
7. **a.** $\overrightarrow{PQ}.\overrightarrow{RQ} = 0$ **b.** 20.41°, 69.59°
8. **a.** $\underset{\sim}{a} \perp \underset{\sim}{b}$ **b.** $\underset{\sim}{a} \parallel \underset{\sim}{b}$ **c.** $\underset{\sim}{a} \perp (\underset{\sim}{b} - \underset{\sim}{c})$ or $|\underset{\sim}{b} - \underset{\sim}{c}| = 0$
   d. $\underset{\sim}{c} \perp (\underset{\sim}{a} - \underset{\sim}{b})$ or $|\underset{\sim}{a} - \underset{\sim}{b}| = 0$
9. Discuss your conclusions with your teacher
10. Discuss your answer with your teacher
11. Discuss your proof with your teacher
12. Discuss your proof with your teacher

Unit 12.3 Activity 1A: Calculating gradients (page 162)

1. **a.** 2 **b.** –6 2. **a.** 0 **b.** –4 3. **a.** 4 **b.** –2
4. **a.** 3 **b.** 3 **c.** 3 5. **a.** –2 **b.** –2 **c.** –2
6. **a.** 0 **b.** 0 **c.** 0 7. **a.** 12 **b.** 27 **c.** 3
8. **a.** 5 **b.** 9 **c.** –11

Unit 12.3 Activity 1B: Variable gradients in practical contexts (page 163)

1. **a.** Moving at constant speed until it passes A, after which it notices an increasing slowing effect until C is passed, after which the slowing effect diminishes until E is reached, at which time the slowing effect disappears.
   b. C
2. **a.** Moving at constant speed until A is passed, after which the car slowly slows down until just after C it rapidly slows. Once past D this slowing effect quickly disappears.
   b. D

3.

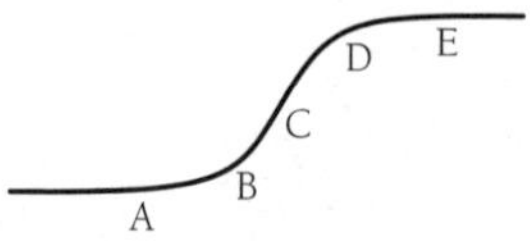

4.

5.

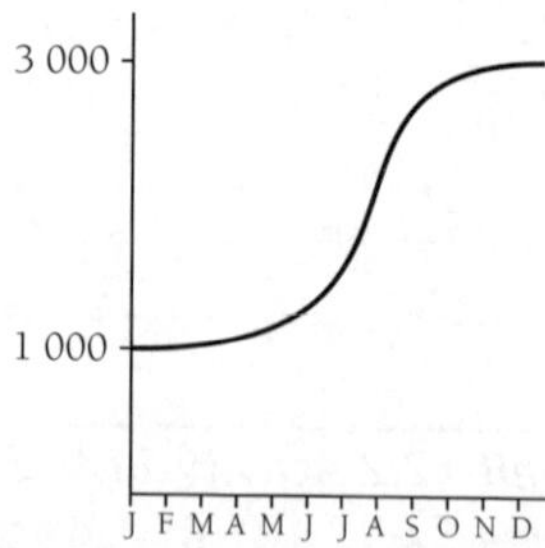

Unit 12.3 Activity 1C: The gradient function (page 165)

1. a.

x	–3	–2	–1	0	1	2	3
y	–27	–8	–1	0	1	8	27
gradient	27	12	3	0	3	12	27

b. gradient = $3x^2$

2. a.

x	–3	–2	–1	0	1	2	3
y	81	16	1	0	1	16	81
gradient	–108	–32	–4	0	4	32	108

b. gradient = $4x^3$

3. **a.** gradient = $5x^4$ **b.** gradient = $4x$ **c.** gradient = $2x - 1$

d. gradient = $2x + 4$ **e.** gradient = 2 **f.** gradient = 0

Unit 12.3 Activity 1D: Sketching graphs of gradient functions (page 167)

1. a.

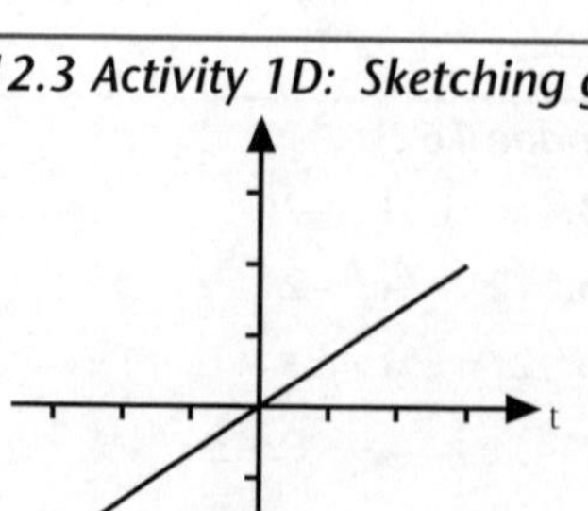

b. The rate of change of distance from the house changes at a steady rate. (***A***)

2. a.

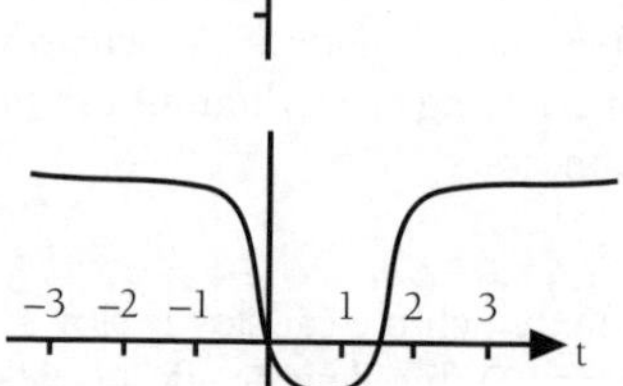

b. The rate at which profits increase is fairly constant until February. It decreases sharply until near the middle of April at which time it quickly recovers to its pre-February value. (***A***)

3. a.

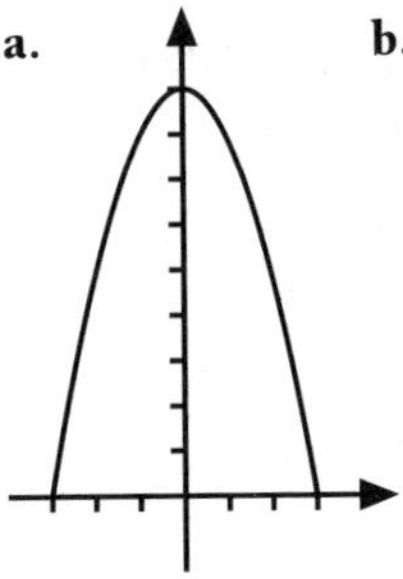

b.

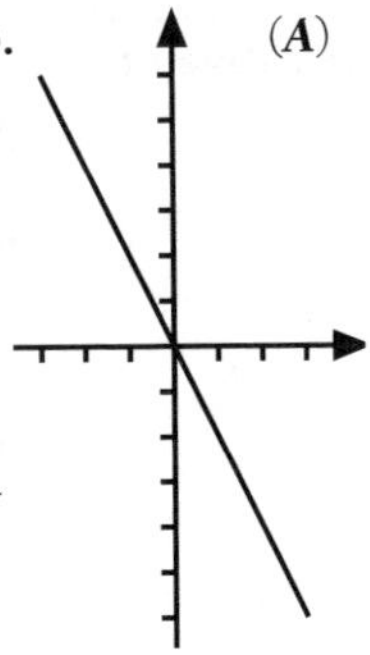

(***A***)

4. a.

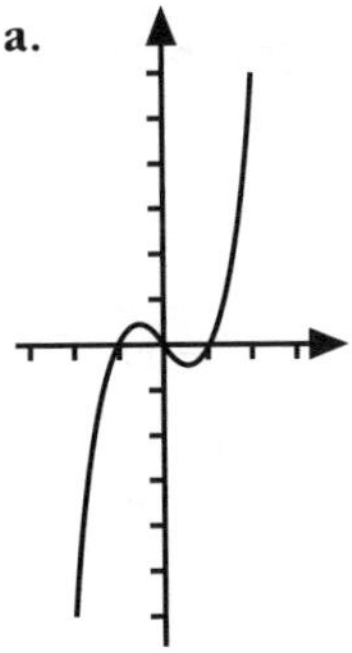

b.

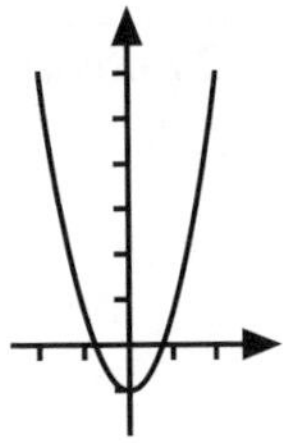

(***A***)

5. a.

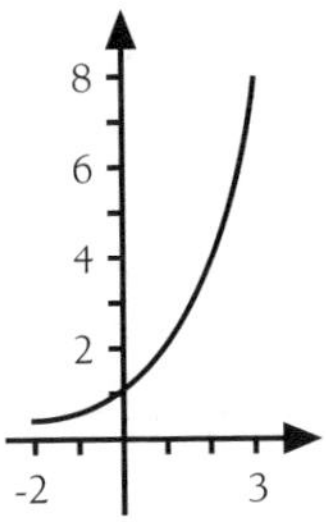

b.

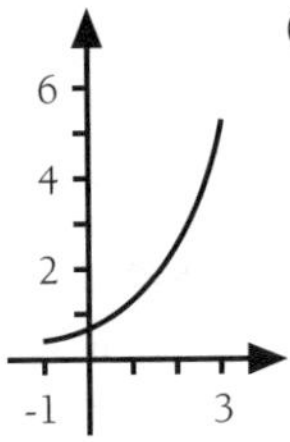

(***A***)

Unit 12.3 Activity 2A: Differentiation (page 171)

1. a. $5x^4$ **b.** $22x^{21}$ **c.** $48x^{47}$ **d.** $21x^6$ **e.** $45x^8$ **f.** $20x$

g. $\frac{2x}{3}$ **h.** $24x^2 - 6x$ **i.** $3x^5$ **j.** $\frac{8x}{3}$ **k.** 0

l. 5 **m.** $4x + 3$ **n.** $70x^9 - \frac{96x^{11}}{5}$ **o.** $2x + 2$ **p.** $8x + 4$

q. 3 **r.** $4x^3 + 6x^2 + 2x$ **s.** $8x - 12$ **t.** $2x + 3$ **u.** $2x + 5$

v. $3x^2 + 6x + 1$ **w.** $5x^4 + 12x^3 + 6x^2$ **x.** $4x^3 + 6x^2 + 6x + 2$

y. $4x + 6$ **z.** $4x^3 - 5x + \frac{3}{2}$ (***A***)

2. a. $12x^2$ **b.** $12x$ **c.** $10x + 3$ **d.** $9t^2 + 8t - 3$

e. $4u + 9$ **f.** $6x$ **g.** 2 **h.** $20x^4 + \frac{1}{7}$

i. $6r^2 + 16r + 6$ **j.** $18t^2 + 16t + 9$ (***A***)

3. a. i. 7 **ii.** 10 **iii.** $\frac{1}{2}$ **b. i.** 3 **ii.** 2 **iii.** 4 **iv.** $3\frac{1}{2}$

c. i. 125 **ii.** 17 **iii.** 0, 1 **d. i.** -6 **ii.** 9 **iii.** 0 **iv.** $-\frac{1}{6}$

e. i. $-15\frac{1}{3}$ **ii.** -9 **iii.** $-2, 4$ **iv.** $-3, 5$ (***A***)

Unit 12.3 Activity 2B: Differentiation with negative indices (page 172)

1. a. $-4x^{-5}$ b. $-5x^{-6}$ c. $-7u^{-8}$ d. $-12u^{-5}$
   e. $-30x^{-7}$ f. $-\frac{10}{3}y^{-6}$ g. $-\frac{3}{2}x^{-3}$ h. $-\frac{4}{5}x^{-5}$
   i. $-\frac{6}{7}x^{-3}$ j. $-6x^{-4} - 12x^{-5}$
2. a. $-5x^{-6}$ b. $-4x^{-5}$ c. $-25x^{-6}$ d. $-4x^{-9}$
   e. $-6x^{-10}$
3. a. $-7x^{-8}$ b. $-4x^{-3}$ c. $3x^2 - 2x^{-3}$ d. $-2x^{-3}$
   e. $-2x^{-3} - x^{-2} - 3x^{-4}$
4. a. $-5x^{-6} - 2x$ b. $-3x^{-2} + 3x^2$ c. $1 - x^{-2} - 2x^{-3}$ d. $2x - 2x^{-3}$
   e. $-12x^{-3} + 4x^{-2}$

Unit 12.3 Activity 3A: Gradients of tangents (page 174)

1. a. 5 b. 0 c. –3 d. 1 (**A**)
2. a. –4 b. –4 c. 3 d. 1.5 (**A**)
3. a. $(1\frac{1}{2}, 2\frac{1}{4})$ b. $(2, 2\frac{2}{3}), (-2, -2\frac{2}{3})$ c. no points d. all points on the line $y = 2x + 3$ (**M**)

Unit 12.3 Activity 3B: Equations of tangents (page 176)

1. a. $y = -5x + 1$ (**M**) b. $y = 8x - 17$ (**M**) c. $y = x + 1$ (**M**)
   d. $y = 4x - 6$ (**M**)
2. a. $8y + x = 26$ b. $11y = x + 100$ c. $7y + x = 37$
3. $8y + x = -89$
4. $y = x - 6$
5. $2y + x = 2$
6. $y = 6x - 8, y = 6x + 16$
7. $\left(-\frac{5}{12}, \frac{113}{48}\right)$
8. a. $y = 5x - 1$ (**A**) b. (2, 4); (4, –8) (**A**)
9. $(1\frac{1}{2}, -3)$ (**A**)
10. $x + 2y + 3 = 0, (-3, 0), (0, -1\frac{1}{2})$ (**A**)
11. $a = 6$; $b = 15$ (**A**)
12. $y = x + 3; y = \frac{1}{2}x + 6$; intersect at (6, 9) and $9 = \frac{1}{4} \times 6^2$ (**A**)

Unit 12.3 Activity 4A: Turning points (page 180)

1. (2, –3) minimum point; increasing $x > 2$; decreasing $x < 2$ (**M**)
2. (–4, 22) maximum point; increasing $x < -4$; decreasing $x > -4$ (**M**)
3. $(\frac{3}{4}, 6\frac{1}{8})$ maximum point; increasing $x < \frac{3}{4}$; decreasing $x > \frac{3}{4}$ (**M**)

4. $(-1, 5\frac{1}{3})$ maximum point; $(-3, 4)$ minimum point; increasing for $-3 < x < -1$; decreasing for $x < -3$, for $x > -1$. (**M**)

5. $(-3, 12\frac{1}{2})$ minimum point; $(-2, 12\frac{2}{3})$ maximum point; increasing for $-3 < x < -2$; decreasing for $x < -3$ and for $x > -2$ (**M**)

6. $(-1, -9)$ minimum point; $(-3, -1)$ maximum point; increasing for $x < -3$, for $x > -1$; decreasing for $-3 < x < -1$ (**M**)

7. $(-2, 18)$ maximum point; $(2, -14)$ minimum point; increasing for $x < -2$, for $x > 2$; decreasing for $-2 < x < 2$ (**M**)

8. $(-3, -52)$ minimum point; $(3, 56)$ maximum point; increasing for $-3 < x < 3$; decreasing for $x < -3$, for $x > 3$ (**M**)

9. $(\frac{-2}{3}, \frac{-4}{9})$ minimum point; $(0, 0)$ maximum point; increasing for $\frac{-2}{3} < x < 0$; decreasing for $x < \frac{-2}{3}$, $x > 0$. (**M**)

10. a. i. stationary ii. decreasing iii. increasing iv. stationary

b. $x = -2\frac{2}{3}$ is minimum, $x = 0$ is maximum (**M**)

11. a. Cubic: x-intercepts, -1, 1; y-intercept 1; x-axis only touched at $x = 1$

b. $3x^2 - 2x - 1$

c. $x = -\frac{1}{3}$, maximum point; $x = 1$, minimum point.

d. $-\frac{1}{3} < x < 1$

e. 1

f. -1

g. $y = -x + 1$ (**M**)

12. $A = 2$, $B = -6$, $C = 3$ (**M**)

Unit 12.3 Activity 4B: Curve sketching (page 182)

1. a.

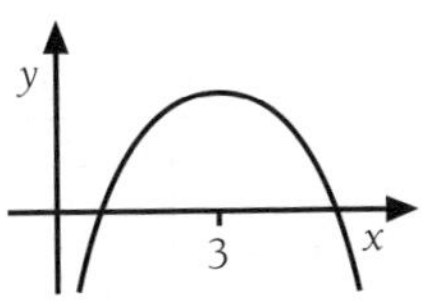

(**M**)

b.

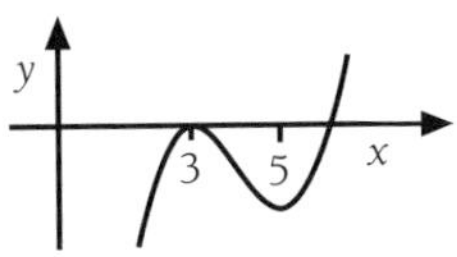

(**M**)

c.

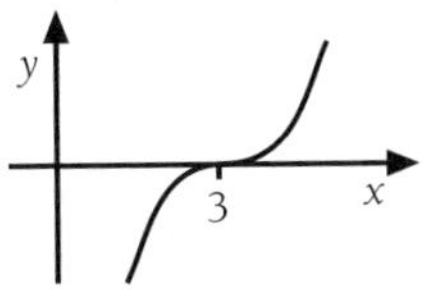

(**M**)

2. $(-2, -11)$ min; $(2, 21)$ max (**E**)

3. $(-2, 0)$ min; $(0, 16)$ max; $(2, 0)$ min (**E**)

4. a. x-intercepts $(0, 0)$, $(2, 0)$, $(-3, 0)$
y-intercept $(0, 0)$ (**A**)

b. $(-1.79, 8.21)$ max
$(1.12, -4.06)$ min (**E**)

c.

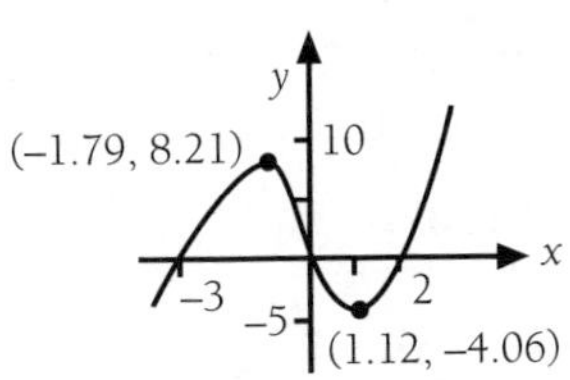

(**M**)

5. a. $(1.155, -6.158)$ min
$(-1.155, 6.158)$ max (**E**)

b. x-intercepts $(0, 0)$, $(\pm 2, 0)$
y-intercept $(0, 0)$ (**A**)

c.

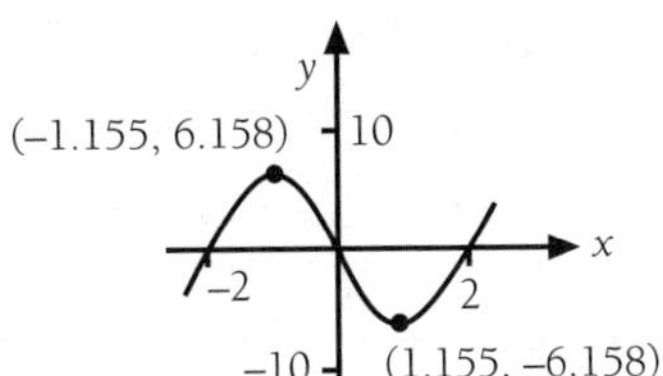

(**M**)

6. a. (–2, 0), (0, 0), (3, 0) intercepts; x-intercepts: set $y = 0$; y-intercepts: set $x = 0$ (***A***)

b. (1.79, 16.42) max; (–1.12, –8.12) min (***A***)

c.

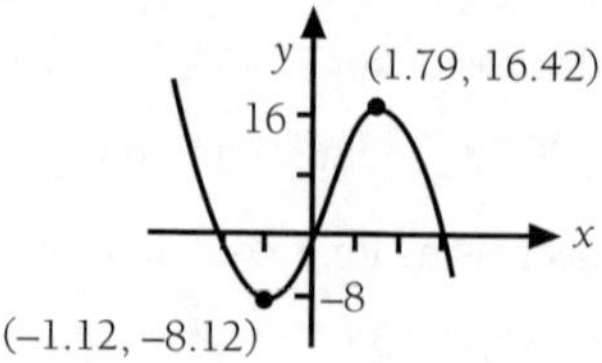

7. (0, 20) max, (2, 4) min (***E***)

8. (0.8, 83.9) max, (4, 0) min (***A***)

9. A (–1, 4); B (2, 4) (***A***)

10. a. Stationary points: (0, 0) stationary point of inflection; (1, –1) minimum.
Reason: $f'(0) = 0$ and $f''(0) = 0$ and $f'''(0) \neq 0$; $f'(1) = 0$ and $f''(1) > 0$. (***E***)

b. x-intercepts: solutions of $3x^4 - 4x^3 = 0 \Rightarrow x = 0$ or $\frac{4}{3}$ (***A***)

c. $(\frac{2}{3}, -0.59)$ by solving $f''(x) = 0$ (***E***)

d.

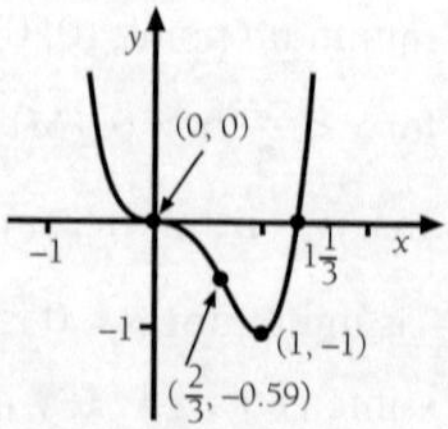

(***M***)

e. Concave down $0 < x < \frac{2}{3}$ (***M***)

11. (0,0) max; (2.1, –20.3) min; (–2.1, –20.3) min
Points of inflection: (–1.2, –11.3), (1.2, –11.3)
Intercepts: (0,0), (3, 0), (–3, 0)

(***E***)

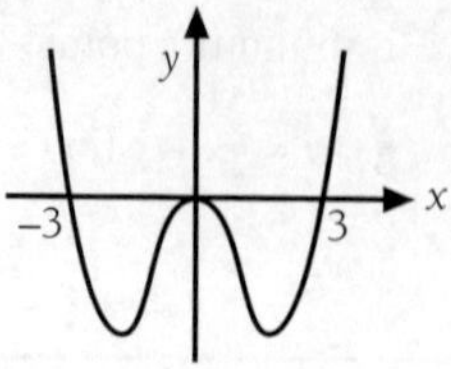

(***E***)

12. a. (–0.67, 0.53) max; (0.67, –0.53) min; $a = -1$; $b = 0$; $c = 1$

b. i. $-0.67 < x < 0.67$

ii. $x > 0$ (***E***)

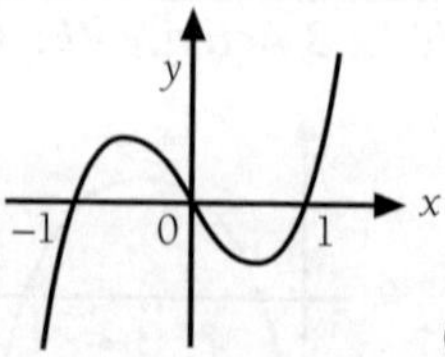

(***M***)

13. a. $p < 0$ (***M***) **b.** $\left(\frac{-1}{2}, \frac{-1}{8}\right)$ min; (0, 0) max; $\left(\frac{1}{2}, \frac{-1}{8}\right)$ min (***E***)

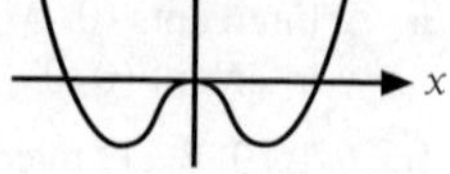

Unit 12.3 Activity 5A: Maxima and minima (page 187)

1. 7 200 m **2. a.** 5 400 litres **b.** After 30 minutes

3. a. 162 km/h **b.** 15 km/h

4. High tide when $t = 5$, with a depth of $216\frac{2}{3}$ cm above mark.

Low tide when $t = 19$, with a depth of $240\frac{2}{3}$ cm below mark.

5. 1 250 m^2 6. $12\frac{1}{2}$ m^2

7. a. $x(240 - 3x)$ m^2 b. 4 800 m^2 when $x = 40$ m (**E**)

8. 1 521 m^2 (**E**) 9. $12\frac{1}{2}$ (**E**) 10. 6 (**E**)

11. a. $x(8 - 2x)(10 - 2x)$ b. x should be about 1.47 cm to give a maximum volume of about 52.5 cm^3 (**E**)

12. Minimum area when the circle has a circumference of $\frac{\pi L}{\pi + 4}$ and the square has side length $\frac{L}{\pi + 4}$. (**E**)

Unit 12.3 Activity 6A: Rates of change (page 191)

1. a. 50 kg b. 1 037 kg (to nearest kg)
   c. 1.046 kg per day d. 2.25 kg per day (**M**)
2. a. 5 m s^{-1} b. 12 m s^{-1} c. 44 m s^{-1} d. 137 m s^{-1} (**M**)
3. a. Show all (t, A) values satisfy the formula. b. 2.984 tonnes
   c. $0.082 - 0.002t$ tonnes per day d. 0.06 tonnes per day e. 41st day (**M**)
4. a. 6 cm b. 0.1 cm per minute c. $25 + t + 0.01t^2$ cm^2
   d. 29.16 cm^2 e. 181.5 cm^2 f. $(5 + 0.1t)^3$
   g. 166.375 cm^3 h. $7.5 + 0.3t + 0.003t^2$ cm^3 per minute (**M**)
5. a. 3 m b. 0.1 m per hour c. 0.2 m per hour
   d. After 20 hours e. $3 + 0.7t + 0.02t^2$ f. 5.28 m^2
   g. $0.7 + 0.04t$ m^2 per hour h. After 107.5 hours (**M**)
6. a. $\frac{db}{dt} = -0.1$ m per hour, $\frac{dh}{dt} = 0.2$ m per hour.
   b. Each hour the base decreases by 0.1 m and the height increases by 0.2 m.
   c. $16\frac{2}{3}$ hours d. After 70 hours e. $14 + 1.2t - 0.02t^2$ f. $1.2 - 0.04t$
   g. 1.2 m^2 per hour h. 15 hours i. $37\frac{1}{2}$ hours. j. 32 m^2 (**M**)

Unit 12.3 Activity 6B: Chain rule (page 196)

1. a. 5 mm b. 15 mm c. 1 mm/s (**A**) d. 30π mm^2/s (**M**)
2. 0.030 cm/sec (or cm s^{-1}) (**M**) 3. a. 12cm s^{-1} (**M**) b. 36 cm^2 s^{-1} (**M**)
4. 0.371 cm s^{-1} (**M**) 5. 360π cm^2 s^{-1}; 2 700π cm^3 s^{-1} (**M**) 6. -0.01 cm s^{-1} (**M**)
7. $\frac{2}{9\pi}$ cm $minute^{-1}$ (**M**) 8. $\sqrt{3}$ ms^{-1} (**M**) 9. $1\frac{2}{13}$ km h^{-1} (**M**)
10. -0.5 radians/sec (**M**)

Unit 12.3 Activity 7A: Differentiating trigonometric functions (page 200)

1. a. $-3 \sin 3x$ (**A**)
b. $8 \cos 4x$ (**A**)
c. $-\sin x + \cos (x + \pi)$ or $-\sin x - \cos x$ (**A**)
d. $10 \sin (1 - 2x)$ (**A**)
e. $6 \cos (2x - \pi)$ or $-6 \cos 2x$ (**A**)
f. $-2 \sin x \cos x$ or $-\sin 2x$ (**A**)
g. $3 \cos (x - \pi) \sin^2 (x - \pi)$ or $-3 \cos x \sin^2 x$ (**A**)
h. $\dfrac{2 \sin t}{\cos^3 t}$ (**A**)
i. $-2 \sin 2x - 12 \cos 3x$ (**A**)
j. $\dfrac{-\sin 2t}{\sqrt{\cos 2t}}$ (**A**)
k. $2 \cos (2x + \frac{1}{6}\pi)$ (**A**)
l. $-n \sin (nx + \pi)$ or $n \sin nx$ (**A**)
m. $\frac{-1}{2} \cos (\frac{1}{4}\pi - \frac{1}{2}x)$ (**A**)
n. $\dfrac{-3 \cos 2t}{\sqrt{\sin^3 2t}}$ (**A**)
o. $4 \cos x (1 + \sin x)^3$ (**A**)
p. $-\sin x \cos (\cos x)$ (**A**)
q. $\cos x$ (**A**)
r. $2 \cos 2x$ (**A**)
s. 0 (**A**)

2. a. 8.46 (2 dp) (**A**)
b. −1.22 (2 dp) (**A**)

3. a. $h = 200$; $A = 200\pi$ (**M**)
b. $m = \cos \dfrac{150\pi}{200} = \dfrac{-1}{\sqrt{2}}$ (**M**)

4. $x = \sqrt{3} + \dfrac{2\pi}{3} \approx 3.83$ (3 sf) $\Rightarrow Q = (3.83, 0)$ (**M**)

5. a. Solve $\dfrac{\pi}{2} = \dfrac{\pi x}{400}$ to get x-intercept at (200, 0); $x = 0$ gives $y = 100$. (**M**)
b. Height $= 50\pi - 100 \approx 57.08$ (4 sf) (**M**)

Investigation (page 201)

1. $h = 300$ m
2. (x int) first at $x = 1.8326 \Rightarrow \overline{OA} = 183.26$ m (2 dp)
3. 104.72 m (2 dp)
4. 16.1° (1 dp)

Unit 12.3 Activity 8A: Differentiation rules and more trigonometric derivatives (page 206)

1. a. $10 \sec^2 2x$ (**A**)
b. $2 \sec (2x - \pi) \tan (2x - \pi)$ (**A**)
c. $-\pi \operatorname{cosec} \pi x \cot \pi x$ (**A**)
d. $2 - \operatorname{cosec}^2 x$ (**A**)
e. $6 \sec^2 x \tan^2 x$ (**A**)
f. $-6 \operatorname{cosec}^2 2x$ (**A**)
g. $-2 \cot x \operatorname{cosec}^2 x$ (**A**)
h. $\sec^2 x - \operatorname{cosec}^2 x$ (**A**)
i. $\dfrac{\cos x}{3x^{\frac{2}{3}}} - \sqrt[3]{x} \sin x$ (**A**)
j. $-\sin 2x \sin x + 2 \cos x \cos 2x$ (**M**)
k. $\dfrac{1 - 3x}{\sqrt{1 - 2x}}$ (**M**)
l. $\sin x(1 + \sec^2 x)$ (**M**)
m. $3(x \cos x + \sin x)$ (**A**)
n. $2(10x + 1)(2x + 1)^3$ (**M**)
o. $\dfrac{-3}{(x - 2)^2}$ (**A**)
p. $\dfrac{8}{(4 - x)^2}$ (**A**)

q. $\dfrac{2x(\sin 2x - x\cos 2x)}{\sin^2 2x}$ (**M**) r. $\dfrac{2-x}{2\sqrt{(1-x)^3}}$ (**M**) s. $\dfrac{\cos x + x\sin x}{\cos^2 x}$ (**A**)

t. $x^2(x\sin x + 3 - 3\cos x)$ (**A**) u. $e^{2x}(2x+1)$ (**A**) v. $1 + \ln(x+3)$ (**A**)

2. a. 0.59 (2 dp) (**A**) b. 0.125 (**A**)

3. a. $2\sec x(x\tan x + 1)$ (**A**) b. $\dfrac{3}{x^2}(x\sec^2 x - \tan x)$ (**A**)

c. $-2(7x+3)(2x+3)^2(1-x)^3$ (**M**) d. $\dfrac{-2(x-1)}{(x-2)^3}$ (**M**) e. $\dfrac{-6}{(2x-3)^4}$ (**A**)

f. $6x^2\sin\sqrt{x} + x^{2.5}\cos\sqrt{x}$ (**M**) g. $\dfrac{2x\sin 2x - x^2}{(\sin 2x)^{\frac{3}{2}}}$ (**M**)

4. $\dfrac{2}{\pi}$ (**A**)

Unit 12.3 Activity 9A: Antidifferentiation (page 211)

1. a. $\dfrac{x^2}{2} + C$ b. $\dfrac{x^4}{4} + C$ c. $\dfrac{3x^5}{5} + C$

d. $\dfrac{x^{12}}{12} + \dfrac{x^{13}}{13} + C$ e. $\dfrac{x^7}{7} - \dfrac{x^6}{6} + \dfrac{x^5}{5} + C$ f. $\dfrac{x^8}{4} + \dfrac{3x^7}{7} + C$

g. $x^4 - \dfrac{x^6}{3} + C$ h. $\dfrac{x^4}{2} - \dfrac{3x^5}{5} + C$ i. $\dfrac{x^3}{3} - \dfrac{3x^2}{2} + C$

j. $2x + C$ k. $3x + C$ l. $\dfrac{x^2}{2} - 3x + C$

m. $\dfrac{x^3}{3} - \dfrac{3x^2}{2} + 4x + C$ n. $\dfrac{x^3}{3} + \dfrac{7x^2}{2} + 12x + C$ o. $\dfrac{4x^3}{3} + 6x^2 + 9x + C$

p. $\dfrac{mx^3}{3} + C$ q. $\dfrac{3}{2}x^4 + 3x^3 + 4x^2 + 12x + C$ r. $\dfrac{4x^5}{5} + \dfrac{20x^3}{3} + 25x + C$

s. $a^2x + C$ t. $\dfrac{x^2}{2b} + C$ u. $\dfrac{cx^2}{2} + \dfrac{x^3}{3d} + k$

v. $\dfrac{Ax^3}{3} + \dfrac{Bx^2}{2} + Cx + D$ w. $\dfrac{x^3}{3A} + \dfrac{x^2}{2B} + \dfrac{x}{C} + D$

x. $\dfrac{A^2x^5}{5} + \dfrac{ABx^4}{2} + \dfrac{(B^2 + 2AC)x^3}{3} + BCx^2 + C^2x + D$ y. $\dfrac{x^5}{5} + \dfrac{Ax^4}{2} + \dfrac{A^2x^3}{3} + C$ (**A**)

2. a. $\dfrac{3x^2}{2} + C$ b. $\dfrac{3x^5}{5} + C$ c. $\dfrac{3x^8}{8} + C$ d. $\dfrac{2x^9}{9} + C$

e. $\dfrac{x^{14}}{14} - \dfrac{x^{11}}{11} + C$ f. $\dfrac{x^8}{4} + \dfrac{4x^6}{3} + C$ g. $\dfrac{x^5}{5} - x^4 + \dfrac{3x^2}{2} + C$

h. $\dfrac{x^4}{4} + x^3 + \dfrac{x^2}{2} + 3x + C$ i. $\dfrac{4x^3}{3} + 10x^2 + 25x + C$ j. $\dfrac{Ex^2}{2} + Dx + C$ (**A**)

3. a. $2x^2 + C$ b. $\dfrac{x^3}{9} + C$ c. $\dfrac{7x^4}{4} + C$ d. $r^2 + C$ e. $\dfrac{8r^9}{9} + C$

f. $\dfrac{5t^7}{7} + C$ g. $8x + C$ h. $ax + C$ i. $t^5x + C$ j. $\dfrac{t^6}{3} + C$

k. $x^2 + 4x + C$ **l.** $\frac{x^4}{4} - x^3 + \frac{5x^2}{2} + 2x + C$ **m.** $\frac{16x^3}{3} + 12x^2 + 9x + C$

n. $\frac{4x^5}{5} + \frac{4x^3}{3} + x + C$ **o.** $\frac{x^4}{4} + x^3 + C$ (**A**)

Unit 12.3 Activity 10A: Finding the constant of integration (page 214)

1. $\frac{x^2}{2}$ (**A**) **2.** $\frac{x^3}{3} + \frac{x^2}{2} + 2$ (**A**) **3.** $x^3 + 3$ (**A**)

4. $x^4 + x^2 - 15$ (**A**) **5.** $\frac{x^4}{2} + x - 3$ (**A**) **6.** $\frac{3x^2}{2} + 2x - 15\frac{1}{2}$ (**A**)

7. $\frac{5x^3}{3} - 95\frac{2}{3}$ (**A**) **8.** $\frac{2x^3}{3} - \frac{x^2}{2} + 3x - 128$ (**A**) **9.** $\frac{3x^2}{2} - \frac{x^3}{3} + 402$ (**A**)

10. $\frac{2x^3}{3} - \frac{3x^2}{2} + 2x + 3\frac{5}{6}$ (**A**) **11.** $\frac{x^2}{2} + 3x - 7$ (**A**) **12.** $21 - r^2$ (**A**)

13. a. $t^2 + t + 12$ **b.** 13.6 seconds (**M**)

14. a. $11\,000 - 60t - t^2$ **b.** After 79.1 seconds (**M**)

15. a. $149 - 5x + \frac{6}{x}$ **b.** After 9.92 km (**M**)

Unit 12.3 Activity 10B: Distance, velocity and acceleration (page 216)

1. a. 80 m **b.** 120 m **c.** 80 m **d.** After 10 minutes
e. $50 - 10t$ m s^{-1} **f.** 20 m s^{-1} **g.** 125 m **h.** –10 m s^{-2}
i. –10 m s^{-2} (**M**)

2. a. 40 m **b.** 25 m **c.** 6 seconds **d.** $30 - 10t$ m s^{-1}
e. –10 m s^{-1} **f.** –30 m s^{-1} **g.** 45 m **h.** –10 m s^{-2} (**M**)

3. a. 80 m s^{-1} **b.** After 25 seconds **c.** $(100t - 2t^2 + 100)$ m
d. 382 m **e.** –4 m s^{-2} (**M**)

4. a. $(50t - \frac{3t^2}{2} + 10)$ cm **b.** 186 cm **c.** 256 cm
d. After $16\frac{2}{3}$ seconds **e.** –3 cm s^{-1} (**M**)

5. a. $(5 + 3t)$ cm s^{-1} **b.** 35 cm s^{-1} **c.** After $13\frac{1}{3}$ seconds (**M**)

6. a. $6t - 3t^2$ **b.** $6 - 6t$ **c.** 0 **d.** 0 **e.** When $t = 0$ or 3 (**M**)

7. a. $\frac{2}{3}$m **b.** –3 m s^{-1} **c.** After 2 seconds **d.** $1\frac{1}{3}$ m (**M**)

Unit 12.3 Activity 11A: Trigonometric integrals (page 221)

1. a. $\frac{3}{4}\tan 4x + c$ (**A**) **b.** $\cos 2x + c$ (**A**) **c.** $12\tan\frac{x}{3} + c$ (**A**)

d. $\sin(x - \frac{\pi}{3}) + c$ (**A**) **e.** $\frac{-1}{3}\cot 3x + c$ (**A**) **f.** $\frac{-3}{2}\operatorname{cosec} 2x + c$ (**A**)

g. $2\sin\frac{x}{2}+\frac{1}{4}\cos 2x+c$ (**A**) **h.** $\frac{3}{4}\sin 4x-\frac{2}{3}\cos 3x+c$ (**A**)

i. $\frac{1}{2}\cos(1-2x)+c$ (**A**) **j.** $2\sin x+c$ (**A**) **k.** $\frac{1}{2}\sec 4x+c$ (**A**)

2. a. 0 (**A**) **b.** 2 (**A**) **c.** $\frac{-2}{3}$ (**A**) **d.** 5.62 (**A**)

e. 0.67 (**A**) **f.** $\frac{2}{3}$ (**A**) **g.** 0.61 (**A**) **h.** $2-\frac{2}{\sqrt{3}}$ (**A**)

i. 3.76 (**M**) **j.** –0.19 (**M**) **k.** 0.078 (**M**) **l.** 1.344 (**A**)

3. L, N, M (**A**)

4. $a=\pi, b=0$ or any a, b such that $|a-b|=n\pi$ where n is an integer (**A**)

5. a. The algebraic sum (taking sign into account) of the distance travelled by that object between $t=\frac{1}{2}$ and $t=2$ seconds. (**A**)

b. $\frac{-1}{\pi}$ metres (**A**)

6. a. $v=2\cos 2t$ (**A**) **b.** max $V=2$ m s^{-1} (**M**) **7.** Mean range $=\frac{605}{\pi}$ m (**E**)

Unit 12.3 Activity 12A: Definite integrals (page 224)

1. a. 6 **b.** $41\frac{1}{3}$ **c.** $7\frac{1}{2}$ **d.** $4\frac{4}{15}$ **e.** $1\frac{1}{3}$ **f.** 2

g. 18 **h.** 9 **i.** $7\frac{1}{2}$ **j.** 21 **k.** $-42\frac{1}{12}$ **l.** $5\frac{13}{24}$

m. $\frac{5}{16}$ **n.** $-8\frac{1}{3}$ **o.** 3.54 (**A**)

2. a. $-55\frac{1}{3}$ (**A**) **b.** $16\frac{1}{12}$ (**A**)

3. a. $a=1.5$ (**E**) **b.** $p=4$ (**E**)

Unit 12.3 Activity 12B: Areas under graphs (page 227)

1. $A_1=12\frac{2}{3}$ $A_2=4\frac{2}{3}$ $A_3=36\frac{2}{3}$ (**A**)

2. a. 6 (**A**) **b.** 10 (**A**) **c.** 39 (**A**) **d.** $1\frac{2}{3}$ (**A**) **e.** $10\frac{2}{3}$ (**A**) **f.** $20\frac{5}{6}$ (**A**)

g. $4\frac{1}{4}$ (**A**) **h.** 2 (**A**) **i.** $7\frac{2}{3}$ (**A**)

Unit 12.3 Activity 12C: Problem solving using definite integrals (page 227)

1. $\frac{1}{3}$ (**M**) **2.** $\frac{4a}{3}$ (**M**) **3.** 1.31 (**M**) **4.** $-\frac{2}{3}$ (**M**)

Investigation (page 229)

Bill's claim is incorrect. 11.1% reduction $\left(\frac{1}{9}\right)$.

Unit 12.3 Activity 12D: Areas (page 230)

1. **a.** 9 (***A***) **b.** $10\frac{2}{3}$ (***A***) **c.** 19.72 (***A***) **d.** 2 (***A***) **e.** 4 (***A***)

f. $24\frac{1}{2}$ (***A***) **g.** $6\frac{1}{2}$ (***A***) **h.** $21\frac{1}{3}$ (***A***) **i.** $5\frac{1}{3}$ (***M***)

2. **a.** $1\frac{2}{3}$ (***M***) **b.** $2\frac{1}{4}$ (***A***) **c.** $5\frac{1}{3}$ (***A***) **d.** Area = $\frac{1}{2} \times 1.5 \times 3 = 2.25$

3. **a.** $1\frac{1}{3}$ (***M***) **b.** $\frac{8\sqrt{2}}{3}$ (***M***) **c.** 16.086 (***E***)

4. $\frac{\pi}{6}$ (***M***) **5.** $49\frac{1}{3}$ m² (***A***) **6.** Proof. (***M***)

7. **a.** $\frac{1}{4}$ (***M***) **b.** $\frac{1}{20}$ (***M***) **c.** $2\frac{2}{3}$ (***M***) **d.** 0.414 (***E***) **e.** 0.586 (***E***)

8. 4 (***M***)

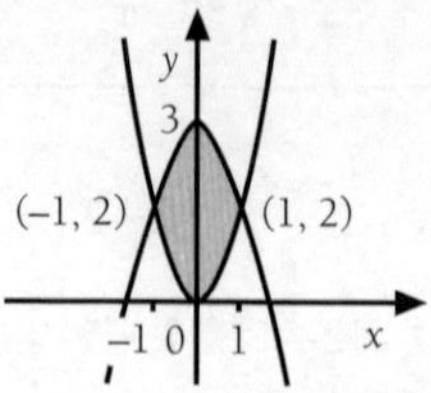

9. $\frac{1}{3}$ (***M***) **10.** $\frac{27}{37}$ (***E***) **11.** **a.** 23.83 (***A***) **b.** 1.272 (***M***)

Unit 12.3 Activity 12E: Integrals using areas (page 232)

1. **a.** **i.** $4\frac{1}{2}$ (***A***) **ii.** Graph is below *x*-axis. (***A***) **b.** **i.** 8 (***A***) **ii.** 0 (***A***)

2. **a.** 1 (***A***) **b.** 1 (***A***) **c.** –1 (***A***) **d.** 0 (***A***) **e.** 0 (***A***)

3. **a.** –5 (***A***) **b.** 0 (***A***) **c.** –5 (***A***) **4.** **a.** 6 (***A***) **b.** 4 (***A***) **c.** 3 (***A***)

5. **a.** $-A_1$ (***A***) **b.** $A_2 - A_1$ (***A***) **c.** $A_2 - A_3$ (***A***)

6. $\int_{-3}^{0} \left(g(x) - f(x)\right) dx + \int_{0}^{5} \left(f(x) - g(x)\right) dx$ (***A***)

7. **a.** 0 (***A***) **b.** $-k$ (***A***) **c.** $-k$ (***A***)

8. **a.**

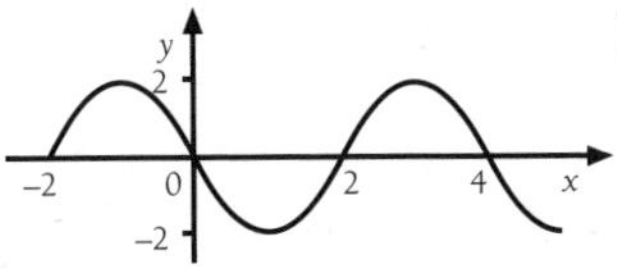

i. $-k$ (**M**) **ii.** k (**M**) **iii.** k (**M**)

b.

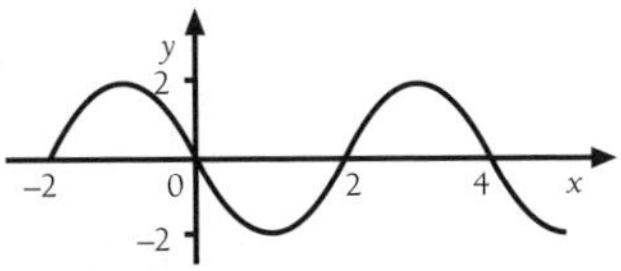

i. $-k$ (**M**) **ii.** 0 (**M**)

c.

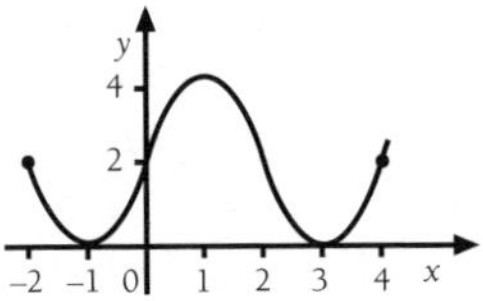

i. $k + 4$ (**M**) **ii.** $12 - k$ (**M**)

d.

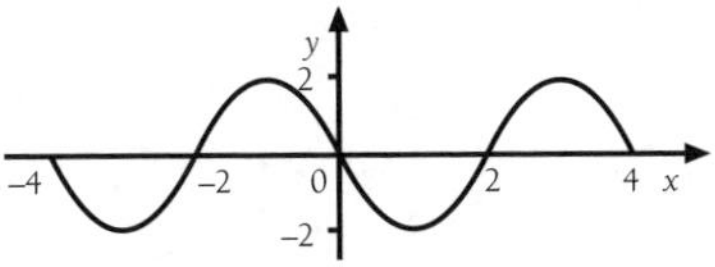

i. $-k$ (**M**) **ii.** 0 (**M**)

9. **D** (***A***) **10.** **A** (***A***)

Glossary/Index

absolute value (|x|) (108, 215, 226): The size of a number or expression, without regard to its sign. Also called the modulus.

acceleration (135, 163, 190, 214): The rate of change of velocity with respect to time.

acute angle (77, 79, 201): An angle of size less than 90°.

addition (+): An arithmetic operation; addition of vectors **(137)**.

all science teachers count (78): A mnemonic used to remember the quadrants in which the different trigonometric functions have positive values.

amplitude (83, 96–102, 122): In the graph of a periodic function, the amplitude is the maximum displacement of a point on the graph from the mean or central position.

anticlockwise (78, 126): The opposite direction to the motion of the hands of a clock.

antiderivative (209–10, 220, 223): Function obtained by reversing the process of differentiating.

antidifferentiation (209–11, 229): also called *integration*. The opposite process to differentiation. Rules for antidifferentiation **(209)**.

applications (30, 35–41, 127): The uses a theory is put to in solving problems.

arc (80): Part of a curve. Arc length formula **(128, 236)**.

area (129, 185–6): Under a curve **(224)**; of a sector **(128)**; of a circle **(80)**.

area of a triangle (128): In any triangle, area = $\frac{1}{2}ab \sin C$.

area under a curve (224): The area between a curve and the *x*-axis is calculated using the definite integral of the equation of the curve. Areas between curves **(228–9)**.

arithmetic progression (AP) (21, 23, 31): *See* arithmetic sequence.

arithmetic sequence (21–3, 41–2): A sequence with a constant difference between adjacent terms.

asymptote (83, 93): Line that a curve gets closer and closer to but never actually touches.

average: The mean (sometimes median or mode).

axis of symmetry (76, 85): Line through the vertex of parabola, dividing it into two congruent parts (two equal halves).

base: The number or letter being raised to a power, eg in 5^4, the base is 5. In a logarithm expression such as $\log_b x$, the base is *b*.

binomial (45): An expression with two terms.

Binomial Theorem (49): A rule for expanding $(a + b)^n$.

bracket rule (200, 204): A specific case of the chain rule, used for differentiating brackets raised to powers.

calculators, use of with: factorials **(48)**; simultaneous equations **(70)**; sequences **(17, 35, 42)**; gradients **(164)**; graphs **(99, 161)**; trigonometric equations **(107–8, 115–20)**; graphs of the trigonometric functions **(89)**; integrals **(220, 229)**.

calculus (169–234): Branch of mathematics involving rates of change, differentiation, integration and associated processes.

Cartesian plane (58): The *x*, *y* plane in which graphs are drawn and points represented by ordered pairs (*x*, *y*).

chain rule (192–6, 199, 205, 240): A rule for differentiating composite functions.

chord: A line segment whose end points are on a curve.

circle (77, 80, 124, 128, 235): Curve with equation of the form $x^2 + y^2 = R^2$. Equation of **(236)**.

circular functions (77, 115): Another name for the trigonometric functions.

coefficient (45): A number that multiplies a variable in a term, eg in the term $3x^5$, the coefficient of x^5 is 3.

collinear (141, 151): Three or more points that lie on the same straight line.

combinations (47–50): Subsets of a set, in which order is not important.

common difference (21–3): The constant difference associated with an arithmetic sequence.

common factors (99, 220): Factors that divide into each term of an expression without remainder.

common ratio (27): The constant value associated with a geometric sequence.

complementary (3, 235): The complement of an event, A, is written A′ and is the set of all elements in the sample space that are outside A.

completing the square (53): A technique used so that a quadratic expression can be written in the form of a perfect square (plus or minus a constant).

compound angle identities (126): A set of six trigonometric identities using sums and differences of two angles.

compress (86, 88): Reduce, by a scale factor.

compression (84): Shortening of the horizontal axis.

concave down (182): Points on a curve where $f''(x) < 0$.

concave up (182): Points of a curve where $f''(x) > 0$.

concurrent: Lines that go through the same point are concurrent.

congruent (76, 94): Identical in shape and size.

consistent (59, 63): A set of equations is said to be consistent if there is at least one solution (there may be infinitely many solutions).

constant (21, 27, 49–50): A number that does not vary.

constant of integration (209, 213–16, 223): A constant that results from integrating.

continuous (122, 224): A continuous function has no holes or breaks in its graph.

cosecant, cosec (93–4, 123–5, 206, 220, 237, 239): The reciprocal of sine.

cosine (cos) (75, 78, 83, 96–7, 100): A trigonometric function.

cosine rule (128, 146, 237–8): A relationship between three sides and one angle of a triangle. In any triangle $a^2 = b^2 + c^2 - 2bc \cos A$.

cotangent, cot (93, 123–5, 206, 220, 237, 239): The reciprocal of tangent.

cube root: A number which when cubed gives the number under the $\sqrt[3]{\ }$ sign.

cubic (182): A polynomial expression in which the highest exponent of the variable is three. A cubic equation is of the form $ax^3 + bx^2 + cx + d = 0$.

cubic functions: An equation of the form y = cubic expression.

cycle (91, 96, 101): One cycle of a periodic graph occurs over every interval of width equal to the period.

data (78): The numbers and information associated with statistics.

decreasing function (177–83): A function whose 'y' values decrease as its 'x' values increase.

definite integral (220, 223–32, 239): An integral evaluated between two limits, resulting in a numerical solution.

degree (182): The highest power of the variable in any of the terms of a polynomial.

degrees, symbol ° (76, 80, 89, 111): A unit of angular measure. A full turn is 360 degrees (360°).

denominator (125, 127, 204): The integer or variable below the horizontal line in a fraction.

derivative (169–70, 192): The rate of change of one variable with respect to the other. Derivative of x^n is nx^{n-1}. This is written $f'(x)$, and is the function derived from $f(x)$ using the formula: $f'(x) = \lim_{h \to 0} \frac{f(x + h) - f(x)}{h}$. Trigonometric derivatives **(199–202, 205, 220)**.

derived function (169): The derivative.

difference (21, 209): The result of subtracting two terms or numbers.

differentiable (169): A function that can be differentiated. A function $f(x)$ is differentiable at $x = a$ if it has a derivative defined at $x = a$.

differential equation: An equation linking variables with one or more of their derivatives.

differentials (192): The two parts dx and dy which make up a derivative.

differentiating (169, 173, 185, 199, 205): Calculating the function $f'(x)$ which is the gradient of the curve $f(x)$.

differentiation (169): Finding a derivative.

direction (135): Line or course of a vector's movement.

distance between two points (150): A formula in co-ordinate geometry.

distributive law (7–8, 126): A law describing how brackets are expanded, $a(b + c) = ab + ac$.

division (÷): An arithmetic operation.

domain (83, 92–4, 169): The set of first elements in the ordered pairs of a relation. Of a trigonometric function **(83, 91, 108, 121)**.

double angle identities (238): The compound angle identities in which the two angles are the same.

elimination (57, 63, 66): A method of solving simultaneous equations.

empty set, Ø (1): A set with no elements in it.

enlarge (88): Make larger, by a scale factor.

enumeration (16): One way of writing a sequence, by listing.

equation: A mathematical sentence involving an equals sign and at least one variable. Trigonometric **(107)**; linear **(55)**; simultaneous **(56–70)**; quadratic **(53)**.

equation of a straight line (236): Gradient intercept form is $y = mx + c$.

estimate: An approximation to an exact value.

even function (92, 94): A function f is even if $f(-x) = f(x)$ for all x in its domain. The graph of an even function is symmetrical about the y-axis.

exact trigonometric values (76, 81): Exact values of the sine, cosine or tangent of certain angles, expressed as fractions, integers or surds.

exact values (79, 95, 111): Trigonometric ratios expressed as fractions or surds which represent the ratios of sides of the special triangles.

expanding (48): Removing brackets from an expression by multiplying.

expansion (24, 45–50): Removing brackets from an expression by multiplying.

exponent: The power to which a base number is raised. For example, in the expression a^x, the exponent is x.

factor: A factor of an expression is a term or expression that divides it without remainder.

factorial (47): n factorial ($n!$) means $n(n-1)(n-2) \dots 3 \cdot 2 \cdot 1$.

factorial notation, symbol ! (47): A shorthand way of representing an important product of whole numbers. $n! = n(n-1)(n-2) \dots 3.2.1$.

factorising (99): Rewriting a mathematical expression as a product of simpler factors.

formula (formulae, plural) (16): An algebraic equation in which the mathematical relationship between different variables is expressed.

fraction: A number expressed as a quotient. The top number is called the numerator, the bottom the denominator. An algebraic fraction or rational expression is a fraction with variables.

free vector (136): A vector with no particular line of action or point of origin but retaining its length, direction and sense of movement.

frequency (101): The frequency of a periodic curve is the number of times the period of the curve occurs in a given unit of time. The frequency is the reciprocal of the period.

function: A relation in which for any value of x there is one *and only one* value of y.

Fundamental Theorem of Calculus (225): $\int_a^b f(x)\,dx = F(b) - F(a)$ where F is an antiderivative of f.

general term (15, 21, 27, 49): A rule for finding any term in a sequence. Also called the nth term.

geometric interpretation (64–5): A representation of formulae or equations using graphs or geometry. Sets of three simultaneous equations in three unknowns can be represented by intersecting planes **(64)**.

geometric progression (27): A geometric sequence.

geometric sequence (GP) (27–34, 35–42): A sequence where each term gives a constant value when divided by the previous term.

gradient (161–3): The ratio of the $\dfrac{\text{vertical change}}{\text{horizontal change}}$ of a straight line. Of a tangent **(173)**.

gradient function (164–7): A function that maps a point onto its gradient; the derivative.

gradient of a curve at P: Alternative description for the gradient of the tangent of a curve at P.

graph (42): A set of ordered pairs plotted on a grid.

graphs: Of parabolas **(178, 185, 229)**; of cubic **(182)**; of trigonometrical functions **(82–9, 91–100)**; of gradient functions **(164)**.

hypotenuse (75, 237): The side of a right-angled triangle opposite the right angle.

identity (123): An equation which is true for all meaningful values of the variables.

inconsistent (59, 65, 70): A set of equations is said to be inconsistent if there is no solution.

increasing function (177–8): A function whose 'y' values increase as its 'x' values increase.

indefinite integral (210, 225, 239): An integral in which the constant of integration cannot be evaluated.

index (indices, plural) (50, 172): Exponent or power (of a base number or variable).

index laws (50): Rules for the multiplication, division, etc of indices (terms written in index form).

inflation (27, 36): A general increase in prices and fall in the purchasing power of money.

integers, I (57, 83, 92, 235): The set of positive and negative whole numbers $\{\ldots, -3, -2, -1, 0, 1, 2, 3, 4, \ldots\}$.

integral (209, 219, 223–8): Another word for antiderivative. And area under a curve **(224–32)**.

integration (209, 219, 239): The process of finding an antiderivative; *see* antidifferentiation.

intercepts (178): Points passed through by a graph on the *x*-axis or *y*-axis.

intersection (58, 64, 228, 235): Curves, lines or planes intersect each other when they have one or more points in common

intersection, symbol $\cap$ (4, 7, 11): The set containing the elements common to two or more sets. The intersection of two sets *A* and *B* is written $A \cap B$.

interval notation (107, 166): A way of representing sets of real numbers in which curved brackets exclude end points of the interval and square brackets include end points of the interval.

inverse (107): A procedure which reverses the effect of another (eg addition and subtraction are inverse operations); the inverse of a function is obtained by exchanging *x* and *y* in the equation of the function.

irrational number (80, 235): A number that cannot be written as the quotient of two integers.

isosceles triangle (75): A triangle with two sides of equal length.

kinematics (189, 213): The study of motion.

Leibniz notation (203): An alternative notation for the derivative $f'(x)$, which is written $\frac{dy}{dx}$.

limit (131): See *limit of a function*. Evaluating limits **(228)**. Values between which a definite integral is defined **(223)**; limit of a sequence **(17, 31)**.

limit of a function $f(x)$ as x approaches a: The value that $f(x)$ approaches as x approaches the value a, from below and above.

line (41): The shortest distance between two points on a plane. Midpoint of a line **(236)**. Gradient of a line **(161)**; intersection with another line **(58)**; intersection with a parabola **(228)**.

line segment (135): Part of a line lying between two points.

linear (55): Pertaining to the straight line. An equation or polynomial in which the highest power of any variable is 1.

listing (1, 16): Also called enumeration.

local maximum (177): A turning point where a function changes from increasing to decreasing.

local minimum (177): A turning point where a function changes from decreasing to increasing.

logarithm (log) (35): The inverse of $y = a^x$ is $y = \log_a x$, a function with domain $(x > 0)$.

logarithm to base 10, $\log_{10} x$ (36): Inverse of $y = 10^x$.

m (162, 175): Commonly used symbol for the gradient of a straight line.

magnitude (135): Size. The length of a vector.

maximum (38, 83, 96, 120): The greatest value of a quantity. See *local maximum*.

mean ($\bar{x}$ or μ) (83, 100): The sum of all the values divided by the total number of values.

midpoint (137, 141): The point halfway between the end-points of a line segment.

minimum (83, 96, 120): The least value of a quantity. See *local minimum*.

minimum turning point (180): A turning point where a function changes from decreasing to increasing.

model (55, 91, 101): A mathematical equation representing a relationship between variables. Modelling with trigonometric functions **(113, 120)**.

multiplying (×): An arithmetic operation.

***n* factorial (47):** *See* factorial notation.

natural numbers, N (3, 47, 235): The set {1, 2, 3, 4, ...}.

nature (70, 97): The number and type of roots of a quadratic equation.

nature of a turning point (178–80): Whether a turning point is a maximum or a minimum.

normal (175): A straight line which is at right angles to a tangent to a curve, and passes through the point of contact of the tangent with the curve.

***n*th partial sum (23, 29):** *See* partial sum.

***n*th term (15, 21, 27):** The member of a sequence associated with n.

numerator (125, 204): The integer or variable above the horizontal line in a fraction.

odd function (92): A function f is odd if $f(-x) = -f(x)$ for all x in its domain. The graph of an odd function has half-turn rotational symmetry about the origin.

operations: Processes applied to numbers, such as squaring, or adding.

optimisation (185): The finding of maximum or minimum values of a variable quantity.

parabola (178, 185, 229): Graph of a quadratic function.

parallel lines (58, 66, 236): Lines with the same gradient.

partial sum (23, 29): The sum of the first n terms of a sequence.

Pascal's Triangle (45): An array of numbers used to evaluate coefficients in binomial expansions.

period (83, 91, 96, 101): The width of the interval of x-values over which a periodic graph repeats itself.

periodic (83): Repeating over regular intervals.

ratio (27, 35, 75, 93): A comparison of two or more quantities.

rational numbers, Q (51, 235): Numbers that can be expressed in the form of a fraction $\frac{a}{b}$, where a and b are integers, and $b \neq 0$.

real numbers, R (15, 83): Numbers that can be represented on a number line. The union of rational and irrational numbers.

rearranging (24, 32, 64, 122): Changing the order of terms in an expression, often so that another variable becomes the subject.

reciprocal (93, 101): The reciprocal of $\frac{a}{b}$ is $\frac{b}{a}$. The reciprocal of x^n is x^{-n}.

reciprocal rule (194): A rule that relates a derivative to its reciprocal.

reflection in the x-axis (96, 99): A transformation of a curve such that each point (x, y) maps onto the point $(x, -y)$.

reflection in the y-axis (77): A transformation of a curve such that each point (x, y) maps onto the point $(-x, y)$.

region (12, 97, 224): Another expression for 'a set of points' on a graph or in a Venn diagram.

related rates of change (192–3): Rates of change expressed using the chain rule as the product of two or more derivatives.

Remainder Theorem (52): The result that when a polynomial $f(x)$ is divided by $(x - a)$ the remainder is $f(a)$.

resultant (137): The vector that results after vectors have been added or subtracted together.

root (228): The inverse of raising a base to the power of n is taking the nth root of the base.

root of an equation (53): A solution to an equation.

rule (11, 16, 60, 96): Formula that expresses a relationship.

satisfy (55, 58, 107): Values which, when substituted, make an equation true, are said to satisfy the equation.

scalar quantity (135): A quantity that has only magnitude and no direction. Also called a constant.

secant, sec (93): The reciprocal of cosine. Graph of sec x **(94)**.

sector (129, 235): A region formed by two radii and an arc. Area of a sector **(128)**.

segment (129): Area enclosed by a chord and arc in a circle.

sequence (15): A function that associates a real number with each natural number. Arithmetic sequences **(21–26)**; Geometric sequences **(27–34)**.

series (19, 30, 236): The sum of the successive terms of another sequence.

sigma notation (Σ) (19): Notation for addition of terms, associated with sequences and series.

simultaneous equations (56–74): Equations, each of which contains the same variables, and which usually have a common solution when solved together.

sine (sin) (75): A trigonometric function.

sine rule (128): In any triangle $\frac{a}{\sin A} = \frac{b}{\sin B} = \frac{c}{\sin C}$.

slope (161–5, 181): Another expression for gradient.

solving (55): Finding the values of the variables that make an equation or inequation true.

speed (v) (135, 163, 190, 214–15): The positive rate of change of distance with respect to time. (Also called velocity.)

spreadsheet (186): A computer program used to perform calculations, draw graphs and fit trend lines using tables of data.

square root: A number which when squared gives the number under the $\sqrt{\ }$ sign.

square root function: The function $f(x) = \sqrt{x}$.

stationary points of inflection (181): Points where a function $f(x)$ is neither increasing nor decreasing, and $f'(x) = 0$..

stationary values (177): Values at which a function is neither increasing or decreasing (the gradient is zero).

statistics: The collection, display and analysis of data.

subject (187): The variable appearing on its own on one side of the equals sign in an equation or formula.

substitution (17, 57, 169, 181): The process of replacing variables with numbers or other variables.

subtend (80): Lie under.

subtracting (–): An arithmetic operation.

sum: The result of adding two or more numbers or like terms. Sum of the first n terms of an arithmetic progression **(23)**; of a geometric progression **(29)**; sum to infinity **(31)**.

surds (52, 79, 82): Irrational numbers that are expressed as square roots, cube roots etc.

table (36): Used to represent some of the ordered pairs of a relation.

tangent (164, 175, 178): A straight line that touches a curve at exactly one point. Gradient of tangent to a circle **(237)**. Tangent to a parabola **(173, 229)**.

tangent (tan) (75, 83, 116): A trigonometric function.

term (15): One member of a sequence or expression.

test failure (181): Test failure occurs when points of inflection cannot be identified by the second derivative test. Test failure occurs when for some x, both $f'(x) = 0$ and $f''(x) = 0$.

third derivative (193): The derivative of the second derivative: $f'''(x)$ or $\frac{d^3y}{dx^3}$.

translation (86, 91, 96–100): A transformation of a curve, denoted by the vector $\begin{pmatrix} a \\ b \end{pmatrix}$, such that each point (x, y) maps onto the point $(x + a, y + b)$.

trend (33): An ongoing pattern in time series data.

trial and error (28, 35): The process of solving a problem by substitution of possible answers until successful.

triangular brackets (< >) (16): Notation used to write a sequence.

trigonometric equation (107–22): An equation involving the sine, cosine or tangent functions.

trigonometric functions (78, 91): The sine, cosine and tangent functions and their transformations. The graphs of the trigonometric functions **(92)**. Integration of trigonometric functions **(219–20)**.

turning point (TP) (177): A point where a function changes from increasing to decreasing (or vice versa) and the gradient of a tangent at the point is zero. Nature of **(178)**; finding turning points **(179)**.

union, symbol $\cup$ **(4, 8, 235):** The union of two (or more) sets is the collection of all outcomes in the sets (or in their intersection). The union of two sets A and B is written $A \cup B$.

unique solution (59, 63, 70): An equation, or set of equations, has a unique solution when there is exactly one solution.

unit vector (145): A vector with magnitude one unit.

variable (55, 63): A letter that represents a number.

vector (135): A quantity with both magnitude and direction which is represented by an arrowed line.

velocity (symbol v) (135, 190, 214): The instantaneous rate of change of distance (displacement) with respect to time.

Venn diagram (3–14): A rectangular diagram used to show relationships between sets (represented as ovals or circles inside the rectangle).

volume of revolution (240): The volume formed when an area is rotated through 360 degrees about an axis.

whole numbers, W (45, 235): The set {0, 1, 2, 3, ...}.

word problems (55, 71): Problems written in words rather than in mathematical form.

***x*-intercept (182):** The point where a graph cuts the x-axis.

***y*-intercept (178):** The point where a graph cuts the y-axis.